The Role of Freshwater Outflow
in Coastal Marine Ecosystems

NATO ASI Series

Advanced Science Institutes Series

A series presenting the results of activities sponsored by the NATO Science Committee, which aims at the dissemination of advanced scientific and technological knowledge, with a view to strengthening links between scientific communities.

The Series is published by an international board of publishers in conjunction with the NATO Scientific Affairs Division

A Life Sciences	Plenum Publishing Corporation
B Physics	London and New York
C Mathematical and Physical Sciences	D. Reidel Publishing Company Dordrecht, Boston and Lancaster
D Behavioural and Social Sciences **E Applied Sciences**	Martinus Nijhoff Publishers Boston, The Hague, Dordrecht and Lancaster
F Computer and Systems Sciences **G Ecological Sciences**	Springer-Verlag Berlin Heidelberg New York Tokyo

Series G: Ecological Sciences Vol. 7

The Role of Freshwater Outflow in Coastal Marine Ecosystems

Edited by

Stig Skreslet
Nordland College, P.O. Box 6003, 8016 Mørkved, Bodø Norway

Springer-Verlag Berlin Heidelberg New York Tokyo
Published in cooperation with NATO Scientific Affairs Division

Proceedings of the NATO Advanced Research Workshop on The Role of Freshwater Outflow in Coastal Marine Ecosystems held in Bodø, Norway, May 21–25, 1985

ISBN 3-540-16089-2 Springer-Verlag Berlin Heidelberg New York Tokyo
ISBN 0-387-16089-2 Springer-Verlag New York Heidelberg Berlin Tokyo

Library of Congress Cataloging in Publication Data. NATO Advanced Research Workshop on the Role of Freshwater Outflow in Coastal Marine Ecosystems (1985 : Bodø, Norway) The role of freshwater outflow in coastal marine ecosystems. (NATO ASI series. Series G, Ecological sciences ; v. 7). Includes index. 1. Marine ecology—Congresses. 2. Coastal ecology—Congresses. 3. Stream ecology—Congresses. I. Skreslet, Stig, 1940-. II. Title. III. Title: Freshwater outflow in coastal marine ecosystems. IV. Series: NATO ASI series. Series G, Ecological sciences ; no. 7. QH541.5.S3N37 1985 575.5'2636 86-6438
ISBN 0-387-16089-2 (U.S.)

This work is subject to copyright. All rights are reserved, whether the whole or part of the material is concerned, specifically those of translating, reprinting, re-use of illustrations, broadcastings, reproduction by photocopying machine or similar means, and storage in data banks. Under § 54 of the German Copyright Law where copies are made for other than private use, a fee is payable to "Verwertungsgesellschaft Wort", Munich.

© Springer-Verlag Berlin Heidelberg 1986
Printed in Germany

Printing: Beltz Offsetdruck, Hemsbach; Bookbinding: J. Schäffer OHG, Grünstadt
2131/3140-543210

PREFACE

Over the last decade or two marine scientists have become more aware of the role of freshwater outflow in coastal waters. Some have raised the question whether or not regulation of river flow may change the biological productivity in coastal marine ecosystems. The idea of organising a workshop to deal with this problem, arose from parallel concern in Norway and Canada, but it was soon recognised that the scientific interests were also shared by research groups in other countries.

After the initial contacts had been made, it was agreed to establish a committee which should assist the workshop director in organising the workshop. The Organising committee consisted of K.F. Drinkwater, Bedford Institute of Oceanography, Canada; S. Skreslet (workshop director), Nordland College, Norway; V.S. Smetacek, Institut fur Meereskunde an der Universitat Kiel, Germany; and H. Svendsen, University of Bergen, Norway.

An Advisory committee was established to provide terms of reference for the Organising committee, with regard to the programme and selection of participants. This committee consisted of J.B.L. Matthews (chairman), Scottish Marine Biological Association, U.K.; T.R. Parsons, University of British Columbia, Canada; R. Sætre, Institute of Marine Research, Norway; M.M. Sinclair, Fisheries and Oceans, Canada; and W.S. Wooster, University of Washington, Canada.

The workshop was organised to fit the requirements of an Advanced Research Workshop funded by the NATO Scientific Affairs Division, which eventually awarded a grant to cover the major part of the expences. However, additional financial support was granted by the Norwegian Fisheries Research Council and the Nordland College Fisheries Fund.

Several colleagues at the Nordland College, both among the administrative and scientific staffs, eventually became involved in one way or the other, and contributed significantly to the success of the workshop. In particular, I am indebted to Ewa Bivand whose professional support, skill and patience was a true relief.

I want to thank all those who have been involved in the workshop, not least the participants whose enthusiasm lifted the programme from scetches on paper to a fruitful reality which points to promising scientific endeavors in the future.

Bodø, 13 November 1985

Stig Skreslet

CONTENTS

OPENING ADDRESS 1
A. Sandberg

INVITED CONTRIBUTIONS

Freshwater outflow in relation to space and time dimensions of complex ecological interactions in coastal waters 3
S. Skreslet

Mixing and exchange processes in estuaries, fjords and shelf waters 13
H. Svendsen

The role of brackish plumes in open shelf waters 47
R.W. Garvine

Laboratory modeling of dynamic processes in fjords and shelf waters 67
T.A. McClimans

Impact of freshwater discharge on production and transfer of materials in the marine environment 85
V.S. Smetacek

A framework for discussion of marine zooplankton production in relation to freshwater runoff 107
J.B.L. Matthews

The dependence of fish larval survival on food and

predator densities   117
W.C. Leggett

Assessment of effects of freshwater runoff variability on fisheries production in coastal waters   139
M. Sinclair, G.L. Bugden, C.L. Tang, J.-C. Therriault and P.A. Yeats

Computer model analysis of pelagic ecosystems in estuarine waters   161
T. Parsons and T.A. Kessler

INSHORE WATERS

The capacity of Lake Melville fjord to store runoff   183
S.A. Akenhead and J. Bobbitt

Inorganic nutrient regeneration in Loch Etive bottom water   195
A. Edwards and B.E. Grantham

Physical exchange and the dynamics of phytoplankton in Scottish sea-lochs   205
P. Tett

Freshwater outflow effects in a coastal, macrotidal ecosystem as revealed by hydrological, chemical and biological variabilities (Bay of Brest, western Europe)   219
B. Quéguiner and P. Tréguer

MARGINAL SEAS

River input of nutrients into the German Bight   231
U.H. Brockmann and K. Eberlein

Effects of freshwater inflow on the distribution, composition and production of plankton in the Dutch coastal waters of the North Sea　241
H.G. Fransz

Freshwater runoff control of the spatio-temporal distribution of phytoplankton in the lower St. Lawrence Estuary (Canada)　251
J.-C. Therriault and M. Levasseur

Biological and physical characteristics of a frontal region associated with the arrival of spring freshwater discharge in the southwestern Gulf of St. Lawrence　261
B. Côté, M. El-Sabh and R. de la Durantaye

SHELF WATERS

Relationships of St. Lawrence River outflow with sea surface temperature and salinity in the northwest Atlantic　271
J.A. Koslow, R.H. Loucks, K.R. Thompson and R.W. Trites

Water retention over Flemish Cap　283
S.A. Akenhead

The Scottish Coastal Current　295
J.H. Simpson and A.E. Hill

Runoff driven coastal flow off British Columbia　309
P.H. LeBlond, B.M. Hickey and R.E. Thomson

The seasonal influence of Chesapeake Bay phytoplankton to the continental shelf　319
H.G. Marshall

River runoff and shrimp abundance in a tropical coastal ecosystem - the example of the Sofala Bank (central Mozambique) 329
A.J. da Silva

Timing and duration of spring blooming south and southwest of Iceland 345
T. Thordardottir

Production, grazing and sedimentation in the Norwegian Coastal Current 361
R. Peinert

Advection of Calanus finmarchicus between habitats in Norwegian coastal waters 375
S. Skreslet and N.Å. Rød

The ecological impact of the East Greenland Current on the north Icelandic waters 389
S.A. Malmberg

GROUP REPORTS

Oceanographic functions of freshwater discharge and consequences of change 405

Interdisciplinary assessment of ecological problems related to freshwater outflow in coastal marine systems 417

SUMMARY CONTRIBUTION

On the role of freshwater outflow on coastal marine ecosystems - a workshop summary 429
K.F. Drinkwater

RECOMMENDATIONS

Recommendations for joint international research cooperation — 439

LIST OF PARTICIPANTS — 443

SUBJECT INDEX — 445

OPENING ADDRESS

Audun Sandberg
Nordland College
Bodø, Norway

Friends and Colleagues,
As head of this College I have a great pleasure to welcome you to Bodø, to Norway and to Nordland College. Situated on the shore of the Vestfjord with the world's largest cod-fisheries it is the right place four your discussion of the impact of freshwater outflow on the productive capacity of the oceans.

The topic you are going to address during these days is indeed a challenging one, and the agenda fully verifies the complexity of the problem - ranging from the macroecology of the world oceans to the living conditions of the shrimp. It is also challenging in a strict scientific sense; as you are never sure of what really effects what, there is a need for analytical designs that handle the problem of control convincingly. I hope this is given due attention, as there will always be explanations behind fluctuations in the fish yield, be it variations in solar activity or volcanic dust rather than man-induced changes in freshwater runoff.

I should also point out to you that man-induced tampering with fresh-waters is not an entirely modern phenomenon. One of my favourites is the ancient Sinhala king who made the following his slogan: "Let not one drop of water reach the sea without making good use of itself". This was 2500 years ago and he implemented this as his political programme with massive use of slaves and elephants.

The oriental despots rose because they controlled the fresh water flow, and they fell because of invasions of "barbarians" from the north, not because of poor fisheries. Even today, more than half of the world's human population is dependent on the controlled rice/water systems of Asia, where fish growing in reservoirs and village tanks is a major source of protein.

However, modern man is more hungry than the Asian peasant, hungry for energy and the good life. And our technological innovative capacity knows no limits. Our multi-year reservoirs can eliminate all spring floods into the North Atlantic. Our engineers can turn around our big rivers that "waste" their waters northwards into the Arctic Seas, all the Yeniseis and McKenzies, and make them run south where every drop can be made good use of.

The challenge for a workshop like this is therefore perhaps not to examine each additional bit of impact on each of the species in the marine ecosystems. The challenge is perhaps more like this: if all the human technological capacity of the human race was applied to our freshwater outflow, what would be the <u>accumulated effect</u> on vital marine ecosystems ? And - challenge number two: should we hunt for optimal sulutions where we can maintain the benefits of freshwater control and still harvest the seas ?

With this, my best wishes for your deliberations. May not all questions remain as open as these at the close of the workshop!

FRESHWATER OUTFLOW IN RELATION TO SPACE AND TIME DIMENSIONS OF COMPLEX ECOLOGICAL INTERACTIONS IN COASTAL WATERS.

Stig Skreslet
Nordland College
Bodø, Norway

ABSTRACT

In marine ecosystems physical and geochemical water properties as well as biomass may be advected by currents over large distances. In addition, several species migrate and exchange biomass between distant biocenoses. Considering also the long generation times of species in temperate regions, and how this relates to transfer of energy in marine food webs, it must be acknowledged that time and space scales in marine ecosystems are very extensive. Ecological processes in temperate shelf waters support reproduction in many important stocks, and are influenced by freshwater outflow from land. However, the role of freshwater input to coastal ecosystems is inadequately understood, which is unfortunate, considering impacts of man's regulation of river flow. To solve apparent conflicts and optimise the future use of coastal zones, new sets of adequate concepts and methods for assessing coastal marine ecosystems, are needed.

INTRODUCTION

The definition of an ecosystem is very much influenced by terrestrial concepts, and is at times very difficult to apply to ecological processes in the sea. The most central criterion of an ecosystem is that most of the energy which is utilised in the foodchains, has been fixed by the system's stocks of autotrophic organisms, i.e. plants or certain bacteria. This is applicable to terrestrial systems where the primary production is rooted in soil, and where a substantial part of the secondary production is performed by herbivorous animals like insects and rodents which have limited mobility. It may also be applicable to lake systems and bodies of landlocked seawater with very restricted flushing of water through the system. In open marine systems which exchange volumes of water, geochemicals and biomass over large distances without being perceptible to the human eye, criteria for defining ecosystems are less obvious. This is a serious obstacle for making adequate conceptual models of marine ecosystems.

FACTORS AFFECTING THE DIMENSIONS OF FOOD WEBS IN THE SEA

Most of the primary production in coastal waters is due to phytoplankton which is advected from one place to the other with currents, as the plants grow and multiply. This is also the case with herbivorous zooplankton, and their predators may be fish populations which transfer biomass over large distances by migration.

Benthic biocenoses in shallow water are often easily located and defined, but their food webs in most cases depend on biomass brought in with currents from elsewhere, maybe from quite distant waters. In temperate waters, most of the zoobenthic species have pelagic early life history stages, dispersing their gametes in water which is advected away. The larvae may live for weeks and months as plankton, before they recruit to biocenoses elsewhere. Thus, benthic biocenoses in coastal waters with pronounced advection of water, are not easily defined as ecosystems.

Most important commercial fish species have planktonic larvae and as they grow, their cruising speed and migration distance usually increases. Thus, the actors on each trophic level in a particular food chain, from primary producer to top carnivore, may all have been distributed over long and perhaps twisted trajectories. The length of each trajectory depends on the life span of the organism, and the velocity of its migration or advection by currents.

Planktonic organisms undertake vertical migrations, both daily and ontogenetic, some of them over large depth ranges. They may therefore be able to alternate between layers of water which is advected in different directions, and thus, over time manage to stay in the same geographic location. This certainly occurs in nature, and is important in relation to biological processes like reproduction and feeding. On the other hand, dispersal gives positive feedback to stock size by expanding the stock's distribution. It increases the stock's potential for exploitation of its environment, and therefore, its stability.

It should be understood that marine ecosystems are not easily defined, due to dispersion of biomass and trophic interactions over large distances in the sea, unless the model is very extensive. This should be kept in mind when dealing with ecological processes in oceans influenced by freshwater outflow.

PHYSICAL AND CHEMICAL INITIATION OF BIOLOGICAL PRODUCTIVITY

Biological production in the sea is initiated where solar energy and plant nutrients are available over time spans which allow plants to multiply. Due to gravity, most of the mineralization of marine biomass occurs in deep water. Production therefore occurs in areas where physical processes mix plant nutrients into water kept within the euphotic layer for a sufficient period of time, either by being brought over shallow bottom or over water with higher density. Both conditions apply to physical processes in estuaries, where haline circulation (Neu 1976) imports seawater from deep layers outside, in exchange for seawater entrained with freshwater outflow.

The input of geochemical matter by rivers, is to a large extent processed in proximal coastal waters (Martin and Whitfield 1983). Several of the dissolved components, like silicate, iron and humic material, are important constituents directly or indirectly involved in marine biological production. Some of the constituents are not found in seawater in concentrations sufficient to support blooming of plants, and may thus be limiting. On the other hand, unpolluted freshwater is usually poor in phosphorous, which seawater is not. The mixing of freshwater and seawater therefore provides a medium well suited to initiate and support plant growth. However important geochemicals transported by freshwater are, my concerns are directed rather exclusively towards the volume of freshwater as such, and its physical relation to ecological processes in the sea.

The effects of freshwater runoff on downstream circulation and mixing, which in the estuary of the St. Lawrence River mainly occurs over a distance of 300 km, has been infered to induce primary production in the outer parts of the estuary and in downstream areas as far downstream as 600 km. The trophic transition from phytoplankton to zooplankton biomass has been interpreted to occur even farther out in the Gulf of St. Lawrence and may be up to 80 days after the freshwater left the river (Fig. 1). This area of maximum zooplankton biomass, the Magdalen Shallows, is also the main spawning area of the southern Gulf of St. Lawrence cod stock. There is some evidence that yearclass strength in this stock is linked to runoff related processes during the autumn prior to the early life history period (Bugden, Hargrave, Sinclair, Tang, Therriault and Yeats 1982). A similar relationship is suggested to be

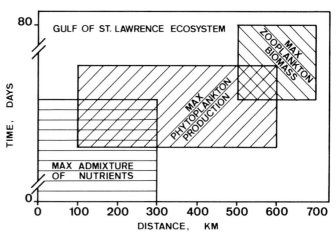

Fig. 1. Approximate dimensions of distance and time in trophic transfer of matter by plankton in the Gulf of St. Lawrence (Data after Steven 1974, 1975).

established along the Norwegian coast where, for a brief period of 13 years, large vernal outflow from southwest Norway was succeeded by the establishment of strong yearclasses of Arcto-Norwegian cod in north Norway the following year (Fig. 2). From this observation it was hypothesised that entrainment in fjords and shelf waters outside southwest Norway stimulates phytoplankton production which supports the reproduction and growth of zooplankton drifting northwards by the Norwegian coastal current to the spawning habitats of Arcto-Norwegian cod (Skreslet 1976, 1981a).

Landings from several commercial stocks in the Gulf of St. Lawrence has been shown to be positively correlated with freshwater outflow, when introducing time lags corresponding to the age at commercial size (Sutcliffe 1972, 1973). A similar relationship is demonstrated by Bakken (1961) in landings of Norwegian sprat.

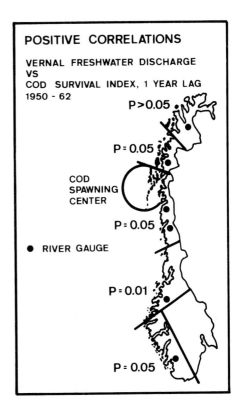

Fig. 2.
Significance levels of positive correlations between freshwater outflow from different rivers in Norway, and ln survival index (survival index = recruitment as proportion of parent stock) of yearclasses of Arcto-Norwegian cod hatched next year (Data after Skreslet 1976).

There is good evidence that production of fish in the sea is coupled to climate (Cushing 1972, 1978, 1982), and that the variations are due to large-scale climatological processes. For instance, Ottestad (1979) has demonstrated a striking relationship between growth in Norwegian pine and catches of spawning Arcto-Norwegian cod 10-11 years later (Fig 3).

Sutcliffe, Drinkwater and Muir (1977) found significant correlations between temperature and landings from several stocks in the Gulf of Maine, but found it difficult to assess the underlying causes for positive and negative correlations. In the Barents Sea, Izhevskii (1961, 1964) observed a positive correlation between Russian trawl catches and seawater temperature 5-6 years in advance. By using temperature as a yearclass index in an algorithm for stock size, he demonstrated a close fit with actual catches (Fig. 4). He concluded that there may be a causal relationship involved, driven by long-period tidal forcing, shifting thermal energy stored in Atlantic water to the North in the Norwegian Sea, during years when maxima in tidal

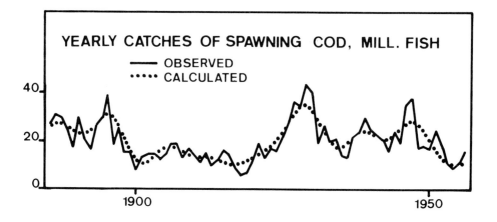

Fig. 3. Observed yearly catches of Arcto-Norwegian cod taken on the spawning grounds in Lofoten, compared with theoretical catches calculated from growth ring width in north Norwegian pine (Redrawn from Ottestad 1979).

waves of different frequency interfere. The warm years were supposed to increase the evaporation, causing heavy snowfall in Scandinavia and large vernal freshwater discharge to Norwegian coastal waters, stimulating planktonic productivity and thus, increased recruitment to fish stocks.

Holistic attempts to relate fish production to freshwater outflow, were already made early in this century by Helland-Hansen and Nansen (1909) who were later supported by Gran (1923). Their hypothesis was, however, challenged by authors who brought forward information which opposed the idea (Sund 1923, Braarud and Klem 1931). More recently, diverse observations demonstrate that there is no simple causal relationships between freshwater outflow and fish production in the sea. Direct correlations between freshwater outflow and fish recruitment may break down when long data series are applied (Rørvik 1979, Sundby 1979) and freshwater outflow may even be negatively correlated with landings from fisheries, i.e. increased outflow is correlated with decreased yearclass sizes, as demonstrated by Sutcliffe et al. (1977) and Sutcliffe, Loucks, Drinkwater and Coote (1983).

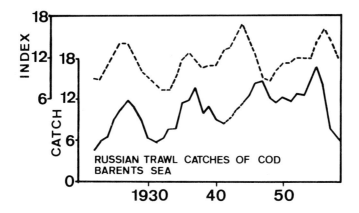

Fig. 4. Stock size indices of Arcto-Norwegian cod calculated with mean annual sea temperature in the Barents Sea as yearclass index, in relation to observed trawl catches in centners (= 100 kg) per trawling hour in the Barents Sea (Redrawn from Izhevskii 1961).

There is compelling evidence that freshwater outflow to the sea is an important factor which links climatic variations with biological productivity in coastal waters. This has caused some scientists (Neu 1976, 1982a, 1982b, Skreslet 1976, 1981b) to advocate that extensive freshwater regulation of river flow by altering the seasonal cycle in hydrological discharge, may be detrimental to biological production and thus, to fish production. Efforts have been made to review the problem (Bugden et al. op. cit., Kaartvedt 1984, Skreslet, Matthews, Leinebø and Sakshaug 1976), but the conclusions of these reviews have been stated in uncertain terms, due to the lack of understanding of the processes involved, as well as to data limitations.

These limitations in understanding have permitted conflicting interpretations, which may have hindered rather than stimulated needed research. The complexity of Nature, featuring processes which interact in ways still inadequately described or perhaps even unknown, should be acknowledged and lead us to accept that all present interpretations are open to question and critical review. The differences of opinion are indicative of the complexity of the task of understanding how marine ecosystems function. Increases in understanding how freshwater outflow influences biological production in coastal water, call for an open mind, willingness to accept conflicting inputs, and sober interpretations of one's own results.

By bringing together scientists who have struggled with this problem, as well as specialists in flanking disciplines who may add fresh insights, the hope arises that not only significant increases in understanding may be achieved, but also key issues of uncertainty may be defined.

REFERENCES

Bakken, E. 1961. Influence of hydrographical and meteorological factors on catch and recruitment strength of the sprat stock in western Norway. Fiskeridir. Skr. Ser. Havunders. 14, 61-71.

Braarud, T. and A. Klem 1931. Hydrographical and chemical investigations in the coastal waters off Møre and in the Romsdalsfjord. Hvalrådets Skr. 1, 1-88.

Bugden, G.L., B.T. Hargrave, M.M. Sinclair, C.C. Tang, J.C. Therriault and P.A. Yeats 1982. Freshwater runoff effects in the marine environment: The Gulf of St. Lawrence example. Can. Tech. Rep. Fish. Aquat. Sci. 1078, 1-89.

Cushing, D.H. 1972. The production cycle and the numbers of marine fish. Symp. Zool. Soc. London 29, 213-232.

Cushing, D.H. 1978. Biological effects of climatic change. Rapp. P. V. Reun. Cons. Int. Explor. Mer 173, 107-116.

Cushing, D.H. 1982. Climate and Fisheries. Academic Press, London.

Gran, H.H. 1923. Snesmeltingen som hovedaarsak til den rike produktion i vort kysthav om vaaren. Samtiden 34, 606-613.

Helland-Hansen, B. and F. Nansen 1909. The Norwegian Sea. Fiskeridir. Skr. Ser. Havunders. 2, (2), 1-390.

Izhevskii, G.K. 1961. Oceanological principles as related to the fishery productivity of the seas. Israel Programme for Scientific Translations, Jerusalem (1964), 186 pp.

Izhevskii, G.K. 1964. Forecasting of oceanological conditions and the reproduction of commercial fish. Israel Programme for Scientific Translations, Jerusalem (1966), 95 pp.

Kaartvedt, S. 1984. Effects on fjords of modified freshwater discharge due to hydroelectric power production. Fisken Havet 1984 (3), 1-104.

Martin, J.M. and M. Whitfield 1983. The significance of the river input of chemical elements to the ocean, pp 265-295 in Wong, Boyle, Bruland, Burton and Goldberg (eds) Trace Metals in the Sea. Plenum, London.

Neu, H.J.A. 1976. Runoff regulation for hydro-power and its effects on the ocean environment. Hydrol. Sci. Bull. 21, 433-444.

Neu, H.J.A. 1982a. Man-made storage of water resources - a liability to the ocean environment? I. Pollut. Bull. 13, 7-12.

Neu, H.J.A. 1982b. Man-made storage of water resources - a liability to the ocean environment? II. Pollut. Bull. 13, 44-47.

Ottestad, P. 1979. The sunspot series and biospheric series regarded as results due to a common cause. Meld. Norg. Landbrukshøgsk. 58 (9), 1-20.

Rørvik, C.J. 1979. Overlevingsindekser for norsk-arktisk torsk. Fisken Havet Ser. B. 1979 (7), 1-13.

Skreslet, S. 1976. Influence of freshwater outflow from Norway on recruitment to the stock of Arcto-Norwegian stock (_Gadus morhua_), pp 133-237 in Skreslet, Leinebø, matthews and Sakshaug (eds), Fresh Water on the Sea. Ass. Norw. Oceanogr., Oslo.

Skreslet, S. 1981a. Informations and opinions on how freshwater outflow to the Norwegian coastal current influences biological production and recruitment to fish stocks in adjacent seas, pp 712-748 in Sætre and Mork (eds) The Norwegian Coastal Current. University of Bergen.

Skreslet, S. 1981b. Importance of natural freshwater outflow to the coastal marine ecosystem of Norway and possible effects of large scale hydroelectric power production on year-class strength in fish stocks. Rapp. P. V. Reun. Cons. Int. Explor. Mer, 178, 79-80.

Skreslet, S., R. Leinebø, J.B.L. Matthews and E. Sakshaug (eds) 1976. Fresh water on the sea. Ass. Norw. Oceanogr, Oslo. 246 pp.

Steven, D.M. 1974. Primary and secondary production in the Gulf of St. Lawrence. Report No 26, Marine Sciences Centre, McGill University, 116 pp.

Steven, D.M. 1975. Biological production in the Gulf of St. Lawrence, pp 229-248 in T.W.M. Cameron (ed.), Energy flow: its biological dimensions. A summary of the IBP in Canada 1964-1974. Royal Society of Canada, Ottawa.

Sund, O. 1924. Snow and survival of cod fry. Nature 113, 163-164.

Sundby, S. 1979. Om sammenhengen mellom ferskvannsavrenningen og en del biologiske parametre. Fisken Havet Ser. B. 1979 (7), 15-26.

Sutcliffe, W.H.Jr. 1972. Some relations of land drainage, nutrients, particulate material, and fish catch in two eastern Canadian bays. J. Fish. Res. Board Can. 29, 357-362.

Sutcliffe, W.H.Jr. 1973. Correlations between seasonal river discharge and local landings of American lobster (Homarus americanus) and Atlantic halibut (Hippoglossus hippoglossus) in the Gulf of St. Lawrence. J. Fish. Res. Board Can. 30, 856-859.

Sutcliffe, W.H.Jr., K. Drinkwater and B.S. Muir 1977. Correlations of fish catch and environmental factors in the Gulf of Maine. J. Fish. Res. Board Can, 34, 19-30.

Sutcliffe, W.H.Jr., R.H. Loucks, K.F. Drinkwater and A.R. Coote 1983. Nutrient flux onto the Labrador shelf from Hudson strait and its biological consequences. Can. J. Fish. Aquat. Sci. 40, 1692-1701.

MIXING AND EXCHANGE PROCESSES IN ESTUARIES, FJORDS AND SHELF WATERS

Harald Svendsen
Geophysical Institute
University of Bergen
Bergen, Norway

ABSTRACT

This paper is a summary of the present state of understanding of mixing and exchange processes in estuaries, fjords and shelf waters. Existing classification methods are evaluated with particular stress put on emphasizing their limitations. Using the equation of turbulent energy transport, a brief presentation of aspects related to mixing processes and sources of turbulent energy is given. Results from model and field studies are applied to estimate the importance of the different mixing processes in estuaries, fjords and shelves, and their influence on exchange processes between the three regimes and adjacent waters. Finally, some thoughts for future work are given.

INTRODUCTION

Turbulent motion is responsible for mixing in the ocean and is among the most complicated problems in oceanography. The study of these small scale processes is subject to large difficulties, especially in coastal waters, i.e. estuaries, fjords and shelf waters, where stratification and topographic effects are strongly manifested.

In discussing mixing processes it is important to bear in mind the distinction between <u>stirring</u> and mixing processes. (Eckard, 1948). Stirring is associated with the larger eddies of the motion and always sharpens the gradients, while mixing is associated with molecular and turbulent diffusion, which act to decrease the gradients.

Horizontal pressure gradients are important in governing the exchange processes between water masses. Since mixing alters the density gradients, a necessary condition for understanding exchange processes is that the large variety of turbulent phenomena is relatively well understood. Knowledge of mixing and exchange processes is crucial when studying ecosystems, due to the close relation between these processes and chemical and biological processes.

It is not possible within the framework of this presentation to review all of the results from the numerous papers which have led to the present state of understanding of mixing and exchange processes in coastal waters. Instead the major intention of the paper is to describe theoretical and experimental results relevant to mixing and exchange processes in estuaries, fjords and shelf waters, and the application of these to explain the observational results.

For those who want a more thorough knowledge about these topics, some books and review paper may be mentioned. A physical introduction to estuaries is given in a book by Dyer (1973). Mixing processes in estuaries are also treated in a book, designed for graduate level courses, by Fisher, List, Koh, Imberger and Brooks (1979). An excellent review of small-scale mixing processes which are relevant for estuaries, fjords and shelf waters has been presented by Turner (1981). Recently, Farmer and Freeland (1983) have reviewed some of the recent theoretical and observational studies of mixing and exchange processes in fjords. Some aspects relevant to mixing in shelf waters are reviewed by Sherman, Imberger and Coros (1978).

CLASSIFICATIONS

The term estuary is used as a common designation of various semiclosed coastal bodies of water from extremely shallow (<10m) tidal mouths of large rivers to large, deep (>1000m) fjords. It is adopted in this presentation not to use the term estuary about fjords. The reasons are, as pointed out above, the great topographic differences between river-estuaries and fjords, and the fact that both physical and dynamical processes in fjords in many respects show substantial difference from the corresponding processes in shallow estuaries.

In order to compare different estuaries and fjords, many attempts of constructing a simple scheme of classification have been proposed. In the next two sections, some of the classification schemes are discussed followed by a short description of different shelves (page 8).

Estuaries

One method of classifying estuaries is geomorphological. Pritchard (1952) introduced the terms coastal plain estuaries and bar-built estuaries to distinguish between estuaries that are drowned river valleys with no restrictions to the ocean and those that have a characteristic bar across their mouths.

There are, however, large individual differences among estuaries within each of these two categories. A finer graduating is obtained by applying a scheme proposed by Pritchard (1956), Cameron and Pritchard (1963), and Bowden (1967), based on hydrodynamical considerations: highly stratified estuaries, partially mixed estuaries and well-mixed estuaries.

Highly-stratified estuary: In the highly stratified estuary also called salt-wedge estuary, the ratio of river flow to tidal flow is large. The river water flows outward above the sea water and, due to shear stress in the interfacial zone, the interface slopes upward downstream (Fig. 1a). Usually salt-water is mixed into the upper freshwater layer by entrainment processes (see page 12) and a weak landward compensating flow occurs in the salt-water wedge. The Coriolis effect causes the interface to slope downward toward the right and the sea surface downward toward the left in the northern hemisphere. (The southwest Pass of the Mississippi is an example of a salt-wedge estuary.)

Partially mixed estuary: A considerable part of the energy of tidal flow in estuaries is dissipated by breaking internal

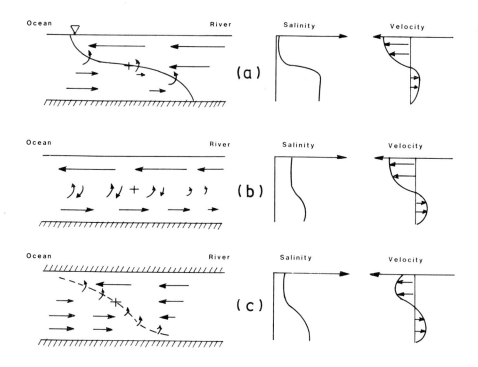

Fig. 1 (a) Highly stratified, (b) Partly stratified and (c) Well mixed estuary. The salinity and velocity distribution are assumed to be representative for the position marked +.

waves and reduced by work against the friction forces on the bottom. The turbulence produced by this process and wind stirring of the surface layer mixes salt water upward and fresh water downward in the estuary. A transition layer with a pronounced vertical salinity gradient develops between nearly homogeneous top and bottom layers (Fig. 1b). In both the top and bottom layers there is a longitudinal gradient of salinity. The salinity in both layers increases and the stratification weakens seaward. In conformity with a highly stratified estuary the lateral slopes are affected by the Coriolis force, and the outflowing and inflowing waters are deflected to opposite sides of the estuary.

In producing a partially mixed estuary it is necessary that the river flow be low compared to the tidal prism, as for instance in Southampton water and James River.

Well mixed estuary: Some estuaries are most of the time vertically homogeneous. The cross-section of such estuaries is usually small, making the tidal velocities large enough to produce sufficient turbulence to mix the entire water column. Contrary to the other estuarine types, the circulation in well mixed estuaries may also occur in the horizontal plane due to the Coriolis force which causes the seaward flow to veer to the right in the northern hemisphere. A compensating landward current occurs on the left due to horizontal mixing and exchange processes (Fig. 1c).

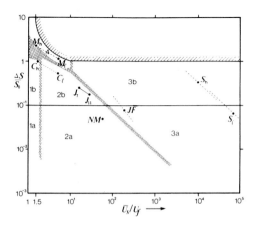

Fig. 2. Classification scheme from Hansen and Rattray (1966). Station code: (M) Mississippi River mouth, (C) Columbia river estuary, (J) James river estuary, (NM) Narrows of the Mersey estuary, (JF) Strait of Juan de Fuca, (S) Silver Bay.

A third method of classifying estuaries, by an analytical approach, was introduced by Hansen and Rattray (1966). This method is based on two dimensionless parameters: a stratification parameter $\Delta S/S_0$ and a circulation parameter U_s/U_f (Fig. 2) where ΔS is the time averaged surface to bottom salinity difference, S_0 is the cross sectional mean salinity, U_s is the time averaged surface velocity, and U_f is the mean velocity based on the freshwater discharged.

In many estuaries, estimation of the two parameters is enticed with considerable uncertainty due to large temporal, cross-sectional and longitudinal variations of the estuary.

As is well known, the velocity and salinity distributions are certainly not independent. This is indirectly assumed in the Hansen and Rattray circulation-stratification diagram. A method of classification that takes into account this relation is suggested by Rydberg (1981). He based the classification on the "bulk-quantities" of mass- and salt-transport as a function of salinity. However, observations which support his ideas have not been undertaken so far.

Fjords

All the classification schemes discussed above also include fjords. Pritchard (1952) has fjords as the third group of estuaries beside coastal plain and bar-built estuaries. In the classification scheme based on stratification, fjords are classified as highly stratified or salt-wedge estuaries. However, since most of the fjords are restricted to high latitudes in mountainous areas, large seasonal variations in both freshwater supply and air temperature occur. As a consequence, the stratification and circulation pattern varies considerably during the year. Another factor that makes it difficult to fit fjords in any of the proposed schemes is that physical and dynamical processes in fjords are governed far more by exhange processes with coastal water than previously assumed, e.g. Svendsen (1977, 1981), Cannon and Holbrook (1981), and Klinck, O'Brien and Svendsen (1981). Stigebrandt (1981) proposed a scheme based on the assumption, introduced by Stommel and Farmer (1953), of two-layer circulation and steady hydraulic control at a constriction

$$F_1^2 + \frac{\rho_1}{\rho_2} F_2^2 = 1 \text{ where } F_1 = \frac{U_1}{(g'H_1)^{\frac{1}{2}}} \text{ and } F_2 = \frac{U_2}{(g'H_2)^{\frac{1}{2}}}$$

are the Froude numbers for the upper and lower layer respectively. U_1, U_2, H_1, H_2 and ρ_1, ρ_2 are the sectionally averaged velocities, depths and densities of the upper and lower layers, respectively, g' is the reduced gravity. This assumption leads to the result that there is an upper limit for the salinity of the brackish water in the estuary depending essentially on the salinity in the sea outside of the fjord constriction, the freshwater runoff and the geometrical dimensions of the fjord constriction. Fig. 3 shows the relative depth $\eta = \frac{H_2}{H_1 + H_2}$ of the

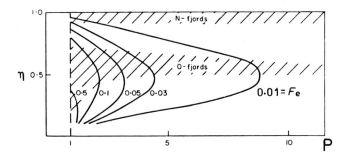

Fig. 3. Classification scheme (fjords) from Stigebrandt (1981). Relative depth of the lower layer (η) is plotted against the mixing parameter (P) for different values of the estuarine Froude number (F_e).

lower layer and the mixing parameter P, defined as the ratio between the flushes of outflowing brackish water and the river water, for different values of an estuarine Froude number F_e:

$$F_e = \frac{Q_f}{g(\frac{\Delta \rho_0}{\rho_0})^{1/2} B_m H_m^{3/2}} \quad \text{where}$$

Q_f is the discharge, $\Delta \rho_0$ is the density difference between fresh and sea water and B_m and H_m is the width and depth respectively at the constriction. Each curve in the figure has a maximum P-value (overmixed).

Although the scheme proposed by Stigebrandt is an important contribution to the understanding of the dynamics of fjords, it is doubtful whether it can be used in any practical application. The most important objections are that a two-layer circulation is a rare occasion in a fjord and that conditions in a fjord are usually highly time dependent.

A classification method which is more suitable for practical applications is a scheme which takes into account both seasonal variations of the hydrography and position in the fjord (head, middle, mouth) (Fig. 4). A geomorphological classification may also be useful, especially in studying exchange processes with adjacent water masses (Fig. 5).

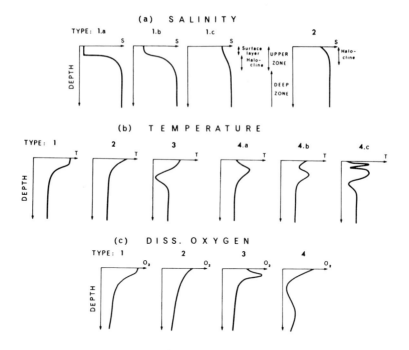

Fig. 4. Classification scheme from Pickard (1961) based on the shape of the profile.

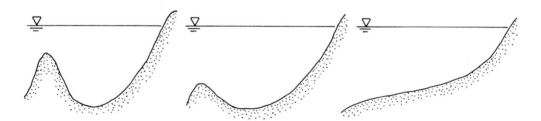

Fig. 5. Sketch of the longitudinal section of a fjord with (a) shallow sill, (b) deep sill and (c) without sill.

Shelves

Continental shelves are usually divided into two domains: the inner and the outer shelf domains (Fig. 6). The inner shelf domain extends from the beach, which is a part of the sea bottom at high water, to a transition zone where water masses lose their coastal water character. From the transition zone to the shelf break, usually at ~200 m, is the outer shelf domain.

The slope of the shelf suddenly increases at the shelf break leading over to the continental slope. The shelf may be hundreds of kilometers wide in some areas, very narrow elsewhere, but rarely absent. Examples of very wide shelves are the North Sea and the Bering Sea shelves. Many shelves have a very complex topography with alternating shallow banks and deep troughs both along and across the shelf.

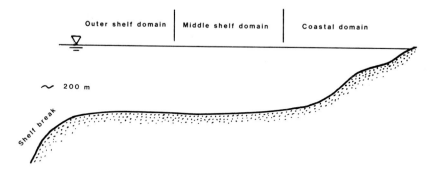

Fig. 6.   Cross section of a continental shelf.

MIXING PROCESSES

Introduction

Mixing processes could be suitable classified by distinguishing between those related to production of turbulent kinetic energy in the interior of a water volume and those related to production of turbulent kinetic energy at the boundaries. The rate of mixing usually decreases with the distance from the energy sources, which causes the mixing intensity to be highly spatially dependent. The turbulent energy level at a particular position is generally due to contributions from different sources. It is possible from the present knowledge of turbulent processes to at least qualitatively identify the different sources. Readers not familiar with this general subject are referred to the book by Turner (1973) and papers by Maxworthy and Browand (1975), Sherman et al. (1978) and Turner (1981).

Vertical mixing

In discussing aspects related to vertical mixing processes it is helpful to examine the well known equations of turbulent energy transport (e.g. Phillips, 1966):

$$\underbrace{\frac{\partial}{\partial t}\overline{c'^2}}_{(1)} + \underbrace{\frac{\partial}{\partial z}\overline{\{w'(\frac{p'}{\rho_0} + c'^2)\}}}_{(2)} = \underbrace{-\frac{\tau}{\rho_0}\frac{\partial u}{\partial z}}_{(3)} - \underbrace{\frac{g}{\rho_0}\overline{\rho'w'}}_{(4)} - \underbrace{\varepsilon}_{(5)} \quad (3.1)$$

where $\overline{c'^2} = \tfrac{1}{2}(\overline{u'^2}+\overline{v'^2}+\overline{w'^2})$ is the turbulent kinetic energy (TKE), and u',v',w',ρ' and p' are the turbulent fluctuations of the velocity components, density and pressure respectively. The over bars denote time averaging, ρ_0 is the mean density of the environment and u is the mean horizontal velocity. The terms in equation (3.1) from left to right represent: (1) temporal change of the TKE, (2) redistribution of the TKE in space by the turbulent motion itself, (3) rate of transfer of the TKE by the mean flow caused by the work of the Reynolds stress against the mean velocity gradient, (4) gain of TKE due to release of potential energy (stable stratification) from the mean density field, and (5) rate of dissipation $\varepsilon = \nu \overline{\nabla^2(u'^2+v'^2+w'^2)}$. Molecular processes enter the equation only through term 5. Term 2 is usually small compared to the other terms. Neglecting it, equation (3.1) becomes:

$$\frac{\partial}{\partial t}\overline{c'^2} = -\frac{\tau}{\rho_0}\frac{\partial u}{\partial z} - \frac{g}{\rho_0}\overline{\rho'w'} - \varepsilon \quad (3.2)$$

<u>Unstratified (well mixed) conditions</u>: The production of turbulence in unstratified embayments is primarily associated with shear flows. Since there is no work due to buoyancy (unstratified), the second term on the right side of equation (3.2) is negligible. Using the so-called "friction velocity" u_* the boundary stress may be wirtten $\tau_0 = \rho_0 u_*^2$ and the energy production term becomes $\frac{\tau_0}{\rho_0}\frac{\partial u}{\partial z} = u_*^2 \frac{\partial u}{\partial z}$. The turbulent kinetic energy in a certain distance from the boudnary is then the difference between the energy transmitted by the Reynolds stress and the energy dissipated covering the distance. τ_0 is transported from the boundary by the Reynolds stress $\tau = -\rho_0 \overline{u'w'}$. The

velocity gradient is related to the frictional velocity by $\frac{\partial u}{\partial z} = \frac{u_*}{kz}$ which by integration gives the logarithmic velocity profile in the boundary layer:

$$u = \frac{u_*}{k} \ln \frac{z}{z_0} \qquad (3.3)$$

where z = distance from boundary, z_0 = roughness length, and k = von Karman constant.

Stratified conditions: A given quantity of turbulent energy produces less transport of mass and momentum (mixing) in stratified fluid compared to unstratified fluid due to the energy consumption related to the work against buoyancy forces (Fig. 7). The relation between the work against buoyancy (gain in

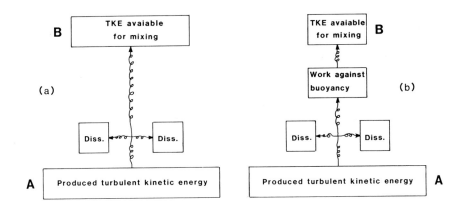

Fig. 7. Produced turbulent kinetic energy in position A and turbulent kinetic energy left for mixing in position B for (a) unstratified (well-mixed) conditions and (b) stratified conditions. (The area of the rectangles represents quantity of energy).

potential energy) and supply of turbulent kinetic energy is expressed by the flux Richardson number (R_f) that is the ratio of the first two terms on the right hand side of equation (3.2).

$$R_f = \frac{g\overline{\rho'w'}}{\tau \frac{\partial u}{\partial z}} \qquad (3.4)$$

Mixing can only take place if $R_f < 1$ (ε = positive in equation (3.2)). If an eddy viscosity coefficient K_e and a coefficient of turbulent diffusion K_d are introduced such that

$$\tau = K_e \frac{\partial u}{\partial z} \quad \text{and} \quad g\overline{\rho' w'} = -K_d \frac{g}{\rho_0} \frac{\partial \rho}{\partial z}, \quad (\rho = \text{mean density}) \qquad (3.5)$$

then

$$R_f = \frac{K_d}{K_e} \frac{-\frac{g}{\rho_0} \frac{\partial \rho}{\partial z}}{(\frac{\partial u}{\partial z})^2} = \frac{K_d}{K_e} \frac{N^2}{(\frac{\partial u}{\partial z})^2} = \frac{K_d}{K_e} R_i \qquad (3.6)$$

where R_i is called the gradient Richardson number which express the relation between the stabilizing buoyancy and the destabilizing velocity shear. N is called the Brunt-Väisälä frequency and measures the dynamic effect of the density gradient. Introducing the coefficients K_e and K_d makes it possible to express small-scale mixing processes based on the conditions at larger scales.

Flow situations where a density interface separates layers with different levels of turbulent energy are a common situation in nature. When a light surface layer flows over a quiescent denser layer, e.g. salt-wedge estuary, the interface is smooth for small velocity shear. However, when the velocity increases, small waves develop at the interface. Such waves may break, and elements of water from the lower layer are injected into the surface layer. This phenomenon is called entrainment. Pure entrainment is rare in nature. Quite often both layers are turbulent, but at different energy levels. Entrainment may then be defined as the net buoyancy flux associated with the net volume transport (e.g. Carstens, 1970).

Horizontal mixing

As seen from equation (3.1), the turbulent energy transport occurs both in the vertical and horizontal planes. In an isotropic turbulence field, the lack of buoyancy forces in the horizontal plane causes a much larger part of the produced turbulent energy to be available for horizontal, rather than vertical, mixing. Horizontal mixing may occur in both lateral directions. However, in a channel-like flow, it is common to distinguish between transverse and longitudinal mixing. Fisher (1976) found the transverse mixing coefficient in a straight rectangular flume to be much larger than the vertical mixing coefficient. Longitudinal mixing coefficients have not yet been estimated satisfactorily due to the difficulty in separating the effect of the transverse shear flow from the longitudinal turbulent fluctuations. Using a time-dependent, one-dimensional advection-diffusion equation Halloway (1981) found that the longitudinal diffusion coefficient must increase seaward and has the form $K=cu_*h$ in the upper regions of the Bay of Fundy to give a good agreement between the model and observations; c is a constant, u_* is the rms friction velocity from the tidal current averaged over the M_2 tide han h is the water depth.

Sources of turbulent energy

Turbulent kinetic energy in estuaries, fjords and coastal waters originates from many different sources. However, based on the present knowledge of mixing processes some of the principal sources can be identified. These are sources associated

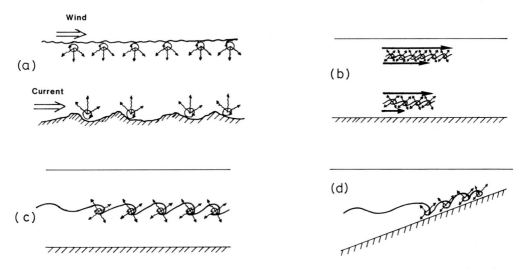

Fig. 8. Sources of turbulent kinetic energy associated with (a) external input of energy, (b) current shear, (c) breaking internal waves and (d) breaking internal waves along a sloping boundary.

with (1) external input of energy from the boundaries (2) internal waves or shear and (3) convective instability (Fig. 8). Input of external energy (1) occurs mainly at the surface and at the bottom. Wind energy supplied to the surface is transferred to turbulent energy by two main mechanisms: shear instabilities in the wind-driven current, and breaking of surface waves. The latter injects turbulent energy at a much smaller scale than the former (e.g. Phillips, 1966). Shear instabilities have been studied in two well known laboratory experiments, those by Kato and Phillips (1969) with a two-layer model, and by Kantha, Phillips and Azad (1977) with a linear density gradient. Also, in the bottom layer turbulent energy is mainly produced by two mechanisms: shear instabilities, and interaction between the flow and bottom roughnesses.

Turbulent energy transferred from internal wave motions (2), long gravity waves and lee waves often constitute a major part of the total turbulent energy level in the interior of a flow. The transfer occurs when the waves become unstable and break either in the interior of the flow or along sloping boundaries. Reviews of these topics have been published by Maxworthy and Browand (1975), and Sherman et al. (1978).

Another source of turbulent energy is related to instabilities of the Kelvin-Helmholtz type. Such instabilities appear in sharp transition regions for both density and velocity. A rule of thumb is that a parallel stratified flow becomes unstable when the minimum gradient Richardson number falls below $1/4$. The topic has been well reviewed by Maxworthy and Browand (1975).

Compared to the energetic processes mentioned earlier, mixing processes related to convective instability (3), thermal convection and double diffusion are of a more modest extent. Thermal convection takes place when the surface is cooled. The vertical extent of convective mixing depends on the strength of the surface cooling and the existing stratification. In periods with a well developed pycnocline, convective mixing is effectively inhibited, while in water masses of weak stratification and intense surface cooling convective mixing may reach great depths.

Double diffusion is an instability which results from the difference in the molecular diffusion rates of salt and heat. When the vertical gradients of both salinity and temperature have the same sign, one of them is "unstable" in the hydrostatic sense (gravitationally unstable). The molecular diffusivity for heat is much larger than that for salt, and allows potential energy to be released from the component that is top heavy (usually the temperature). See Turner (1974) and Sherman et al. (1978 for reviews of this topic.

MIXING PROCESSES IN ESTUARIES, FJORDS AND COASTAL WATERS

Estuaries

Except during periods of strong wind, mixing in estuaries is mainly caused by turbulence resulting from bottom friction and internal shear. Turbulence produced in the bottom layer dominates well mixed estuaries, while turbulence produced by internal shear dominates highly stratified estuaries. In partially mixed estuaries both can be important. As the buoyancy in weakly stratified estuaries is insignificant, the turbulent energy available for mixing is the difference between that produced at the boundaries and that dissipated (see equation (3.2)). Are the theoretical and experimental results on production of turbulence at a boundary applicable to estuaries? An indication may be obtained by comparing the observed mean velocity profiles from the bottom layer in estuaries and the logarithmic mean velocity profile expressed by equation (3.3). Based on numerous field experiments, it is found that the logarithmic mean velocity profile to a great extent is representative for estuary bottom layer. However, both the von Karman constant (k) and the roughness parameter (z_0) vary considerably. Observed z_0 values of one to two orders higher than those obtained in the laboratory experiments by Nicuradse (1933) are not a rare event. There are several sources for this variability. Topographic effects and oscillating movements related to highfrequency surface waves (Grant and Madsen, 1979) may contribute to a considerable increase of z_0.

Unfortunately, very few field studies in estuaries have procured data suitable to detailed studies of the Reynolds shear stress field. However, results from some field studies, e.g. Smith and McLean (1977), indicate that the mean Reynolds shear stress is, in conformity with the mean logarithmic velocity field, fairly well described by classical turbulent boundary-layer theory. It should be pointed out, however, that the Reynolds stress field is very sensitive, much more so than the velocity field, to topographic influence.

Except near the bed region, the mean velocity profile differs more and more from the "logarithmic" with increasing stratification. As in the case of an atmospheric boundary layer (Monin and Yaglom, 1977), Anwar (1983) found the mean velocity profile (Fig. 9) to be well described by a log-linear profile in a stratified flow in Great Ouse Estuary:

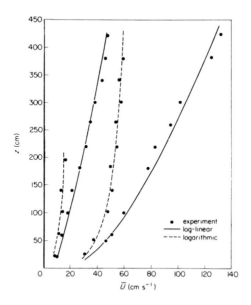

Fig. 9. Typical mean velocity profiles measured in stratified flow (solid line) and well-mixed estuarine flow (broken line). From Anwar (1983).

$$u = \frac{u_*}{k} (\ln \frac{z}{z_0} + \beta \frac{z-z_0}{L}) \quad \text{(See e.g. Webb, 1970, and Monin and Yaglom, 1977)} \quad (4.1)$$

u_*, k, z and z_0 are the same as in equation (3.3). β is a dimensionless constant and L is the well known Monin-Obukhov length (stability parameter). Near the bottom the term representing the production of potential energy (second term in the bracket) is negligible.

Two main sources of turbulent energy production dominate in highly stratified estuaries, i.e. the sources related to the bottom shear layer and the free shear layer. However, the loss of energy by work against buoyancy may cause the total

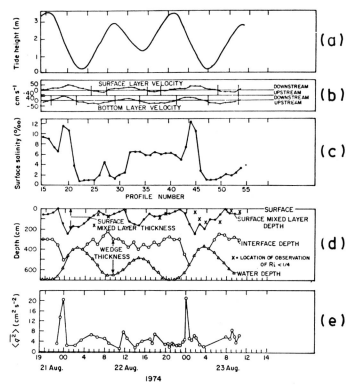

Fig. 10. (a) Tide curve for the Duwamish Estuary; (b) tidal currents in the upper and lower layers; (c) surface salinity; (d) surface mixed layer depth, interface depth and water depth with the position of observed occurrences of subcritical gradient Richardson number; (e) average turbulent kinetic energy over the frame extent. From Partch and Smith (1978).

turbulent energy available for mixing to be less than that in a well mixed estuary under equivalent external energy input.

Like the space averaged logarithmic and log-linear profiles mentioned above, most existing models of circulation and mixing-processes in estuaries are at least tidally averaged. However, although tidal averaging to a great extent describes the mean velocity distribution in estuaries, it is well known that mixing processes in estuaries are highly time dependent. The basic physical processes can therefore be hidden in the averaging.

Contrary to intuition, Partch and Smith (1978) found in a salt-wedge estuary-type (Duwamish River) that maximum turbulence was produced on the maximum ebb current but during a period of minimum mean shear (Fig. 10). As the internal Froude number was near critical at that time, the authors concluded that the maximum value was associated with breaking of long internal waves. The variations in the degree of vertical mixing have also been treated in other field studies (e.g. Ingram, 1983). Energy inputs at peirods much less than tidal periods are also common in estuaries (e.g. Dyer, 1982). (Relevant review papers include Bowden (1977), Fischer et al. (1979), Garner, Nowell and Smith (1980), and Turner (1981)).

Fjords

Increasing attention has been devoted to studies on mixing processes in fjords in recent years. An excellent review of the recent theoretical and observational studies is found in an article by Farmer and Freeland (1983).

As in shallow estuaries, the generation of turbulence in fjords arises mainly from mechanical input. Wind stress acting on the surface, fresh water supply, and tides are the main energy sources. Other sources are internal gravity currents, thermal convection and double diffusion.

Mixing due to wind action and fresh water supply are limited to the surface layer. As most fjords are situated in high latitudes, both energy sources have a marked seasonal variation. Superposed on the long-term variation are periodic and intermittent variations of much shorter time scales, e.g. sea breeze in the summer season. The fresh water has a dual effect as one of the current components of the surface-layer shear flow, and as a stabilizing effect which counteracts deepening of the surface layer. Besides mixing the wind stress either accelerates or retards the surface layer flow.

Significant mixing may take place through the front between river water (plume) entering a surrounding brackish layer. Based on a simple theory for the energetics of frontal entrainment McClimans (1979) found that the salinity of the plume depends on $\frac{W}{B}$ and $\frac{B}{H}$, where W is the width of a seaward narrows, B is the river width and H is the river thickness.

The complicated interaction between the effects of the wind, fresh water and tides makes it very difficult to study mixing processes related to these energy sources. Of the recent model studies on the thickness variation of the upper layer, most are related to the early contribution by Stommel and Farmer (1952, 1953), and the extension of their work proposed by B. Kullenberg (1955). Two such models are by Long (1975) and Stigebrandt (1981). Assumptions of steady state and hydraulic control are applied in both models. Although hydraulic control (densimetric Froude numbers = 1) is never observed in fjords, the models give some valuable insight into the upper-layer processes. For small discharge in a wide fjord, Long found the layer thickness (h) to be given by

$$h = \frac{AK\sigma^3}{b_0 Q_f}$$

while Stigebrandt's model separates the effect of the hydraulic control and wind mixing

$$h = \frac{\gamma w^3 A}{Q_f g \beta S_2} + \phi \frac{Q_f^{2/3}}{(g \beta S B^2)^{1/3}}$$

A is the surface area. B is the fjord breadth. b_0 is the buoyancy of sea water. K is the mixing efficienty parameter. σ is the rms turbulence velocity. w is the wind speed. γ, β are constants. Q_f is the discharge. ϕ is a contant. S_2 is the salinity of the lower layer at the mouth of the fjord. S is the salinity of the brackish water.

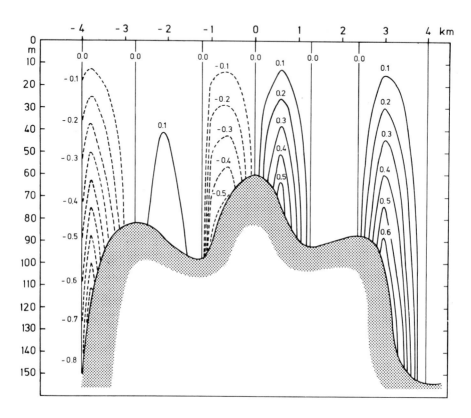

Fig. 11. Plot of the vertical force in the vicinity of the shallowest point of the fjord sill. This vertical force is a result of the surface tide and is the generating mechanism of the internal tide. Arbitrary scale. From Cushman-Roisin and Svendsen (1983).

The presence of internal tides in sill fjords is well known. In brief, the barotropic tide passing over a sill induces a vertical velocity which displaces the isopycnals in an oscillatory manner. This generates a vertical buoyancy force (Fig. 11) which causes the internal tides. In general, internal tides can be generated not only over sills but also wherever tidally driven fluid oscillates over irregular topography. The wave generation occurs mainly on the flanks of the sill (Cushman-Roisin and Svendsen, 1983), and the energy propagates along rays (Fig. 12). Under certain conditions, breaking of the internal waves may take place, and turbulent energy is released. Weather breaking shall take place along the bottom depends on the ratio of the slope of the rays to the slope of the bottom (Cacchione and Wunsch, 1974).

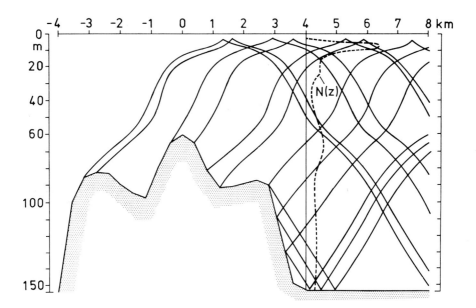

Fig. 12. Rays emanating from the sill during the case of January 17,1977 (Winter cond.). The Brunt-Väisälä frequency profile is plotted along the mooring line, on the inside of the sill (dashed line). From Cushman-Roisin and Svendsen (1983).

The rate of turbulence generation associated with sill processes depends on the sill topography and the tidal amplitude, and these vary from fjord to fjord. Over long sills, vorticity production may cause considerable mixing. Intense tidal mixing without wave generation ($F_i>1$) was observed above the sill (sill length 30 km) in Puget Sound (Gyer and Cannon, 1982).

A special manifestation of internal gravity waves and lee waves is intense mixing near obstacles. Some of the numerous theoretical and experimental studies on lee waves which may be relevant to fjords are reviewed by Farmer and Freeland (1983). Theoretical studies of two-layer flows show that a transition from super-critical to sub-critical conditions causes a hydraulic jump downstream of the obstacle. It is expected that hydraulic jumps cause intense mixing. High entrainment rates associated with hydraulic jumps are observed in laboratory experiments, and the excellent series of echo-sounding of the growth and collapse of lee-waves in Knight Inlet gives evidence of strong mixing (Farmer and Smith, 1980).

Thermal convection and double-diffusion may become important in fjords with strong cooling in winter time. In some of the Arctic fjords, vertical exchanges due to surface cooling reach the bottom (see e.g. Sælen, 1950). Winter cooling produces destabilizing vertical temperature gradients while the salinity gradients are stabilizing, and the conditions necessary for double diffusion are met.

Shelves

The main source of turbulence responsible for mixing in shelf waters is generated by wind stress at the surface and by tidal currents. Many shelf areas are characterized by large variations, both in space and time, of mixing intensity. Seasonal variations of wind mixing may cause large variations in the stratification in shelf waters. In the South Atlantic Bight continental shelf, homogeneous conditions exist in the midshelf zone in the fall and winter season due to intensified wind mixing while increased runoff and decreased wind mixing cause stratified conditions in the spring and summer seasons (Atkinson, Lee, Blanton and Chandler, 1983) (Fig. 13). Many shelf areas undergo such a yearly cycle between well-mixed, partly stratified and even strongly stratified conditions due to varying wind strength.

As in estuaries and fjords, mixing due to tidal action in shelf waters has a marked periodic character related to the tidal cycle. In shelf areas with small variations between spring and neap tides, mixing due to tidal currents show small variations between each tidal cycle. However, some areas have a very pronounced spring-neap tidal cycle. An extreme example is the south Australian Gulf where there is almost no mixing at neap tide while at spring tide the water masses are exposed to strong tidal mixing (Provis and Lennon, 1983).

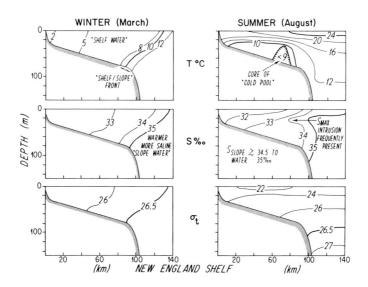

Fig. 13. Schematic cross-shelf temperature, salinity and density (sigma-t) sections for the northern Mid-Atlantic Bight for winter (left) and summer (right). From Allen et al. (1983).

Mixing processes of great significance in shallow shelf seas besides wind stirring of the top-layer and tidal mixing, are free shear layer mixing, up-welling and mixing related to lateral excursion of frontal edges (Garvine and Monk, 1974 ; Mork, 1981) and to internal tides generated at the continental shelf break (Rattray, 1960).

EXCHANGE PROCESSES

While mixing processes are central to vertical exchange within estuaries and fjords, residual flows are crucial for exchange processes between estuaries, fjords and adjacent waters. In shelf waters both mixing processes (across fronts or by eddies) and residual flows can be important.

Estuary-shelf waters

The "residual" flow responsible for the exchange of water between estuaries and adjacent shelves consits of several components. The classical semiclosed (open seaward) estuarine circulation with seaward flow in the surface layer and a landward near bottom flow is well known. This current component is driven by the fresh water runoff. Other important components are the "residual" flows induced by density gradients, local wind forcing and Stoke's drift related to tidal motions and possibly other oscillating current components.

Gravitational flow was earlier assumed to be the dominant component and was subject to great attention (e.g. Pritchard, 1956). Since the mid-seventies, the importance of circulation induced by local wind forcing (Weisberg, 1976; Elliot, 1978; Wang, 1979a) and Stoke's drift (Uncles and Jordan, 1979; Dyke, 1980) has received increasing attention. Results from an investigation in Severn Estuary (Uncles and Jordan, 1979) show the lower-layer landward Stoke's drift to be an important part of the residual transport of water generating a compensating seaward Eulerian residual transport near the surface. The residual transport is proportional to the tidal range.

Clear evidence of how strongly the locally induced circulation in an estuary may affect the shelf circulation was obtained by Norcross and Stanley (1967). Using bottom drifters they observed that drifters released 70 km seaward of the mouth of the Chesapeake Bay were consistently drawn toward the mouth of the Bay. Similar results were found by Pape and Garvine (1982) in an investigation of the subtidal circulation in Delaware Bay and adjacent shelf water.

The exchange of water induced by local forcing may at times be largely overshadowed by water exchange due to non-local forcing. Elliot (1978) found a significant portion of the sea-level and current fluctuations in the Potomac estuary to be induced by non-local effects. Wang (1979b) and Wang and Elliot (1978) showed subtidal fluctuations (>10 days) in Chesapeake Bay to originate from coastal sea-level changes generated by alongshore winds through the Ekman flux.

Fjord-shelf

In fjords the semiclosed circulation, characteristic of shallow estuaries and the related exchange processes with adjacent waters, are limited for most of the year to the upper layer. However, the distance seaward from the fjord mouth to where recirculation is felt is much less than that found in shelf areas adjacent to shallow estuaries. In deep silled fjords where an intermediate layer between the upper layer and the sill is present, it has recently been apparent that processes in adjacent shelf areas may have a major impact. Based on the results from field studies of fjord-shelf interaction at the west coast of Norway (Svendsen, 1977 and 1981) and at the northwest coast of the U.S.A. (e.g. Holbrook, Muench and Cannon, 1980), a numerical two-layer model study was made by Klinck, O'Brien and Svendsen (1981). Both the field and the model studies show that the geostrophic alongshore current of the wind-forced coastal circulation have a strong effect on the circulation within the fjord. The geostrophic currents control the free surface and pycnocline displacement at the fjord mouth, thereby strongly affecting fjord circulation. This mechanism is an alternative to the classical idea of hydraulic control at the mouth proposed by Stommel and Farmer (1952). The geostrophic control at the mouth of the fjord on the low-frequency response in the fjord is in agreement with the results from an analytic, linearized, continuously stratified model by Proehl and Rattray (1984).

Replacement of deep water in a fjord takes place when the density of the water above and seaward of the sill exceeds the density of the water in the fjord (water masses below the sill-level). Such events are related to upwelling caused by wind-driven surface Ekman transport in adjacent water and internal waves along the coast passing the fjord mouth. An excellent review of this topic is given by Gade and Edwards (1980).

Shelf-ocean

The exchange of water on a continental shelf takes place both in alongshore and across-shelf directions. Wind- and density-driven circulations appear to be responsible for most of the water exchange on continental shelves. The density-driven circulation arises from influx of fresh water from the coast and thus has for most shelves a pronounced seasonal character. The wind may have a seasonally prevailing direction that causes a seasonal variation in the transverse circulation related to up- or down-welling events. Up- and down-welling events may constitute a significant part of the exchange of shelf water with adjacent ocean water on narrow shelves. Transient wind events of a time scale of a few days have a major influence on the variations of the shelf circulations (see e.g. Csanady, 1982; Allen, 1980; Winaut, 1980).

Besides wind and density gradients, the deep ocean circulation boarding the shelf water at the shelf break is a possible driving force for the circulation of the shelf. During the winter months, on some continental shelf breaks (e.g. the northeast shelf of North America), a sharp transition zone/front is formed separating saline, warm ocean water from cold, less saline shelf water (Fig. 13). On other shelves (e.g. in northwest Europe), such a pronounced front does not appear to occur possibly due to the fact that this shelf is considerably wider than its counterpart along northeast North America and the fronts may tend to form on the shelf rather than at the shelf break (for other possible explanations, see Bowden, 1981).

Fronts on shallow continental shelves arise from spatial differences in the effect of vertical mixing due to tidal friction and wind stress (Csanady, 1978; Simpson, Allen and Morris, 1978). Simpson and Hunter (1974) suggest that the marked frontal structure occuring on the shallow European continental shelf during the summer months is produced by variations in the level of tidal mixing. They showed that the locations of the fronts is essentially determined by the parameter h/u^3, where h is the water depth and u the amplitude of the tidal stream. Simpson et al. (1978) conclude that although windstress does make a significant contribution to the total mixing it is of lesser importance in determining the locations of fronts because it is more spatially uniform than tidal stirring. The shape of the front and cross-frontal mixing is affected by variable wind stress (Csanady, 1978, 1984; Hsueh and Cushman-Roisin, 1983). Frontal waves may be triggered by variable wind stress and governed by baroclinity and bottom topography (Mork, 1981). An important mechanism of lateral exchange between shelf and adjacent ocean water is frontal wave instability and eddy shedding. This mechanism is probably the origin of pockets of coastal water within adjacent ocean water (Mork, 1981).

FUTURE WORK

It is well documented that most of the processes in coastal water are highly timedependent. However, since many of these processes still are poorly understood and difficult to model, most existing models are based on averaging. Important nonlinear processes are thus left out. Emphasis should therefore be placed on developing methods of studying the scale and intensity of these nonlinear turbulent motions and the mechanism by which the turbulent energy is produced and distributed.

Since the hydrographic conditions at any point in coastal waters are mainly results of both mixing and advective processes, study of mixing should be combined with simultaneous studies of advective processes.

The interaction of both estuaries and fjords with adjacent waters (shelfwaters) is complex and still poorly understood. Recent investigations have shown that the interaction is far more important than previously assumed and merits much greater attention in future studies.

On shelves the dynamics of the coupling between the outer shelf and offshore currents needs additional study. The same applies to alongshore variations of low frequency mean currents as well as the effects of density fronts and topographic irregularities on cross-shelf transport.

Modelling of coastal water processes will undobtedly receive increasing attention. New information of the kind outlined above will be of great help in developing improved numerical models with better representation of frictional and diffusive processes.

REFERENCES

Allen, J.S., 1980: Models of wind-driven currents on the continental shelf, Ann. Rev. Fluid Mech., 12, 389-433.

Allen, J.S., R.C. Beardsley, J.O. Blanton, W.C. Boicourt, B. Butman, L.K. Coachman, A. Huyer, T.H. Kinder, T.C. Royer, J.D. Schumacher, R.L. Smith, W. Sturges and C.D. Winant, 1983: Physical oceanography of continental shelves. Rev. Geophys. Space Phys., 21, 1149-1181.

Anwar, H.O., 1983: Turbulent measurements in stratified and well-mixed estuarine flows. Estuarine, Coastal and Shelf Science, 17, 243-260.

Atkinson, L., T.N. Lee, J. Blanton and W.S. Chandler, 1983: Climatology of the southeastern U.S. continental shelf waters. J. Geophys. Res., 88, 4705-4718.

Bowden, K.F., 1967: Stability effects on turbulent mixing in tidal currents. In Boundary Layers and Turbulence. The Physics of Fluids Supplement, 278-280.

Bowden, K.F., 1977: Mixing processes in estuaries. In Estuarine Transport Processes, ed. B. Kjerfve, Univ. of South Carolina Press, 11-36.

Bowden, K.F., 1981: Summing-up. Phil. Trans. R. Soc. Land. A 302, 683-689.

Cacchione, D. and C. Wunsch, 1974: Experimental study of internal waves over a slope. J. Fluid Mech., 66(2), 223-239.

Cameron, W.M. and D.W. Pritchard, 1963: Estuaries, 306-324. In The Sea, Vol. 2, ed. M.N. Hill. Interscience, New York.

Cannon, G.A. and J.R. Holbrook, 1981: Wind-induced seasonal interaction between coastal and fjord circulation. In Norwegian Coastal Current, 131-151, ed. R. Sætre and M. Mork, Univ. of Bergen.

Carstens, T., 1970: Turbulent diffusion and entrainment in two-layer flow. Proc. Amer. Soc. Civil Eng., WWI, 97-104.

Csanady, G.T., 1978: Wind effects on surface to bottom fronts. J. Geophys. Res., 83, 4633-4640.

Csanady, G.T., 1982: Circulation in the coastal ocean. D. Reidel Publ. Comp. Dordrecht, Holland, 279 pp.

Csanady, G.T., 1984: The influence of wind stress and river run-off on a shelf-sea front. J. Phys. Oceanogr., 14, 1383-1392.

Cushman-Roisin, B. and H. Svendsen, 1983: Internal gravity waves in sill fjords: vertical modes, ray theory and comparison with observations. In Coastal Oceanography, ed. H.G. Gade, A. Edwards and H. Svendsen.

Dyer, K.R., 1973: Estuaries: A physical Introduction. J. Wiley, Chichester, Sussex.

Dyer, K.R., 1982: Mixing caused by lateral internal seiching within a partially mixed estuary. Est., Coastal and Shelf Sci., 15.

Dyke, P.P.G., 1980: On the Stokes drift induced by tidal motions in a wide estuary. Estuarine and Coastal Marine Science, 11, 17-25.

Eckard, C., 1948: An analysis of the stirring and mixing processes in incompressible fluids. J. Mar. Res., 7, 265-275.

Elliot, A.J., 1978: Observations of the meteorologically induced circulation in the Potomac estuary. Estuarine and Coastal Marine Science, 6, 285-299.

Farmer, D.F. and H.J. Freeland, 1983: The physical oceanography of fjords. Progress in Oceanography, ed. M.V. Angel and J.J. O'Brien.

Farmer, D.E. and J.D. Smith, 1980: Tidal interaction of stratified flow with a sill in Knight Inlet. Deep-Sea Res. 27A, 239-254.

Fischer, H.B., 1976: Mixing and dispersion in estuaries. Ann. Rev. Fluid Mech. 8, 107-133.

Fischer, H.B., E.J. List, R.C.Y. Koh, J. Imberger and N.H. Brooks, 1979: Mixing in Inland and Coastal Waters. Acad.Press, New York.

Gade, H.G. and A. Edwards, 1980: Deep-water renewal in Fjords. In Fjord Oceanography. Ed. H.J. Freeland, D.M. Farmer and C.D. Levings. Plenum Press, New York, 453-489.

Cardner, G.B., A.R.M. Nowell and J.D. Smith, 1980: Turbulent processes in estuaries. In Estuarine and Wetland Processes, eds. P. Hamilton and K.B. MacDonald. Plenum Press, New York, 1-34.

Garvine, R.W. and J.D. Monk, 1974: Frontal structure of a river plume. J. Geophys. Res., 79, 2251-2259.

Grant, W.D. and O.S. Madsen, 1979: Combined wave and current interaction with a rough bottom. J. Geophys. Res., 84, 1797-1808.

Gyer, W.R. and G.A. Cannon, 1982: Sill processes related to deep water renewal in a fjord. J. Geophys. Res., 87, 7985-7996.

Halloway, P.E., 1981: Longitudinal mixing in the upper reaches of the Bay of Fundy. Estuarine, Coastal and Shelf Science, 13, 495-515.

Hansen, D.V. and M. Rattray, Jr., 1966: New dimensions in estuary classification. Limnology and Oceanography, 11, 319-326.

Holbrook, J.R., R.D. Muench and G.A. Cannon, 1980: Seasonal observations of low frequency atmospheric forcing in the Strait of Juan de Fuca. In Fjord Oceanography, eds. H.J. Freeland, D.M. Farmer and C.D. Levings. Plenum Press, New York, 305-317.

Hsueh, Y. and B. Cushman-Roisin, 1983: On the formation of surface to bottom fronts over steep topography. J. Geophys. Res., 88, 743-750.

Ingram, R.G., 1983: Salt entrainment and mixing processes in an under-ice river plume. In Coastal Oceanography, eds. H.G. Gade, A. Edwards and H. Svendsen. Plenum press, New York, 551-564.

Kantha, L.H., O.M. Phillips and R.S. Azad, 1977: On turbulent entrainment at a stable density interface. J. Fluid Mech., 79, 753-768.

Kato, H. and O.M. Phillips, 1969: On the penetration of a turbulent layer into a stratified fluid. J. Fluid Mech., 37, 643-655.

Klinck, J.M., J.J. O'Brien and H. Svendsen, 1981: A simple model of fjord and coastal circulation interaction. J. Phys. Oceanogr., 11, 1612-1626.

Kullenberg, B., 1955: Restriction of the underflow in a transition. Tellus, 7, 215-217.

Long, R.R., 1975: Circulation and density distribution in a deep, strongly stratified, two-layer estuary. J. Fluid Mech., 71, 529-540.

Maxworthy, T. and F.K. Browand, 1975: Experiments in rotating and stratified flows: oceanographic applications. Ann. Rev. Fluid Mech., 7, 273-305.

McClimans, T.A., 1979: On the energetics of river plume entrainment. Geophys. Astrophys. Fluid Dynamics, 13(1), 67-82.

Monin, A.S. and A.M. Yaglom, 1977: Statistical fluid mechanics: Mechanics of Turbulence, 3rd ed. MIT Press, Cambridge MA.

Mork, M., 1981: Circulation penomena and frontal dynamics of the Norwegian Coastal Current. Phil. Trans. Soc. Land., 302, 635-647. Printed in Great Britain.

Nicuradse, J., 1933: Laws of flow in rough pipes. N.A.C.A.Tech. Memo. 1292, 62.

Norcross, J.J. and E.M. Stanley, 1967: Inferred surface and bottom drift, June 1963 through October 1964. Circulation of shelf waters off the Chesapeake Bight. Prof. Pap. Environ. Sci. Serv. Admin., 3(2), 11-42.

Pape, E.H. and R.W. Garvine, 1982: The Subtidal Circulation in Delaware Bay and adjacent shelf waters. J. Geophys. Res., 87, 7955-7970.

Partch, E.N. and J.D. Smith, 1978: Time dependent mixing in a salt wedge estuary. Estuarine and Coastal Marine Science, 6, 3-19.

Phillips, O.M., 1966: The dynamics of the upper ocean, 261 pp, Cambridge Univ. Press, New York.

Pickard, G.L., 1961: Oceanographic features of inlets in the British Columbia mainland coast. J. Fish. Res. Bd. Can., 1816, 907-999.

Pritchard, D.W., 1952: Estuarine hydrography. Adv. Geophys., 1, 243-280.

Pritchard, D.W., 1956: The dynamic structure of a coastal plain estuary. J. Mar. Res., 15, 33-42.

Proehl, J.A. and M. Rattray, Jr., 1984: Low-frequency response of wide deep estuaries to non-local atmospheric forcing. J. Phys. Oceanogr., 14, 904-921.

Provis, D.G. and G.W. Lennon, 1983: Eddy viscosity and tidal cycles in a shallow sea. Estuarine, Coastal and Shelf Sci., 16, 351-361.

Rattray, M., Jr., 1960: On the coastal generation of internal tides, Tellus, 12, 54-62.

Rydberg, L., 1981: A proposal for classification of estuaries. In The Norwegian Coastal Current, eds. R. Sætre and M. Mork. Univ. of Bergen, Norway. 215-228 pp.

Sherman, F.S., J. Imberger and G.M. Coros, 1978: Turbulence and mixing in stably stratified waters. Ann. Rev. Fluid Mech., 10, 267-288.

Simpson, J.H. and J.R. Hunter, 1974: Fronts in the Irish Sea. Nature, 250, 404-406.

Simpson, J.H., C.M. Allen and N.C.G. Morris, 1978: Fronts on the continental shelf. J. Geophys. Res., 83, 4607-4614.

Smith, J.D. and S.R. McLean, 1977: Spatially average flow over a wavy boundary. J. Geophys. Res, 82, 1735-1746.

Stigebrandt, A., 1981: A mechanism governing the estuarine circulation in deep, strongly stratified fjords. Estuarine, Coastal and Shelf Science, 13, 197-211.

Stommel, H. and H.G. Farmer, 1952: Abrupt change in width in a two layer open channel flow. J. Mar. Res., 11, 205-214.

Stommel, H. and H.G. Farmer, 1953: Control of salinity in an estuary by a transition. J. Mar.Res., 12, 13-20.

Svendsen, H., 1977: A study of the circulation in a sill fjord on the west coast of Norway. Mar.Sci.Comm., 3, 151-209.

Svendsen, H., 1981: A study of circulation and exchange processes in the Ryfylkefjords. Report No. 55, Geophys. Inst., Univ. of Bergen.

Sælen, O.H., 1950: The hydrography of some fjords in northern Norway. Tromsø Museums Årshefter, Nat.Hist.Avd. No. 38, Vol. 70.

Turner, J.S., 1973: Buoyancy effects in fluids. Cambridge Univ. Press. 367 pp.

Turner, J.S., 1974: Double-diffusive phenomena. Ann Rev. Fluid Mech., 6, 37-56.

Turner, J.S., 1981: Small scale mixing processes. In Evolution of Physical Oceanography, eds. B. Warren and C. Wunsch. MIT Press, Cambridge MA, 236-262.

Uncles, R.J. and M.B. Jordan, 1979: Residual fluxes of water and salt at two stations in the Severn estuary. Estuarine and Coastal Marine Science, 9, 287-302.

Wang, D.P., 1979a: Subtidal sea level variations in the Chesapeake Bay and relations to atmospehric forcing. J. Phys. Oceanogr., 9, 413.421.

Wand, D.P., 1979b: Wind driven circulation in the Chesapeake Bay, Winter 1975. J. Phys. Oceanogr., 9, 564-572.

Wang, D.P. and A.J. Elliot, 1978: Nontidal variability in the Chesapeake Bay and Potomac River: Evidence for nonlocal forcing. J. Phys. Oceanogr., 8, 225-232.

Webb, E.K., 1970: Profile relationships: the log-linear range and extension to strong stability. Quarterly Journal of the Royal Meteorological Society, 96, 67-90.

Weisberg, R.H., 1976: The nontidal flow in the Providence River of Narragansett Bay: A stochastic approach to estuarine circulation. J. Phys. Oceanogr., 6, 721-734.

Winant, C.D., 1980: Coastal circulation and wind-induced currents. Ann. Rev. Fluid Mech., 12, 271-301.

THE ROLE OF BRACKISH PLUMES IN OPEN SHELF WATERS

Richard W. Garvine
College of Marine Studies
University of Delaware
Newark, DE   19716   USA

ABSTRACT

The coupled circulation between estuaries and the adjacent continental shelf has many important impacts on coastal ecology. An especially important component of this coupled circulation results from continuation onto the shelf of the estuarine gravitational circulation. Surface outflow of lighter water from an estuary will usually produce brackish plumes on the shelf, while bottom inflow of heavier water can be drawn from a significant portion of the shelf. Observations of the near field of brackish plumes are reviewed first. These plumes divide naturally into two groups, river plumes which contain nearly fresh water and estuary plumes containing a mixture of fresh and coastal water. Models of brackish plume dynamics are then reviewed. Finally, observations and a model of the landward flowing bottom water are summarized. A clear need exists for more detailed observations and more sophisticated models of both brackish plumes and bottom return flow on continental shelves.

1. Introduction

Research into the physical dynamics of estuaries and continental shelves has a history of many decades, but only within the past decade or so has it become clear that the interaction or coupling between these two regimes is both complicated and crucial for the ecologies of both. Simple continuation of estuarine gravitational circulation onto the adjacent shelf implies the existence of brackish water plumes there with their own particular dynamics. I review observations of these in Section 2 and modeling aspects in Section 3. Of no less likely ecological importance, however, is the shelf bottom water landward flow required to supply the well known landward bottom drift in an estuary. I discuss this relatively unknown phenomenon in Section 4.

2. Brackish plumes in shelf waters

It is useful to divide the spatial characteristics of brackish plumes in shelf waters into two fields, a near field where the salinity contrast with shelf water is both readily measurable and clearly connected to its riverine or estuarine source, and a far field where substantial mixing with shelf water has created a diffuse pattern unassociated with any particular coastal source. In this paper I will concentrate attention on the near field because of its more immediate impact on both coastal circulation and ecology. As Csanady (1984) has shown with a diagnostic, analytic model, the circulation induced in the far field by a coastal source of brackish water is weak compared to typical mean shelf currents. This is true especially where the mean flow is alongshelf in the direction of Kelvin wave propagation and where the brackish water is cooler as well as fresher than outer shelf or slope water, such that the density contrast is reduced. These conditions obtain, for example, in the Middle Atlantic Bight of the U.S. east coast. Nevertheless, the full ecological impact of brackish water in the far field is likely to be important, despite the absence of strong induced changes in circulation.

In the remainder of this section I will review the known physical properties of brackish plumes in two categories: river plumes and estuary plumes. By the term river plume I refer to a brackish plume in shelf waters within which nearly fresh water is present so that the plume itself is the region where the primary mixing between upland drainage and coastal sea water occurs. In contrast, by the term estuary plume I refer to a brackish plume where nearly fresh water is absent because the primary mixing region lies upstream within the estuary itself. Examples of river plumes are those of the Mississippi River at each of its three passes in its delta, the Amazon River, the Fraser River, and the Connecticut River. Examples of estuary plumes include those of the Columbia River, the St. Lawrence River, and Chesapeake Bay.

Hansen and Rattray (1966) defined a parameter P which has proved useful both in classifying estuaries as well as in providing an indicator of whether or not a river plume, as defined here, will be present. P is defined as the ratio of the mean downstream velocity of freshwater in a river or estuary to the root-mean-square tidal current there. A method for estimating P using river discharge and tidal prism data is given in Garvine (1974). Clearly, the larger P, the less mixing will occur before the brackish water reaches the shelf, but how large must P be to correspond to a river plume? Figure 1 shows data from the Connecticut

River that provide a clear answer for at least that estuary. There $x_{0.1}$, the distance from the mouth (positive upstream) where the surface salinity reaches 0.1 times that of coastal seawater, is plotted vs. P for both low (ebb) and high (flood) slack water. At $P \approx 3/4$ the curve for low slack water crosses the P axis, so that then and for higher P values at least 90% of the mixing between fresh water and coastal seawater occurs seaward of the mouth; that is, a river plume is present. The corresponding value for high slack water is $P \approx 2$. Thus, one may postulate the general rule that for $P \geq 1$ a river plume will be present while small values of P should correspond to an estuarine plume, that is, the outflow of previously mixed water. This rule seems reasonable on physical grounds, since for $P=1$ the mean freshwater velocity and root-mean-square tidal current are just equal.

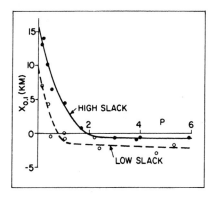

Figure 1. Distance from the mouth $x_{0.1}$ (positive upstream) where surface salinity reaches 0.1 times the coastal sea water value vs. P, the Hansen and Rattray (1966) parameter for the Connecticut River estuary.

A second parameter useful in characterizing the near fields of brackish plumes is the Kelvin number,

$$K = w/r_i$$

where w is the width of the river or estuary mouth and r_i is the internal Rossby radius,

$$r_i = (g'D_m)^{1/2}/f$$

with g' the reduced gravity of the brackish layer, D_m its mean depth at the mouth, and f the local Coriolis parameter. K is a measure of the degree to which earth rotation or Coriolis effects shape the plume dynamics. For $K \ll 1$ the circulation directly associated with the plume can hardly be distinguished from a nonrotating system (f=0), for $K \sim 1$ Coriolis effects are important, while for $K \gg 1$ they are dominant and so likely to produce flows very near geostrophic balance. As examples, $K \approx 0.3$ for the Connecticut River plume, indicating that Coriolis effects should be modest, while $K \approx 3.0$ for the Chesapeake plume, indicating that Coriolis effects should be large.

In the remainder of this section I will focus successively on the physical characteristics of river plumes and estuary plumes.

2a. River plumes

For illustrative purposes I will review properties of the Connecticut River plume. In general, it is similar to other river plumes as described by Wright and Coleman (1971) for the Mississippi, Ingram (1981) for the Great Whale, and Grancini and Cescon (1973) for the Po. The Connecticut River is one of the four largest in mean fresh water discharge, 560 m³/s, draining into the Middle Atlantic Bight of the U.S. east coast. Its mouth lies at the eastern end of Long Island Sound a short distance from the continental shelf south of New England. Its mean value of P is about unity, so that it usually forms a river plume, in the sense defined above, and its Kelvin number K is about 0.3, so that Coriolis effects are weak, but not negligible. Tidal currents in eastern Long Island Sound are of semi-diurnal period, have typical speeds of 1 m/s, and are directed alongshore roughly normal to the direction of the lower river valley. The resulting plume is thus deflected eastward with the ebbing tidal currents and then westward with the flood tidal currents roughly twice each day. Figure 2 shows nearly synoptic maps of the salinity distribution at 0.5 m depth at low (ebb) slack and the succeeding high (flood) slack water during a period of high discharge (1130 m³/s or P=5.9). The salinity varies from nearly zero to about 25‰, the level of the coastal sea water at that season. The shape of the plume is not narrow and jet-like, but rather broad and suggestive of the lateral spreading of buoyant, brackish water over heavier coastal sea water. For both tidal stages the off shore boundary toward which the tidal currents had been impinging shows a frontal character with very high horizontal gradients. This boundary was especially well resolved during the high slack observations. Figure 3 shows an aerial photograph of the plume boundary at high slack water. Plume water lies to the right and is lighter in tone. A line of foam clearly marks the offshore frontal boundary. The sharpness of this front and smoothness of the plume shape contrast with the diffuse nature of buoyant plumes rising vertically in the gravity field, such as those form coal-fired power plants. Note that Figure 2 indicates that frontal boundaries are present only on the offshore, not the onshore, side of the plumes. In this sense ebb and flood plumes are symmetric. Nevertheless, the bulk of the flood plume lies nearer shore than that of the ebb plume, a feature consistent with the action of Coriolis force which tends to deflect fluid to the right of its motion in the northern hemisphere. The modest degree of this asymmetry is consistent with the modest size of the plume Kelvin number, K=0.3.

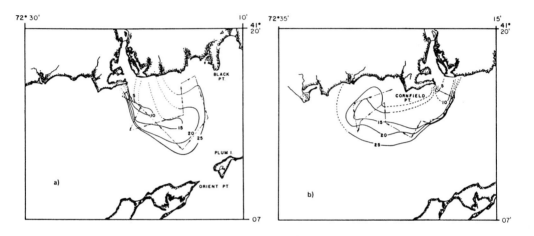

Figure 2. Isohalines at 0.5 m depth for the Connecticut River plume on April 21, 1972 at a) low (ebb) slack water and the succeeding high (flood) slack water. Line with arrows shows the ship track. From Garvine (1974).

Figure 3. Aerial photograph showing the frontal boundary of the Connecticut River plume, April 26, 1972. River water is on the right of the foam line.

Figures 4 and 5 show the distribution of plume properties in vertical plane sections across the Connecticut plume and along its centerline, respectively, at low slack water. The brackish water is confined to the upper 2 m by the very high static stability (Figure 6). Not surprisingly, most of the reduction in plume density (σ_t) results from the reduced salinity of the discharge. The plume terminates abruptly offshore at the frontal boundary. (Isolines have been drawn in the neighborhood of the front with slopes consistent with those found by Garvine and Monk (1974) in detailed observations there; see Figure 8.) Note the tendency of the plume to deepen toward offshore, a consequence of the basic plume dynamics, as discussed in Section 3. In contrast the plume thins along its centerline away from the mouth (Figure 5), a result of plume lateral spreading. Beyond about 5 nautical miles (9 km) from the mouth the plume structure weakens and deepens, indicating local mixing with the underlying salt water. Figure 6 shows the distributions along the same axis of N, the Brunt-Vaisala frequency at the pycnocline beneath the plume, and c, the long internal wave phase speed $(g'D)^{1/2}$, where g' is the local reduced gravity of the plume layer and D the pycnocline depth. The stability is very high with $N \approx 0.4 s^{-1}$, consistent with the large extent over which plume spreading is possible with very little attendant mixing across the pycnocline. The phase speed c declines almost monotonically from about 60 cm/s near the mouth to about 20 cm/s. Because the plume current speed q generally exceeds c, as I discuss in Section 3, large currents may be anticipated. This feature, high current speeds generated by the gravitational spreading of the buoyant layer, is a hallmark of river and estuarine plumes generally.

Figure 7 shows the current field associated with the Connecticut plume at one half hour before low slack water as determined by tracking a large array of drifters and drogues. The subsurface (4 m depth) velocity field, shown by the dashed lines on a 2 km grid, simply reflects the barotropic ebb tidal current field of eastern Long Island Sound. The current is nearly slack inshore but increases offshore to about 75 cm/s toward east, since the tidal current changes later in the deeper water offshore. The surface velocity field (solid lines) is nearly the same as that at depth for that region offshore of the front where coastal seawater is present, but inshore of the front the surface flow is dramatically different, moving toward the front which itself is moving offshore, in concert with the spreading action of the buoyant water pool. The speed q

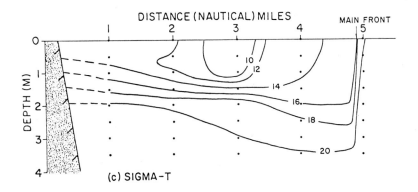

Figure 4. Density (σ_t) transect across the Connecticut River plume for April 13, 1973. From Garvine (1974).

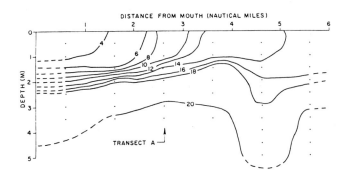

Figure 5. Isopycnals (σ_t) along the axis for the Connecticut River plume. Transect A marks the section shown in Figure 4.

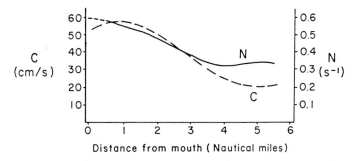

Figure 6. Brunt-Vaisala frequency N at the pycnocline base and internal wave phase speed c for the centerline section of Figure 5.

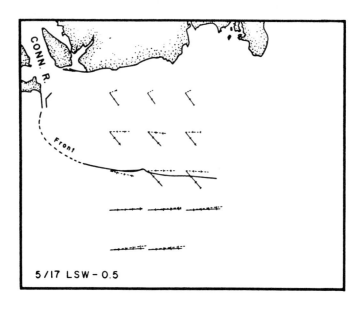

Figure 7. Eulerian velocity field interpolated on a 2 km grid from drogue and drifter velocity observations by Garvine (1977) at one-half hour before low (ebb) slack water on May 17, 1975. Solid vectors denote surface current and dashed vectors 4 m depth current. Tick marks indicate 25 cm/s intervals.

there is about 50 cm/s, while c is about 20 cm/s, so that the Froude number $F=q/c$ well exceeds unity; that is, the flow is supercritical, faster than an internal gravity wave. Sequential aerial photographs showed that the frontal propagation speed is less than q, so that as the surface plume water reaches the front it has to sink, forming a local surface convergence circulation. Indeed, such frontal convergence readily explains the frontal trapping of foam shown in Figure 3 and other floating material the these fronts.

Detailed observations near the frontal boundary of the Connecticut plume (Garvine and Monk, 1974) revealed an intense circulation and horizontal gradient field there. Figure 8 shows a vertical section of σ_t normal to the front. The outcropping isopycnals for $14 \leq \sigma_t \leq 17$ reach an asymptotic depth beneath the plume of about 1 m in a horizontal distance of about 50 m. Thus, the frontal zone scale (50 m) is about 10^2 smaller than the plume scale (10 km, Figure 4). Not all the outcropping isopycnals descend monotonically from the surface. Some show a distinct tongue shape where plume water evidently is entrained downward into the coastal water. This pattern is consistent with that observed in a laboratory study of density currents by Britter and Simpson (1978), and is coherent with the observed surface convergence. The latter is shown especially well in Figure 9. Here the current normal and relative to the front is shown as a profile in depth observed 30 m from the surface front on the brackish water side. Above 1 m depth the flow is toward the front, but below it reverses and soon reaches a level of roughly 50 cm/s. The latter is simply the negative of the normal propagation speed of the front itself relative to the heavier coastal seawater, and appears here as a simple consequence of maintaining the vessel at a fixed distance from the front during the observations.

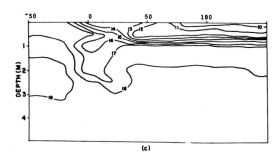

Figure 8. Isopycnals in a vertical section normal to the frontal boundary of the Connecticut River plume for March 29, 1973. From Garvine and Monk (1974).

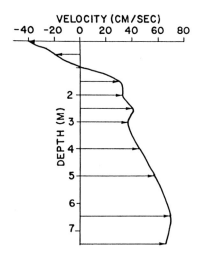

Figure 9. Horizontal velocity vs. depth for the current component locally normal to the frontal boundary at 30 m from the surface front within the Connecticut River plume. Positive values denote flow away from the front. From Garvine and Monk (1974).

2b. Estuary plumes

Because of their greater size, estuary plumes are more difficult to observe and hence are less known. Since the primary mixing lies within the estuary itself, the resulting brackish water discharge usually has salinity values considerably above fresh water and P values that are small. For example, at long term mean values of fresh water inflow the values of P for the four major estuaries of the Middle Atlantic Bight of the U.S. east coast, the Connecticut River, Chesapeake Bay, the Hudson River, and Delaware Bay, are, respectively, 1.0, 2.4×10^{-2} 1.4×10^{-2}, and 2.3×10^{-3}. The Connecticut forms a river plume, as illustrated above, Chesapeake Bay and the Hudson River form estuary plumes, while Delaware Bay lacks any distinctive plume.

Because of their typically larger scale estuary plumes tend to have associated Kelvin number K larger than river plumes. That of the Chesapeake plume is about 3.0, compared to only 0.3 for the Connecticut, so that Coriolis effects are prominent. The deflecting tendency of Coriolis force is mirrored in the tendency for the estuary plume to turn right or anti-cyclonically in the northern hemisphere after discharging from the mouth. After this turning motion these plumes usually move alongshore as a coastal current with the brackish water forming a thin, buoyant layer banked against the coast on its right side and often forming a frontal boundary in its left or offshore side. Figure 10 from Boicourt (1973) shows the surface salinity field of the Chesapeake plume after a period of very high discharge. Both the turning region near the mouth and the subsequent coastal current are clearly represented. The plume width is about 30 km and its length over 100 km, significant portions of the shelf there.

Figure 11 shows a typical salinity cross-section of the Hudson River plume from Bowman and Iverson (1978). Here the sharp halocline shows the tendency of such plumes to "lean" against the right hand coastline, here the New Jersey shore, as well as to terminate seaward at a front similar to those associated with river plumes. Bowman and Iverson noted that where the isohalines outcropped a sharp surface front was present where surface convergence had trapped much floating debris.

Most of the physical characteristics of estuary plumes, especially the current field, frontal boundaries, and mixing dynamics, are insufficiently known. Field observations are particularly needed.

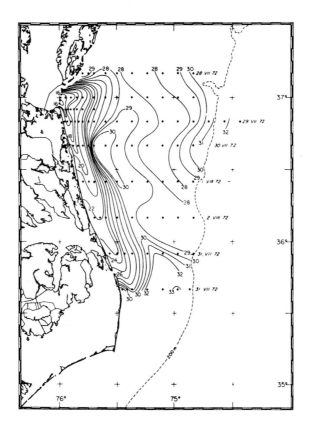

Figure 10. Surface salinity distribution for the Chesapeake Bay plume from Boicourt (1973) July 28-August 2, 1972. The Bay mouth is at the upper left. Sampling stations are marked by solid circles.

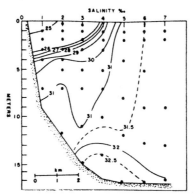

Figure 11. Vertical section across the Hudson River Plume showing isohalines for August 19, 1976. From Bowman and Iverson (1978).

A series of comprehensive observations will be conducted for the Chesapeake plume during 1985-1987 by the MECCAS program funded by the U.S. National Science Foundation and should add greatly to our understanding of these important elements of shelf circulation and ecology.

3. Models of brackish plume dynamics

To explore the essentials of plume dynamics I will assume for simplicity that the plume itself is relatively shallow compared to the shelf water depth and that it has uniform density and thus uniform reduced gravity.

Consider the simple case of a quasi-steady state for an established estuarine plume. I neglect vertical friction, arguing that the highly stratified pycnocline below reduces it, despite the potentially large velocity shear there, and that surface wind stress is a complicating but not essential element. Lateral friction is also neglected. In vector form then:

$$(\vec{q} \cdot \vec{\nabla})\vec{q} + f(\vec{k} \times \vec{q}) + \vec{\nabla}(g'D) = 0. \quad (1)$$
$$Q^2/L \qquad\quad fQ \qquad\quad g'D/L$$

Here $\vec{q}$ is the horizontal velocity vector, ∇ the horizontal gradient operator, f the Coriolis parameter, $\vec{k}$ the local vertical unit vector, g' the reduced gravity of plume water, and D the local plume depth. Scaling estimates have been introduced below each term with Q the typical current speed and L the horizontal scale of variation. From the mouth to at least as far as plume turning is notable the first (advection) and third (pressure gradient) terms should be of the same order. This implies that

$$Q \sim (g'D)^{1/2} \equiv c$$

or, in other words, the Froude number $F \equiv |\vec{q}|/c \sim 1$. For example, for the Chesapeake plume where $g' \approx 0.05$ m^2/s and $D \approx 5$ m, we would have $Q \sim c \approx 0.5$ m/s, relatively fast flow.

The relative importance of the Coriolis force may be estimated by its ratio to the pressure gradient force in Eq. (1):

$$\frac{fQ}{g'D/L} \sim \frac{fL}{c} = \frac{L}{r_i} \equiv K$$

where r_i is again the internal Rossby radius c/f and K the Kelvin number. Thus, the Kelvin number appears as a direct measure of the importance of earth rotation, as discussed in Section 2. In general, for estuary plumes K will be of order unity, while for river plumes it will usually be small.

Conservation of mass for plume water gives

$$\nabla \cdot (D\vec{q}) = 0 \qquad (2)$$

Thus, Eqs. (1) and (2) form a system of three scalar equations for the dependent variables D and the two components of $\vec{q}$.

If Eqs. (1) and (2) are manipulated, two scalar properties may be found to be invariant following the motion, the potential vorticity P and the total energy per unit mass or Bernoulli function B given respectively by

$$P = (f + \zeta)/D,$$
$$B = c^2 + q^2/2,$$

where $\zeta = \vec{k} \cdot (\nabla \times \vec{q})$, the relative vorticity, and $q^2 = \vec{q} \cdot \vec{q}$. When combined with Eq. (2), conservation of P and B form an alternative set of governing equations. Furthermore, should P be itself uniform across the outflow layer at the mouth, it will remain so throughout the plume; this would obtain, for example, if the plume flow originates in a slow moving region upstream within the estuary where the depth of the upper layer is constant at D_0. Gill (1977) has analyzed such a flow in a channel undergoing rotation. When the normalized discharge volume flux $Vf/g'D_0^2$ is small (V is the dimensional brackish layer discharge), as it is for most estuaries, Gill's results show that B also varies little across the exiting plume so that B will also be nearly uniform in the plume and equal to $g'D_0$. Thus, as the plume spreads and thins over the shelf, $c^2 = g'D$ decreases while q^2 tends toward $g'D_0$. However, before this limit is reached the local Froude number q/c will become sufficiently large that vertical shear instability will likely result in rapid mixing with shelf water below and thus in local destruction of the plume as a distinct layer. Perhaps in this manner much of the plume or near field water eventually becomes part of the far field, as defined in Section 2. Such action may be illustrated in Figure 5 where the isopycnals spread abruptly beginning at 4 nautical miles from the river mouth.

I have applied the ideas above to the restricted case of the steady state flow of a river plume ($f=0=K$) discharging onto a shelf where a constant alongshore current u_a interacts with it (Garvine, 1982). Plume and ambient water interact in the model through contact at the offshore frontal boundary which is modeled as a discontinuity by jump conditions similar in concept to a hydraulic jump. The shape of the front as well as the plume flow field are determined by the model equations which can be easily solved using the method of characteristics, if the flow is everywhere supercritical ($F>1$), as observations suggest. Figure 12 shows the resulting plume depth field beyond the exit channel or mouth. The plume ultimately must turn to parallel the ambient alongshore current, i.e., parallel to the x-axis. As it does so it spreads laterally and thins toward downstream. At given depth in the plume the pressure is proportional to D; thus, the decline in D downstream results in acceleration of the flow such that $B=c^2+q^2/2$ is conserved. Meanwhile lower pressure is felt inshore, thus supplying the pressure gradient force needed to turn the flow parallel to the shore. For given x, the plume actually reaches its greatest depth offshore at the front and is shallowest inshore. This action appears to explain the similar configuration of the Connecticut River plume cross-section shown in Figure 4. For estuary plumes, or more generally, for plumes with order one or larger Kelvin number, the action of Coriolis force tends to deflect or turn them alongshore as well. For sufficiently large K, the plume depth could well be greatest inshore rather than at the front, as in Figure 12. Clearly it would be useful to explore models even of this simple type for application to estuary as well as river plumes.

4. Bottom water return flow

While river and estuary plumes on the continental shelf are a well known consequence of the outflow of land drainage, the compensating bottom water return flow from the shelf into estuaries may well prove to have as significant, but differing, ecological results. For most estuaries the landward, bottom volume flux is nearly as large as the seaward, surface flux, both many times greater than the fresh water inflow rate. For estuarine ecology the landward flow has obvious significance in that material and biota of shelf origin are transported far up-estuary. For continental shelf ecology the landward flux needed to supply this is likely to be significant as well and may be a critical pathway in the life cycle of many shelf species.

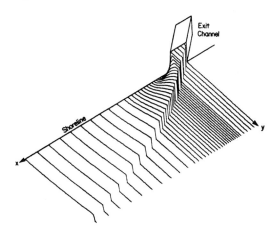

Figure 12. Isometric projection of a river plume interface depth (D) field drawn positive upward for an ambient alongshore current moving toward positive x. The frontal boundary corresponds to the abrupt fall in D at the offshore side. From Garvine (1982).

Bumpus (1965) found the earliest direct evidence of the large domain on the shelf subject to bottom landward motion toward major estuaries. He deployed more than 75,000 sea bed drifters off the east coast of the U.S. Landward of the 60 m isobath in the Middle Atlantic Bight he found that the bottom drift was generally toward shore at speeds of about 1 km/day and often converged on the mouths of the major estuaries. Norcross and Stanley (1967) found a similar landward bottom flow toward Chesapeake Bay with may sea bed drifters recovered there which had been deployed as far as 70 km seaward on the shelf. Later Pape and Garvine (1982) found a similar bottom inflow into Delaware Bay again using sea bed drifters. Figure 13 shows the Lagrangian mean velocity field for the bottom flow deduced from averages over eight of their deployments of a total of nearly 4000 drifters. Note the clear convergence of the bottom shelf flow on the mouth from as far as 40 km seaward. Once inside the estuary the bottom flow moved landward and laterally toward each shore, implying a bottom horizontal divergence. The shelf return flow averaged about 1 km/day, the same level reported by Bumpus (1965). The velocity field of Figure 13, however, was subject to significant wind-induced variations on meteorological time scales. When winds alongshore toward the northeast were dominant following a deployment, Pape and Garvine found significantly higher rates of drifter recoveries; conversely, for winds alongshore toward the southwest they found lower rates. These results imply that bottom transport landward is augmented or reduced by the action of bottom Ekman transport induced by the coastal barrier to surface Ekman transport. The long term average, as in Figure 13, nevertheless reflects the persistent action of the estuarine landward flow as a sink for adjacent shelf bottom water.

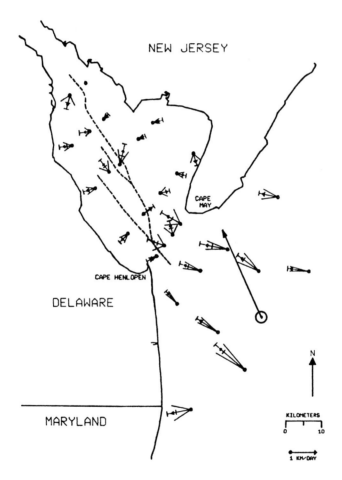

Figure 13. Vector map of bottom current mean velocities deduced from all returns of sea bed drifters deployed at the base of each vector. Brackets on the vector length indicate 95% confidence intervals on speed while lines to each side indicate the same for direction. Dashed lines inside Delaware Bay mark the main channels. From Paper and Garvine (1982). The large open circle marks the mooring location for the record shown in Figure 14.

To assess characteristics of the bottom return flow in more detail than sea bed drifters permit we moored a current meter 5 m above the bottom in 20 m depth at a distance of 20 km southeast of the mouth of Delaware Bay. Figure 14 shows the resulting progressive vector diagram (or virtual displacement) for July 1 to September 1, 1983. The motion has three principal components: tidal excursions of a few km in length which tend to be elliptical, a few wind driven displacements, mostly alongshore (northeast-southwest) of a few tens of km extent, and a persistent drift toward the estuary mouth (direction of large arrow). The latter dominates the mean velocity whose speed is 3.6 km/day. Thus, while the sea bed drifters, which glide along the bottom, showed a residual drift toward the mouth, so does the flow 5 m above bottom. Consequently, the landward transport of near bottom shelf water appears to be large and persistent.

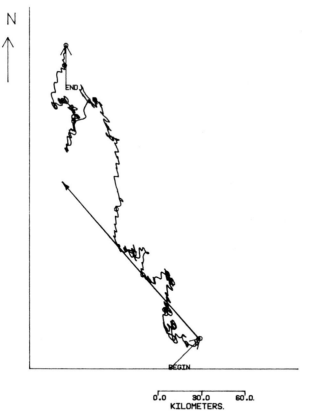

Figure 14. Progressive vector diagram from a two month record of current obtained 5 m above bottom on the shelf southeast of Delaware Bay in 20 m water depth. The large arrow points toward the mouth of the estuary. See Figure 13 for mooring location.

At this writing the only model of such a return flow is that of Beardsley and Hart (1978). They postulate a linear, steady bottom layer flow dynamics where the vorticity balance was between vortex line shortening and bottom friction as the estuarine bottom layer, modeled as a point sink, drew the bottom water shoreward into shallower water. Their model predicted horizontal scales for the region of landward flow on the shelf of several tens of km, the same scale found in the observations cited above. Their model thus may well represent at least a lowest order description of this process.

Clearly little of substance or detail is known of the bottom return flow on the shelf. Its likely impact on both shelf and estuarine ecology argues for greatly increased observations, especially long term current measurements, and further modeling efforts.

5. Concluding remarks

Brackish water plumes in shelf water and their mass continuity counter part, landward flowing shelf bottom water, are both powerful components of the more general coupled circulation between estuaries and the continental shelf. Because of the very low frequencies at which they operate, including the climatological mean state, they contribute greatly to net displacement of water and thus are both certain to exert critical impacts on both estuarine and shelf ecology. Thus, both these components will be rewarding to study through the combined approach of field observations and models, including laboratory as well as mathematical types. Over the next decade we can expect to see major advances in this important subfield of coastal oceanography.

Acknowledgments

This work was supported by the National Science Foundation through grant OCE-8315207. Ms. Ann K. Masse performed the analysis of the current meter record shown in Figure 14.

References

Beardsley, R. C. and J. Hart, 1978. A simple theoretical model for the flow of an estuary onto a continental shelf. J. Geophys. Res., 83, 873-883.

Boicourt, W. C., 1973. The circulation of water on the continental shelf from Chesapeake Bay to Cape Hatteras. Ph.D. Thesis, The Johns Hopkins University, 183 pp.

Bowman, M. J. and R. L. Iverson, 1978. Estuarine and plume fronts. In Oceanic Fronts in Coastal Processes, Chap. 10 (ed., M. J. Bowman and W. E. Esaias), Springer-Verlag, Berlin, 114 pp.

Britter, R. E. and J. E. Simpson, 1978. Experiments on the dynamics of a gravity current head. J. Fluid Mech., 88, 223-240.

Bumpus, D. F., 1965. Residual drift along the bottom on the continental shelf in the Middle Atlantic Bight area. Limnol. & Oceanogr., Suppl. to 10, R50-R53.

Csanady, G. T., 1984. Circulation induced by river inflow in well mixed water over a sloping continental shelf. J. Phys. Oceanogr., 14, 1703-1711.

Garvine, R. W., 1982. A steady state model for buoyant surface plume hydrodynamics in coastal waters. Tellus, 34, 293-306.

Garvine, R. W., 1974. Physical features of the Connecticut River outflow during high discharge. J. Geophys. Res., 79, 831-846.

Garvine, R. W., 1977. Observations of the motion field of the Connecticut River plume. J. Geophys. Res., 82, 441-454.

Garvine, R. W. and J. D. Monk, 1974. Frontal structure of a river plume. J. Geophys. Res., 79, 2251-2259.

Gill, A. E., 1977. The hydraulics of rotating-channel flow. J. Fluid Mech., 80, 641-671.

Grancini, G. and B. Cescon, 1973. Dispersal processes of freshwater in the Po River coastal area. Limnol. & Oceanogr., 18, 705-710.

Hansen, D. V. and M. Rattray, Jr., 1966. New dimensions in estuary classification. Limnol. & Oceanogr., 11, 319-326.

Ingram, R. G., 1981. Characteristics of the Great Whale River plume. J. Geophys. Res., 86, 2017-2023.

Norcross, J. J. and E. M. Stanley, 1967. Inferred surface and bottom drift, June 1963 through October, 1965, circulation of shelf waters off the Chesapeake Bight. Prof. Papers Environ. Sci. Serv. Admin., U. S. Govt., 3(2), 11-42.

Pape, E. H., III and R. W. Garvine, 1982. The subtidal circulation in Delaware Bay and adjacent shelf waters. J. Geophys. Res., 87, 7955-7970.

Wright, L. D. and J. M. Coleman, 1971. Effluent expansion and interfacial mixing in the presence of a salt wedge, Mississippi River delta. J. Geophys. Res., 76, 8649-8661.

LABORATORY MODELING OF DYNAMIC PROCESSES IN FJORDS AND SHELF WATERS

T.A. McClimans
Norwegian Hydrotechnical Laboratory
7034 Trondheim, Norway

ABSTRACT

Some newer experience with laboratory models of the dynamics of stratified flows in fjords and coastal regions is presented. Brief comments on historical developments are given together with theoretical considerations of modeling. Some laboratory results are compared with field measurements. The effects of the earth's rotation on large scale coastal flows are highlighted. Of particular interest for the evaluation of the marine environment are large scale flow meanders and eddies which lead to a significant local variability.

INTRODUCTION

The following physical processes have been successfully simulated in reduced scale laboratory models:

- long waves (tides, seiches, Kelvin waves)
- horizontal density flows (stratified flows)
- frontal mixing
- wind mixing
- topographic steering
- dispersion
- effects due to the rotation of the earth

The most complicated flow dynamics are concerned with turbulent, rotating, stratified flows, for which homogeneous flows are special cases.

A comparison of model results with field measurements shows that small scale hydraulic (laboratory) models can produce both qualitative and quantitative results for diagnostics and prognostics for, e.g., regulation of freshwater discharge and topographic changes. Laboratory models give further insight in cause-effect relationships and are therefore valuable for establishing theories and calibrating analytical or numerical computer models.

Hydraulic laboratory work can be divided into demonstrations, experiments and models. <u>Demonstrations</u> are intended to illustrate qualitative aspects of complicated flow processes. To this category belongs Mortimer's (1951) wind-induced internal waves in a three layer basin. <u>Experiments</u> are performed to quantify cause-effect relationships and provide important empirical data together with a definition of the range of validity of mathematical formulations. To this category belongs Keulegan's (1955) investigation of frictional forces in layered flows. <u>Models</u> are built to evaluate complicated physical systems and proposed changes. To this category belongs Tully's (1949) model study of Alberni Inlet.

The present discussion is limited to the laboratory simulation of the dynamics of stratified flows with special emphasis on the effects of the earth's rotation in fjords and shelf waters. Von Arx (1962) gives a general description of the theory of laboratory modeling of large scale oceanic flows. The present discussion can in part be considered an update on this earlier work, much of it appearing also in McClimans (1983). A more detailed discussion on the simulation of the energetics of mixing and diffusive processes is given in McClimans & Mathisen (1979).

DYNAMICS OF FJORD AND SHELF WATERS

River runoff, wind, tides and variations in the seaward density field all contribute to the dynamics of stratified flows in fjord and coastal waters. All contribute to variations in surface slope (barotropic mode) as well as variations of the local density structure (baroclinic mode). To simplify these concepts, it is common practice in hydraulics to represent these continually stratified flows as two or three layer flows. In the following we will consider two particular situations: flow in a fjord (two and three layer) and flow of river (or coastal) water along a coast.

Fig 1 depicts a longitudinal section of a fjord. Symbols to be used are noted on the figure. In cross-section, fjord hydrography often reveals the three-layer structure as shown in Fig 2. In time, the river outflow is absorbed into the brackish layer due to horizontal and vertical mixing and loses its identity.

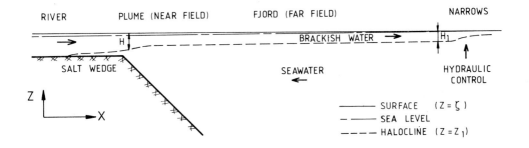

Figure 1. Longitudinal section of river flow through a fjord (with nomenclature). Vertical coordinate exaggerated.

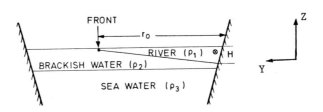

Figure 2. Cross section of a river flowing through a fjord (with nomenclature). Vertical coordinate exaggerated.

River flow or flow of brackish water from a fjord to an open sea will be affected by the earth's rotation as shown in plan view in Fig 3. Here the time scales are large compared to large rivers flowing to narrow fjords. In cross section, this flow resembles the river flow in Fig 2. Details of these dynamics will be discussed in the following sections. In Fig 3, there are shown two dynamical length scales, the inertial radius r_i which describes the path of a frictionless jet in a rotating system, and the Rossby deformation radius r_o, which describes the extent of buoyant plume spreading as the flow is diverted to the right (in the northern hemisphere). These will be discussed later.

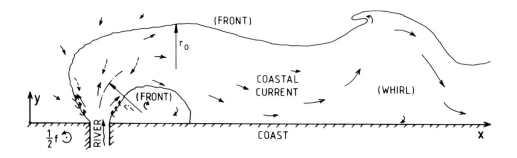

Figure 3. River outflow forming a coastal current (with nomenclature).

MODEL THEORY

The dynamic simulation of stratified flows in the sea must follow the rules of model theory as well as the experience which goes under the heading of scale effects or model effects.

Simulating large geophysical flows in the laboratory usually requires the use of distorted hydraulic models in which the vertical scale is exaggerated vis-á-vis the horizontal scale. In particular, rivers with their large width (B) to depth (H) aspect ratio are difficult to simulate geometrically within a reasonable (economic) size. This raises the question of the validity of testing the dynamics of stratified flows in distorted laboratory models.

For simplicity, the model laws for simulating flows in a distorted laboratory model are developed for a two-fluid system by extending Stommel & Farmer's (1952) equations to include rotation. These laws hold for the more general case of continuous time-varying stratification. The essential feature is that buoyant spreading is a process for which the pressure p is approximately hydrostatic, except in the region of frontal convergence where the vertical accelerations are significant and turbulent mixing is quite vigorous. Thus, with surface elevation ζ, vertical coordinate z positive upwards, density ρ and acceleration of gravity g,

$$p(z) = \int_z^\zeta g\rho dz' \qquad (1)$$

except in the narrow region near a front. In this region the mass entrainment velocity and the side frictional stress τ_s are large.

The laterally and vertically integrated equation of longitudinal (x) momentum for a stationary, incompressible river plume in a homogeneous, deep, quiescent sea can be written, for the river plume,

$$(u^2 hb)_x = -ghb\zeta_x - gh^2 b\, \rho_{1x}/2\rho_2 + fvbh - 2\tau_s h/\rho_2 - \tau_i b/\rho_2 \qquad (2)$$

and for the deeper layer below the density jump at z_1,

$$(\rho_1 h + \rho_2 z_1)_x = 0 \qquad (3)$$

Here, u(x) is the (integrated) longitudinal velocity, h(x) is the thickness of the plume, b(x) is the width of the plume, f is the socalled Coriolis parameter representing the earth's rotation, v(x) is the lateral velocity, τ_i is the interfacial shear stress below the plume and the subscript x denotes differentiation w.r.t. the longitudinal coordinate. Although the Coriolis term fvbh is small in the x-direction, it is large in the lateral y-direction and must be retained in this analysis of model laws.

The effects of wind and surface films are not taken into account. Further, a small term involving the acceleration of the entrained fluid is left out and the Boussinesq approximation is used. The former may be included in the stress terms without loss of accuracy.

In the following, the stress terms are replaced by a quadratic velocity drag formula

$$\tau/\rho = C_D u^2 \tag{4}$$

which has been often used for densimetric flows. Here a distinction will be made between C_{D_i} for the interface and C_{D_s} for the vigorous frontal region (or C_{D_w} for wall friction).

Combining (2) - (4) gives

$$(u^2 hb)_x = -\frac{hb}{2}(g\zeta_x + g'h_x) + fvbh - 2C_{D_s} u^2 h - C_{D_i} u^2 b \tag{5}$$

for which $g' = g(\rho_2 - \rho_1)/\rho_2$ is the socalled reduced gravity. The terms ζ_x and h_x are independent by virtue of the horizontal gradient in g'. When entrainment is negligible, ρ_1 is reasonably constant and the two terms may be combined. Such a system is referred to as a reduced gravity model and behaves much like a one-layer flow.

Scaling laws for the dynamic simulation and the interpretation of laboratory results are derived from the dimensionless equations of motion. This requires a choice of scaling factors. For distorted models two length scales are necessary. The following scaling factors will be applied to (5)

$$\left.\begin{array}{ll} H & = \text{vertical length} \\ B & = \text{horizontal length} \\ U & = \text{horizontal velocity} \\ f^{-1} & = \text{time scale (T)} \end{array}\right\} \tag{6}$$

Grouping all physical quantities in curly brackets, (5) may be expressed in dimensionless form as

$$(u^2 hb)_x = -\left\{\frac{gH}{U^2}\right\}\frac{hb}{2}\zeta_x - \left\{\frac{g'H}{U^2}\right\}\frac{hb}{2}h_x + \left\{\frac{fB}{U}\right\} vbh$$
$$- 2\{C_{D_s}\} u^2 h - \left\{\frac{B}{H} C_{D_i}\right\} u^2 b \tag{7}$$

in which all quantities outside the curly brackets are now dimensionless. A distorted laboratory model simulates the natural flow only when the assumptions leading to (5) are fulfilled, and only when the products in curly brackets are the same in the model as they are in nature.

The dimensionless groups in curly brackets contain, from left to right,

$$\text{Froude number } Fr = \frac{U}{(gH)^{\frac{1}{2}}} \qquad (8)$$

$$\text{Densimetric Froude number } F = \frac{U}{(g'H)^{\frac{1}{2}}} \qquad (9)$$

$$\text{Rossby number } Ro = \frac{U}{fB} \qquad (10)$$

The drag coefficients have been defined in (4). They depend on both F and the Reynolds number

$$Re = \frac{UH}{\nu} \qquad (11)$$

where ν is the kinematic viscosity.

Normally, C_{D_i} increases with F, but decreases with increasing Re (Abraham & Eysink, 1971). Since F is simulated in Froude models, the last term in (7) can be simulated at lower (model) Re only by reducing B/H. This implies distortion as a necessary (but not sufficient) condition for simulating interfacial friction. In general, the distortion necessary to simulate friction from a near-critical Re to the large Re in nature is from 10 to 100. This is too rough an estimate for the simulation of the friction forces if interfacial friction is a major factor in the force balance. In the near field of the river plume, however, the advective acceleration term on the left hand side of (7) is balanced by and large by pressure terms. For the large scale coastal currents the pressure term is balanced by and large by the Coriolis term (geostrophic balance). It is therefore important that the model does not exaggerate the friction term to the point that it becomes a dominant term in the balance. Thus, at near critical Re ~ $O(10^3)$ the distortion in the laboratory should be greater than, say, 10.

The lateral friction is not reduced by distortion. Thus, an increase of C_{D_s} in the model may affect the dynamics of the plume, rendering a combined Froude-Reynolds law. This would greatly limit the possibility of scaling model results to natural flows. Fortunately, the momentum exchange along the front occurs primarily in large wave instabilities and vortices which derive their energy from the spreading, buoyant flow. Although the details of these three-dimensional processes are not simulated in a distorted model, the energetics of the process has been argued to be independent of distortion provided $B/H \gg 1$ in the laboratory (McClimans, 1979). The importance of horizontal momentum exchange in distorted Froude models was treated in McClimans & Gjerp (1979).

The exact simulation of side friction is not necessary since it represents a smaller contribution to the dynamics. It is essential, however, that it is not greatly amplified. Here wall roughness is an important factor. Fortunately, the boundary layer along smooth model walls is often of the same (scaled) size as the boundary layers along a natural coastline.

By choosing the time scale as f^{-1} and simulating the density, the set of restrictions imposed by (8)-(10) are consistent. Such a model may be termed a Froude-Rossby model. It should be mentioned in this context that nonstationary flows also are consistent with these laws when $T = f^{-1}$. However, for time dependent flows there is a further restriction on the distortion, imposed by vertical accelerations.

A case in point is the dimensionless form of the horizontal component of vorticity for irrotational flows

$$\frac{\partial u}{\partial z} - \left\{ \frac{H^2}{B^2} \right\} \frac{\partial w}{\partial x} = 0 \qquad (12)$$

in which w is the vertical velocity, scaled to the horizontal velocity by UH/B. Clearly, this relation is valid in the model only when the ratio of H/B is identical to that in nature, or when $\partial w/\partial x = \partial u/\partial z = 0$. Surface gravity waves cannot be simulated with distortion.

Long shallow water waves like tides and seiches have negligible vertical accelerations in nature. In most cases these waves will be reflected along the sloping bottom. Miche (1944) developed the following criteria for total reflection of a wave from a bottom with slope α

$$\frac{a}{\lambda} < \left(\frac{\alpha}{2\pi^3}\right)^{\frac{1}{2}} \sin^2 \alpha \tag{13}$$

where a is the wave amplitude and λ is the wave length. Nagashima (1971) used laboratory experiments to show that this relation holds also for interfacial waves. Since both a/λ and α are proportional to the distortion of a model, it is clear that the limits for wave reflection are greater in a distorted model than in nature. Thus for gently sloping bottoms for which (13) is not fulfilled, distortion cannot be allowed. This is normally not a problem for Norwegian bathymetry.

SCALING LAWS

Denoting the ratio of natural quantities to laboratory quantities by the index r, the model requirement can be expressed as

$$(Fr)_r = (F)_r = (Ro)_r = (C_D)_r = 1 \tag{14}$$

Choosing the primary similarity requirement $(F)_r = 1$ leads to

$$U_r = (g'_r H_r)^{\frac{1}{2}} \tag{15}$$

The kinematic relation for horizontal currents yields a time scale

$$T_r = B_r/U_r = B_r/(g'_r H_r)^{\frac{1}{2}} \tag{16}$$

This constitutes Froude's scaling law for densimetric flows. If both Fr and F are constraints, then $g'_r = 1$. If Ro is also a constraint (for rotational flows) then $f_r = T_r^{-1}$.

Scaling laws for all hydrodynamic quantities can be derived from the above. Take, for example, river discharge Q:

$$Q_r = U_r B_r H_r = B_r^4 f_r^3 / g'_r \qquad (17)$$

showing the strong influence rotation has on the interpretation of model results of large scale flows. For non-rotating flows, f has to be replaced by $(g'_r H_r)^{1/2}/B_r$.

Returning to the length scales shown in Fig 3, it is instructive to look at their ratio. The inertial radius is

$$r_i = u/f \qquad (18)$$

and the Rossby deformation radius is

$$r_o = (g'h)^{1/2}/f \qquad (19)$$

Thus, their ratio is the densimetric Froude number

$$F = r_i/r_o \qquad (20)$$

Further

$$Ro = \frac{r_i}{B} \qquad (21)$$

$$F/Ro = B/r_o \qquad (22)$$

These ratios are important for interpreting the large scale flow simulations for which geostrophic balance dominates.

SCALE EFFECTS

Effects not modeled by (8) - (10) give spurious results which are interpreted as scale effects. They are different in the laboratory than in nature. The simulation of friction has already been discussed.

It is important for momentum exchange as well as mass exchange that the flow is turbulent. This is one of the limiting conditions beyond which the simulation is not valid. For momentum exchange in jets at large F, it is well accepted that fully turbulent flow occurs when Re>

1500. For the case of F~1 there is a large reserve of potential energy to create turbulence, and the critical number appears to be between 500 and 1000. For the frontal regions, there appears to be a highly energetic horizontal exchange. For the region below the plume, the gravitational stability damps out the turbulence below a critical value (Keulegan, 1955)

$$\theta = \frac{(\nu g')^{1/3}}{u} < 0.178 \qquad (23)$$

θ is sometimes referred to as the Keulegan number.

The exchange of mass is also highly dependent on turbulence. In this paper the main emphasis is on the dynamics, so a further discussion of mass exchange will not be pursued.

Surface tension gradients pose an important limiting factor for river plume studies. After some experience with spurious surface currents induced by surface tension reducing agents, current practice is to avoid surface tension effects by dimensioning the experiments so that surface films have no dynamic significance. This of course leads to a larger apparatus.

Using the Flemming-Revelle formula and laboratory experience, it can be shown (McClimans & Sægrov, 1982) that the lower limit of layer thickness for the simulation of river plumes is 1 cm for saline and 4 cm for thermal stratification. Thermal plume studies require an apparatus which is 4 times larger than that for fresh water plumes in seawater.

The requirement $g'_r = 1$ limits the applicability of the laboratory results. In the absence of surface forcing (atmosphere and tide) and with $g' \ll g$, there is very weak coupling between the surface slope and the internal density field. With entrainment, however, g' is not a constant and there is a strong argument for $g'_r = 1$. It should therefore be mentioned that newer laboratory studies of densimetric coastal currents over a wide range of g' show no reason for concern (Griffiths & Linden, 1981).

The question of applying the laboratory results to another g' is closely connected to the possibility of interpreting H to another value, since the two occur together in the scaling. For most of the densimetric surface flows, the bottom plays a minor role and the flows can be considered as quasi-two-dimensional. Thus, for these flows, H_r can be interpreted for a reasonable range of distortions other than that chosen for the model simulation. This holds particularly for deep fjords.

SOME NEWER LABORATORY RESULTS

The laboratory model of Alberni Inlet was mentioned in the introduction. Since that pioneering study there have been published several laboratory simulations of stratified fjord and coastal flows, e.g. Long (1953), Rattray & Lincoln (1955), von Arx (1962), Simmons (1969), Ingebrigtsen (1979), McClimans & Mathisen (1979) and Vinger, et al. (1981). In the following, some newer results obtained in the 5 m diameter rotating basin at NHL in Trondheim, Norway will be shown and compared with field observations to show the types of stratified flows occurring in fjords and shelf seas.

A cross-section of the laboratory facility is shown in Fig 4. The axis of alignment is supported both at the ceiling and floor and the weight is supported by an air cushion under a skirt submerged in a pool of water. The alignment is kept to within 2×10^{-5} rad of vertical and the rotation rate is constant to within ±0.4%. Water and electricity supply systems are also shown. Flow speed and density measurements are recorded on board. Photographs of dye and confetti streaks make up an important part of the data.

Experimental setups for simple fjord and coast geometries are shown in Figs 5 and 6. The distortions of these flows are, respectively 10 and 100. The flows were run at Reynolds numbers of several thousands to assure turbulence. Results of the river flowing in the fjord are shown in Fig 7 with and without rotation. A comparison with field observations, shown in Fig 8, reveals the importance of rotation. The path of the river jet both in nature and in the rotating basin may be described very well by the inertial radius $r_i \simeq 3$ km.

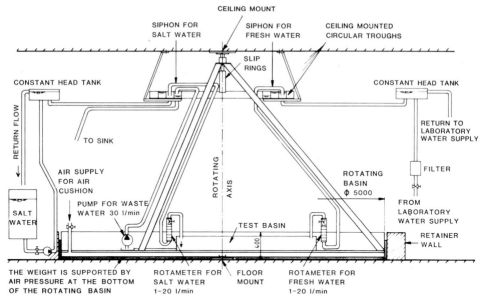

Figure 4. Cross section of the 5 m diameter rotating basin at NHL. Three beams connect the basin to the ceiling mount.

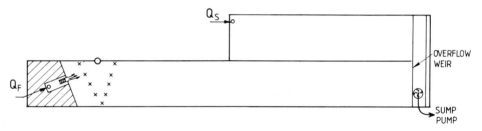

Figure 5. Plan view of a rotating fjord model. x-position of conductivity probes; O-position of camera mounted 2 m above model; Q_f - river water supply; Q_s - seawater supply at bottom.

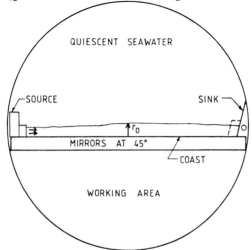

Figure 6. Plan view of a rotating coast model.

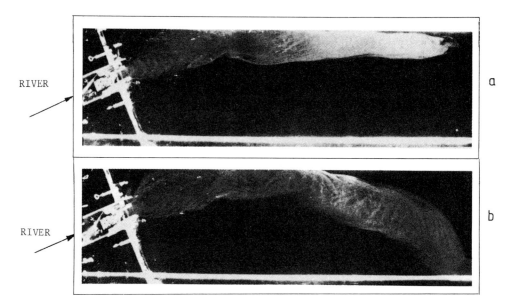

Figure 7. River plume simulation (a) without rotation and (b) with rotation.

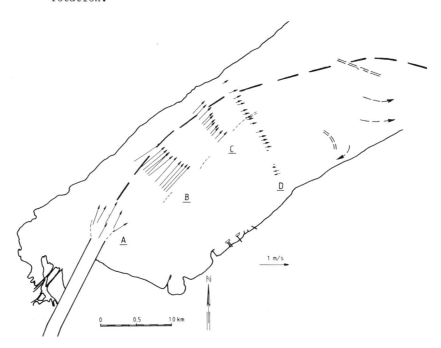

Figure 8. Trajectories of the Orkla River plume:
→ field observations ; — — model results.

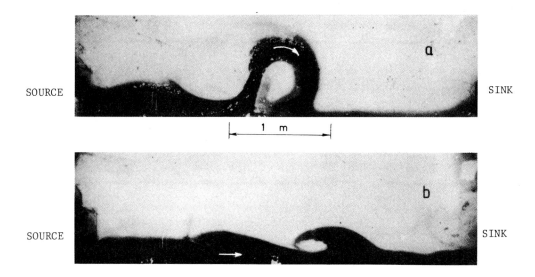

Figure 9. Laboratory simulation of baroclinic coastal currents.
(a) Source F = 0.6, (b) Source F = 1.2.

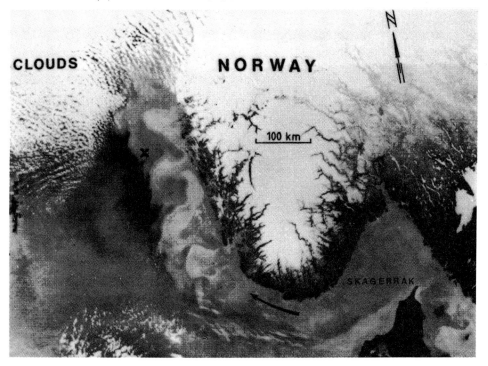

Figure 10. NOAA 6 Thermal image of the Norwegian Coastal Currents (13 April, 1981).

Results for the buoyant coastal current are shown in Fig 9 for initial densimetric Froude numbers greater than and less than 1. Wave growth and formation of meanders in which the coastal waters attain a large offshore penetration, as seen often in satellite thermal images like that in Fig 10, implies that F<1 for the Norwegian Coastal Current (Carstens, et al., 1984). Wave growth rates observed in the laboratory are compared with theory in Fig 11 (McClimans, et al., 1985).

Insitu field observations confirm the large scale variability implied by the laboratory model studies. This must be taken into account in the evaluation of single point measurements from cruises.

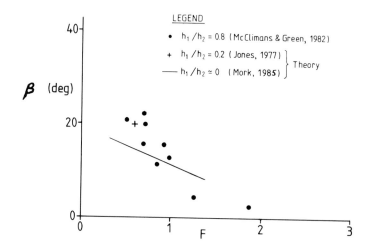

Figure 11. Angle (β) of offshore growth of northward propagating meanders. Comparison of laboratory model with theory. (McClimans, et al., 1985).

CONCLUSIONS

Quantitative estimates for stratified flows in fjords and shelf seas can be obtained in small scale laboratory Froude-Rossby models. Results from laboratory simulations reveal a large scale variability of meandering flows like those observed in satellite thermal images and insitu field measurements. Interpretation of local observations is difficult without a knowledge of the synoptic flow field.

ACKNOWLEDGEMENTS

This work has been supported by the Fund of License Fees.

REFERENCES

Abraham, G. and Eysink, W.D. (1971) Magnitude of interfacial shear in exchange flow, J. Hyd.Res. 9:125-151.

von Arx, W.S. (1962) Introduction to physical oceanography. Addison - Wesley.

Carstens, T., McClimans, T.A. and Nilsen, J.H. (1984) Satellite imagery of boundary currents. In Remote sensing of shelf sea hydrodynamics. Ed. J.C.J. Nihoul. Elsevier.

Griffiths, R.W. and Linden, P.F. (1981) The stability of buoyancy driven coastal currents. Dyn. of Atmos. and Oceans 5:281-306.

Ingebrigtsen, J. (1979) Laboratory model of a coastal current. Thesis, Geophysical Inst., Univ. of Bergen.

Jones, S. (1977) Instabilities and wave interactions in a rotating two-layer fluid. PhD thesis, St. Johns College, Cambridge.

Keulegan, G. (1955) Seventh progress report on model laws for density currents. Interface mixing in arrested saline wedges. NBS Report 4142.

Long, R.R. (Editor) (1953) Fluid models in geophysics. Proceedings of the first symposium on the use of models in geophysical fluid dynamics. Johns Hopkins University, Baltimore, U.S. Government Printing Office.

McClimans, T.A. (1979) On the energetics of river plume entrainment. Geophys. Astrophys. Fluid Dynamics, 13:67-81.

McClimans, T.A. (1983) Laboratory simulation of river plumes and coastal currents. ASME symposium on modeling of environmental flow systems, Boston. FED 8:3-9.

McClimans, T.A. and Gjerp, S.A. (1979) Numerical study of distortion in a Froude model. Proc. 16th Coastal Eng. Conference III: 2887-2904.

McClimans, T.A. and Green, T. (1982) Phase speed and growth of whirls in a baroclinic coastal current. River and Harbour Laboratory Report STF60 A82108.

McClimans, T.A. and Mathisen, J.P. (1979) Stationary wind mixing in a deep-silled, narrow fjord: a comparison of laboratory observations with a numerical model. Mar.Sci.Comm. 5:127-174.

McClimans, T.A. and Sægrov, S. (1982) River plume studies in distorted Froude models. J.H. Res. 20:15-27.

McClimans, T.A., Vinger, Å. and Mork, M. (1985) The role of Froude number in models of baroclinic coastal currents. Ocean Modelling 62:14-17.

Miche, M. (1944) Mouvements ondulatorires de la mer. Annales des Ponts et Chaussées Nr. 7. (In French.)

Mork, M. (1985) Studies of meso-scale variability in the Norwegian Coastal Current. (In preparation.)

Mortimer, C.H. (1951) The use of models in the study of water movements in lakes. Proc. Int. Ass. Limnology 11:254-260.

Nagashima, H. (1971) Reflection and breaking of internal waves on a sloping beach. J. Ocean.Soc. Jap. 27:1-6.

Rattray, M., Jr. and Lincoln, J.H. (1955) Operating characteristics of an oceanographic model of Puget Sound. Trans. Amer. Geophys. Union 36: 251-261.

Simmons, H.B. (1969) Use of models in resolving tidal problems. J. Hyd. Div., ASCE. 95.125-146.

Stommel, H. and Farmer, H.G. (1952) On the nature of estuarine circulation. Part I. WHOI Ref. No. 52-88.

Tully, J.P. (1949) Oceanography and prediction of pulp mill pollution in Alberni Inlet. Fish. Res. Bd. Can. Bull. No. 88.

Vinger, Å., McClimans, T.A. and Tryggestad, S. (1981) Laboratory observations of instabilities in a straight coastal current. In The Norwegian Coastal Current II: 553-582. Eds. R. Sætre and M. Mork. University of Bergen.

IMPACT OF FRESHWATER DISCHARGE ON PRODUCTION AND TRANSFER OF MATERIALS IN THE MARINE ENVIRONMENT

V. S. Smetacek
Institut für Meereskunde an der Universität Kiel
D-2300 Kiel 1, FRG

ABSTRACT

Selected case studies are presented to illustrate how seasonality of freshwater discharge, in conjunction with hydrography/topography of the receiving environment, can influence both production and transfer of material as well as species composition of the plankton. Spatial and temporal environmental patterns (zones or phases with specific combinations of physico-chemical properties) that arise along the salinity gradient are colonized by characteristic organism types. It is argued that these organisms possess survival strategies (benthic or pelagic resting stages) that enable maintenance of resident populations within such regions. Some pelagic "life history types" are presented and it is shown that behavioural features of these organisms can have considerable impact on production as well as vertical and horizontal transfer of material introduced by freshwater discharge.

INTRODUCTION

Freshwater discharge has a twofold effect on the marine system: a) it introduces allochthonous dissolved and particulate materials and b) it drastically modifies the structure and dynamics of the physical environment. Because of the wide range of variability associated with freshwater discharge (depending on its magnitude and seasonality, the drainage basin and the hydrography/topography of the receiving marine environment) its impact on biological processes varies accordingly. Further, such environments are inherently heterogeneous, both spatially and temporally, because of the environmental gradient continuously maintained by the admixture of fresh and marine waters. Thus, any comparative analysis of production and transfer of materials along such gradients should be based on the spatial and temporal scales relevant to these processes.

The importance of considering scales of variability in estuarine ecosystems has been stressed by Lewis and Platt (1982). Medium to small scale zones in physico-chemical and hence also biological properties are created along salinity gradients. Because these environmental patterns are subject to short-term variation and also horizontal displacement, adequate sampling and assessment of freshwater influenced environments are rendered difficult (Seliger et al.1981, Sinclair et al. 1981, Therriault and Levasseur this vol.). Classification of such environments can be based on topography/hydrography (Svendsen this vol.), average properties such as nutrient and production levels (Boynton et al. 1982) or length of the phytoplankton growth cycle (Sinclair et al. 1981). However, all these aspects are interrelated and the patterns of interaction are highly complex. Therefore, more variables will have to be considered and their specific scales of variability addressed when constructing conceptual frameworks for assessing the influence of freshwater discharge on marine ecosystems.

In this review I deal primarily with the longer scales associated with the biology of organisms adapted to environments subject to fresh water discharge and show how the patterns of production and transfer of this material to benthic or pelagic food webs can be regulated hereby.

SELECTED CASE STUDIES

In the following, some contrasting locality-specific situations are presented to illustrate how freshwater discharge, in conjunction with seasonality and topography of the receiving environment, can influence production and transfer of organic matter in estuarine and marine systems.

Semi-enclosed water bodies

San Francisco Bay: This shallow bay consists of a southern and a northern arm. Major rivers enter the northern estuary but the southern arm receives only little freshwater directly. The northern Bay supports large diatom blooms during periods of moderate discharge in summer but low biomass (mainly of flagellates) during high discharge in winter (Cloern et al. 1983). The conclusion is that at high flushing rates, the normal bloom diatom species are unable to maintain themselves in the estuary, whereas moderate rates enable build-up of high biomasses. The dominant diatoms in the estuary, Thalassiosira eccentrica (re-identified as T. decipiens, J. Cloern, pers. comm.) and Skeletonema costatum are covered with mineral grains which greatly increase specific gravity of the cells. Cloern et al. (1983) suggest that the same physical processes that maintain suspended sediment maxima in well-mixed estuaries - "density-selective retention of particles within an estuarine circulation cell" - are also responsible for concomitant maintenance of large diatom stocks.

During a drought year no diatom bloom was observed, although favourable growth conditions in terms of light and nutrients prevailed. Cloern et al. (1983) attributed the absence of the bloom in the drought year to a change in the physical structure of the estuary which prevented accumulation of a large diatom biomass. However, Nichols (1985) suggests that increased benthic grazing in this region of the estuary during the drought year might have been responsible for preventing build-up of bloom biomass. The drought year also significantly affected the stock of mysids (that feed on diatoms) which form a major food source of juvenile striped bass. Abundance of the latter during the drought year was the lowest recorded to date. Circumstantial evidence thus indicates that perturbation of the seasonal discharge can have far-reaching repercussions on the food chain. However, the presence of such a causal relationship needs to be confirmed.

Southern San Francisco Bay is a shallow, well-mixed bay that maintains unusually low phytoplankton stocks (mainly of nanoflagellates) during much of the year in spite of high nutrient levels (Cloern 1982). However, during periods of increased river runoff in the spring, low-salinity water enters the southern Bay from the northern arm and stratification develops. Large diatom blooms then accumulate in the surface layer and dissipate after some weeks. Because of the shallowness of the bay, stratification does not significantly improve light availability, hence this is not the reason for bloom development. Heavy zooplankton grazing is also not responsible. Rather, Cloern (1982) suggests that filtration by the large aerobic benthic population is the primary factor controlling the size of the phytoplankton stock in the Bay. Stratification segregates the surface layer from the benthos, permitting build-up of a diatom bloom. Thus, stratification enhances production and retention of nutrients in the bay; the fate of nutrients introduced by freshwater discharge here can be controlled by benthic filter feeders which is quite a different situation from that in other deeper environments (Officer et al. 1982).

Chesapeake Bay is yet another example of how stratification-circulation patterns can influence spatial distribution of phytoplankton populations within an estuarine system. As this is a large body of water, local wind mixing over shallower areas introduces an additional complication. Further, the bay supports large populations of zooplankton (dominated by the typical estuarine copepod genus Acartia) and pelagic regeneration is an important source of nutrients over much of the growth season (Boynton et al. 1982). Seliger et al. (1981), on the basis of a detailed sampling programme, have shown that different regions of the bay are separated from one another by distinct fronts which favour either the growth of diatoms or dinoflagellates. Significant differences in biomass levels occur between shallower areas and the main channel; water is retained over longer periods in the well-mixed, former regions and the enhanced nutrient supply from benthic regeneration permits accumulation of large phytoplankton stocks. Apparently, benthic grazing does not outstrip growth rates of the phytoplankton in the bay as it does in some of the river mouths such as that of the Potomac (Cohen et al. 1983). Species behaviour in terms of vertical movement - sinking and buoyancy or swimming - is selected for by specific environmental features that result from interaction between hydrography and topography of the Bay. Tyler and Seliger (1981) have traced the seasonal movement of a population of Prorocentrum marie-lebouriae within the bay: The vegetative population growing in the surface layer is advected down the Bay; this population forms resting stages downstream and sinks out of the surface layer thereby entering the inflowing water of the saline wedge which transports the cells back upstream. The resting cells are eventually deposited on the bottom in the upper reaches of the bay from where the next year's population is apparently seeded. Similar behaviour, but at shorter time scales and without deposition of benthic resting stages has been reported for Skeletonema costatum populations in the Bay as well (Seliger et al. 1981). The geographical positions of the fronts change seasonally and interannually, which is at least partly due to corresponding changes in freshwater flux.

An interesting, albeit unusual example of how the magnitude of freshwater discharge can influence species composition of the phytoplankton is described for the Potomac Estuary by Cohen (in press). During a winter with high discharge rates the dinoflagellate Katodinium rotundatum bloomed in the stratified water column; in a winter with low discharge, the water column was mixed and a diatom bloom of Chaetoceros spp. developed. The dinoflagellate bloom persisted for a period of 4-5 months indicating that the cells maintained their position actively, possibly by vertical migration. The eventual fate of the bloom was sedimentation. It apparently arose from benthic cysts present in the area where the bloom occurred. This example illustrates how a particular species can maintain a given position within an estuary throughout its life cycle.

Western Kiel Bight: Although not an estuary in the strict sense, the Bight is part of the Belt Sea (the transition zone between the North and Baltic Seas) which is characterized by typical estuarine water exchange and is hence salinity-stratified over much of the year. The Belt Sea in general is much more productive than the Baltic proper. The reason for its high productivity is to be sought in the hydrography/topography of the area. Most of the Belt Sea is shallow and the low-salinity surface layer is only 10 - 15 m deep which greatly extends the length of the growth season relative to the open Baltic at the same latitude. Not only does the spring bloom (of typical bloom diatoms) begin much earlier (by about 6 weeks) but pronounced autumn blooms of Ceratium spp. and diatoms occur as late as October and November respectively (Smetacek et al. 1984). The spring diatom bloom is terminated by nutrient depletion before the sparse overwintering zooplankton has significantly increased biomass and hence grazing pressure. The bulk of biomass produced by the spring bloom sediments and is rapidly utilized by the benthos (Graf et al. 1982). Nanoflagellates are of importance only in the late spring and are heavily grazed by zooplankton that attain a biomass maximum in this period. Summer phytoplankton is dominated by medium and large dinoflagellates and diatoms. Sedimentation rates are at their annual minimum when zooplankton stocks are at their maximum (late spring and summer) indicating that copepod grazing retards sinking losses (Smetacek 1984). Zooplankton stocks decline to low levels by late summer.

The autumn phytoplankton blooms arise when surface salinity increases (declining Baltic outflow) and vertical mixing due to winds increases; this results in admixture of summer-stagnant bottom water rich in nutrients to the surface layer in which a fairly large population of Ceratium spp. is already present; stratification, albeit weaker than in summer, is nevertheless still maintained. Nutrient input triggers a massive Ceratium bloom; mass mortality and subsequent sinking is the usual fate of this bloom. Noji et al. (in press) speculate that cell mortality occurs because the ceratia are unable to complete their sexual cycle in the turbulent autumn water column. This exlanation assumes that (as in the case of many diatoms) a sexual phase is obligatory after a certain number of vegetative cell divisions. Mass sedimentation of this bloom biomass also elicits a strong response from the benthos as in spring (Czytrich et al. in press). Sedimentation of the autumn diatom bloom, largely as intact cells, has also been observed (Noji et al. in press). The Belt Sea is a nutrient sink because rapid sedimentation is the fate of phytoplankton blooms - whether diatoms or dinoflagellates (Smetacek 1980). The seasonal cycle of biomass and composition of the zooplankton (dominated by small copepods but not Acartia) varies much less than that of the phytoplankton indicating that the former are not tightly geared to the latter (Smetacek 1985a). The annual cycle of Kiel Bight plankton closely resembles that of St. George's Bay in Canada, which also has salinity stratification and a fairly long residence time of its water (Hargrave et al. 1985).

Norwegian Fjords: The wide variation in annual patterns of the plankton in different Norwegian fjords can be attributed to corresponding differences in hydrography/topography in relation to freshwater runoff. Such differences can occur independently both in the composition and seasonal cycle of phytoplankton (Sakshaug 1976, Dahl 1978, Eilertsen et al. 1981) as well as zooplankton (Fosshagen 1979, Matthews and Heimdal 1979). In most fjords a typical batch-culture type spring diatom bloom occurs before the peak in seasonal runoff (Sakshaug 1976); as this bloom is generally not heavily grazed, it is likely that following nutrient depletion, sedimentation to the benthos is the usual fate of much of its biomass. Heavy sedimentation of the spring bloom has been reported from Lindaspollene (Wassmann 1983) and Kosterfjorden (Båmstedt 1985). In Trondheimsfjord a second bloom of diatoms of a different species composition to the spring bloom develops in the low-salinity layer during the peak runoff season in early summer. In contrast to the batch-culture type spring bloom, this second bloom has been compared to a "chemostat system driven by the continuous supply of nutrients from freshwater and entrained water" (Sakshaug 1976). As large zooplankton stocks have developed by this time, it is likely that much of the biomass of this second bloom is utilized within the pelagic systems of the lower Fjord and the coastal current on the shelf. However, this remains to be verified.

River plumes

River plumes overlying the shelf or open ocean invariably support higher production and carry higher pelagic biomass than neighbouring coastal waters. Major rivers such as the Zaire and the Amazon influence large areas of the ocean. Thus, during the season of peak runoff (winter) of the Amazon, the influence of its plume in terms of higher plankton biomass is felt as far away as Barbados (Kidd and Sander 1979). As in the case of estuaries, the bulk of the nutrients introduced by rivers is taken up by neritic bloom diatoms that tend to settle out of the surface layer following nutrient depletion. This has been shown for the Zaire plume (van Bennekom and Berger 1984, Cadee 1984), the Po plume (Revelante and Gilmartin 1976), the Fraser plume (Parsons and Kessler this vol.) and the Hudson plume (Malone 1984). However, seasonality plays an important role as demonstrated in the case of the Hudson River plume where diatoms dominated during conditions of weak stratification during spring to be replaced by nanoflagellates during strong summer stratification. The bulk of the diatom bloom settled out whereas the nanoplankton was largely utilized by zooplankton within the surface layer (Malone and Chervin 1979). The authors suggest that diatoms are seeded from benthic sources below the plume whereas the nanoplankton originates from estuarine runoff. This would further suggest that seasonality in the estuarine system affects that of the plume.

The case of the Nile plume is the most striking example of how river regulation can affect the ecology of the receiving basin. Before erection of the High Dam, massive diatom blooms that grew on nutrients introduced by the Nile occurred in the river plume in autumn - the season of maximum runoff (Dowidar 1984). The annual zooplankton maximum also coincided with this bloom. A large sardine population used to appear during this season to feed in the plume. During the period when the reservoir was being filled (1966-77) the sardine catch dropped to extremely low levels. However, since 1979 the discharge rates have been increased, albeit well below those of the pre-dam period. The sardine catch in 1979 and 1980 increased significantly to about a third of the earlier harvest although the diatoms had apparently been replaced by dinoflagellates (Dowidar 1984). However, this is yet to be confirmed.

Coastal currents

The Norwegian coastal current (NCC) is the best-known of the several coastal currents driven by freshwater runoff in the Northern Hemisphere. It has a considerable impact on the ecology of the Norwegian shelf. Thus the growth season is prolonged by about 20 % in comparison to adjoining oceanic waters because salinity stratification permits earlier initiation of the spring bloom (Rey 1981). A similar situation has been described for the Icelandic coastal current (Thordardottir this vol.). The spring bloom in the NCC again consists of typical neritic diatoms although the gelatinous, colonial flagellate Phaeocystis pouchetii is also important (Smayda 1980). Nanoflagellates at low biomass levels dominate during the summer when grazing pressure, exerted by a large Calanus population (also by euphausiids and pteropods), is at its highest for the year (Peinert this vol.). Lower grazing pressure during the spring bloom, when production and standing stock of phytoplankton are at their annual maximum, indicates that settling out of the diatoms is also important in the NCC (Peinert this vol.). Winter nutrient levels in the NCC are lower than in offshore waters (Føyn and Rey 1981) and there is as yet little information on how much organic matter is exported from the fjords. NCC water is hence not enriched relative to the Norwegian Sea.

The most important impact of the NCC on the marine environment is likely to be on the zooplankton and fish (Skreslet 1976, 1981). The prolongation of the growth season means that short-lived organisms such as copepods can have more generations and long-lived ones such as fish grow larger each year than their oceanic counterparts. Colebrook (1985), on the basis of long-term observations in the North Atlantic, suggests that the size of the overwintering copepod populations is the single most important factor determining the maximum size of their summer stocks. If this is indeed true, then large-scale circulation patterns such as the NCC will have a profound influence on the positioning of overwintering stocks in relation to their summer grazing waters and ultimately control the food supply of fish. However, there is as yet little direct evidence in support of these hypotheses.

These examples show that environments subject to freshwater discharge, in spite of continuous or periodic flushing, nevertheless maintain resident populations of characteristic phyto- and zooplankton species. Adaptation to the immediate environment, as reflected in features such as temperature or salinity tolerance or mode of nutrition, is obviously a primary requirement for such resident species. In a locally growing pelagic population, the relationship between the rates of reproduction + recruitment (from upriver sources) and the loss rate due to flushing will determine its spatial persistence (Smayda 1983). However, resident populations are not always growing but are invariably seasonal in their appearance (Round 1981). They must therefore possess mechanisms that either retain non-growing stages in the region of residence or re-introduce the species by means of a migratory stage. Increasing evidence now shows that plankton species distribution may well reflect adaptation not only to the immediate growth environment but also to larger temporal and spatial scales in the seasonal and regional range effected by means of life cycles involving periodic spatial displacement of specific stages (Fryxell 1983, Steidinger and Walker 1984, Smetacek 1985). Whereas total system biomass is ultimately constrained by light and nutrient availability, the pathways by which this biomass is eventually distributed in the environment can well be governed by behavioural features of the life cycles of the dominant organisms involved. Thus, spatial and temporal environmental patterns created by freshwater discharge can select for organisms with particular "life history types" which in turn will significantly influence cycling and deposition of matter in the marine environment. In the following section, much of which is still speculation, I shall give some examples of such "life history types" and their relevance for the theme of this review. This is a speculative and by no means exhaustive presentation: its main purpose is merely to draw attention to what I feel is an important but sorely neglected aspect of marine ecosystems.

PLANKTONIC LIFE HISTORY TYPES

Tychopelagic diatoms: In the turbulent upper reaches of estuaries, where mixing occurs down to the bottom, large, heavily silicified centric diatoms often predominate. Blooms of these diatoms (Actinocyclus, Coscinodiscus, Thalassiosira, Biddulphia spp.) occur in temperate and tropical estuaries (Smayda 1983). These are generally tychopelagic forms that are equally capable of living on the sea bed. Drebes (1974) reports that centric diatoms characteristic of shallow, turbulent water are frequently covered with mineral particles, presumably attached to microfibrils or a mucilaginous coating. Cloern et al. (1983) and Smetacek (1985) suggest that these mineral particles increase specific gravity and hence permit rapid resettling of the cells. Heavy silicification will have a similar effect. Apparently, the environment in the upper reaches of estuaries (such as in the San Francisco Bay example) selects for phytoplankton with high potential settling rates. This community will be of significance in retarding the transport of both dissolved silica as well as mineral particles to the sea. Production of mucus by the cells will aid in binding smaller mineral particles brought in by rivers and will play a fundamental role in formation of mud sediments in estuaries (Degens and Ittekot 1984). By reducing turbidity, these communities will also effect production further downstream (Parsons and Kessler this vol.) Where the water is less turbid and benthic algae can flourish, mucous secretion by diatom carpets becomes important in stabilizing sediments and reducing resuspension (Holland et al. 1974). Macrophytes can also significantly influence nutrient chemistry of surface water and, by reducing turbulence, increase deposition of suspended particles (Kemp et al. 1984). The fate of the tychopelagic phytoplankton blooms has only been studied in some localities. As development and position of this bloom within the estuary are a direct function of the discharge rate (Cloern et al. 1983), the eventual fate of its biomass will also be dependent on the same factors. It appears that the rich benthic fauna typical of these regions consumes much of the biomass of these blooms. Bivalves in particular are reported to significantly deplete phytoplankton stocks in estuarine environments (Dame et al. 1980, Cloern 1982, Cohen et al. 1983). Crustaceans such as mysids and shrimp are also abundant in shallow, turbulent environments whereas holozooplankton is of relatively lesser importance.

Pelagic bloom diatoms: The bulk of nutrients introduced to the marine environment by river discharge is converted into organic matter by smaller-celled diatom genera such as Skeletonema, Thalassiosira, Chaetoceros, Thalassionema, Nitzschia etc.. These genera are also typical of temperate and boreal spring and subtropical upwelling blooms. They exhibit high growth rates in turbulent, nutrient-rich waters and have a high tolerance to salinity changes (Margalef 1978, Smayda 1983). Apart from these properties, bloom diatoms share another common feature: they tend to sink rapidly out of the surface layer on nutrient depletion, thus transferring energy, essential elements and opal to the sediments. The geochemical and biological significance of this process has only recently been appreciated (Walsh 1983, Smetacek 1984, 1985b). This behaviour of bloom diatoms has been equated with mass mortality in the past, and there is little doubt that the vast majority of these cells do indeed die in the process. However, mass sinking is suggested to primarily represent the transition from a growing stage in surface waters to a resting stage positioned in deeper water or on the sea-bed in the life histories of the respective species (Smetacek 1985b). Such a seeding strategy is necessary as the species must colonize a new environment before the onset of each growth season. In estuaries and freshwater plumes, seeding is effected during mixing of the fresh surface layer with the deeper saline layer. That seeding by benthic resting stages or by suspended cells can be of considerable significance in initiation of these blooms has been shown by Garrison (1979) for Monterey Bay in California. In a different type of environment - the lower St. Lawrence Estuary - Therriault and Levasseur (this vol.) argue that seeding of the fresh surface layer by entrainment of seed cells from the saline deeper layer is of great importance in initiating the spring phytoplankton bloom. Under conditions of increased runoff in spring, stratification is more stable and mixing between surface and deeper layers restricted. The phytoplankton bloom in the surface layer does not take place until June although the growth environment is apparently conducive to bloom formation well before this month. Retardation of the bloom has important ecological implications because nutrients introduced by the St. Lawrence are transported to and presumably utilized on the adjoining shelf instead of in the estuary.

Species-specific differences in diatom sinking behaviour and frustule dissolution rates have been suggested to be of adaptive significance for specific environments (Smetacek 1985b). Thus, it is conjectured that the frustules of oceanic bloom species dissolve more rapidly than those of neritic ones. The frustule acts as ballast during passive sinking; jettisoning this ballast will permit positioning of a sub-surface "seed bed" of small resting spores which will be of obvious advantage in oceanic species. Such large populations of sub-surface seed beds, particularly of Chaetoceros species have been described by Karsten (1905) and more recently by Peinert (1985). In rivers without deltas such as the Zaire, oceanic water is entrained into the freshwater plume and seeding will be effected by pelagic resting stages. Van Bennekom and Berger (1984) report that the bulk of silica incorporated into frustules in the Zaire river plume is dissolved above 500 m depth in the Angola basin. In contrast, the thick frustules of tychopelagic species (Melosira sp.) sediment out to the bottom at the river mouth (Cadee 1984). It would be interesting to study the fate of silica in river plumes with differing diatom species composition. The mass sinking behaviour of diatom blooms has far-reaching consequences for the cycling of matter in the environment. Not only are nutrients introduced by runoff incorporated into organic matter and transferred to the sediments but terrigenous materials of low sinking rates (clay) are co-precipitated out as well.

Red-tide dinoflagellates are characteristic of stratified waters in frontal regions and hence are also commonly associated with saline fronts (Walker 1984). Because of their motility, dinoflagellates exploit their environment more selectively than diatoms which is reflected in the compactness of their populations and the regularity with which some species appear in particular areas. Although they rarely form blooms as extensive as those of diatoms, they are nevertheless of great importance because of the toxicity of many species. Many red-tide dinoflagellates have benthic resting cysts (Dale 1983) which play a fundamental role in the population dynamics of the species (Anderson and Wall 1978). The importance of life cycle events (encystment and germination) in maintaining resident populations in given regions has been established for several species (Tyler et al. 1982, Anderson et al. 1983). Further examples of how dinoflagellate life history types enable gearing of resident populations to the annual cycles of their environment have been described above in the section on Chesapeake Bay. Therriault et al. (1985) have shown how the annual appearance and size of a Protogonyaulax tamarensis population resident in the St. Lawrence Estuary is governed by freshwater runoff.

The genus Ceratium, although not toxic, can also attain nuisance proportions in some areas subject to freshwater discharge such as the New York Bight (Mahoney and Steimle 1979), the German Bight (Hickel 1982) and the Belt Sea of the Baltic (Edler 1984). In the German Bight massive summer blooms of C. furca have been observed in the boundaries of the Elbe River plume (Hickel 1982). C. tripos winter blooms have occurred in the other environments. In all cases, these Ceratium blooms were associated with severe oxygen depletion in deeper water and at the sediment surface. The Kiel Bight studies show that sedimentation of these blooms is not part of the life history strategy (marine ceratia are not known to have benthic resting stages) but is instead preceded by mass mortality.

Zooplankton: Estuarine zooplankton assemblages tend to differ from the more typically marine ones. For instance, heterotrophic nanoflagellates and ciliates are reported to play a more important role in estuarine than in marine waters (Burkill 1982). These small heterotrophs aid in recycling and hence retaining nutrients within the surface layer and eventually provide food for crustacean zooplankters further downstream (Parsons and Kessler this vol.).

In shallow estuaries the genus _Acartia_ tends to be the dominant copepod, whereas in deeper estuaries and fjords typically marine, albeit coastal, zooplankton assemblages dominated by genera such as _Pseudocalanus_, _Paracalanus_ and _Centropages_ are more the rule. It appears that differences in behaviour and life history strategies in relation to hydrography/topography rather than properties of the growth environment select between these two groups of copepods. Thus, neither food quality nor temperature or salinity appear to be the decisive factor (Schlieper 1958, Hernroth and Ackefors 1977, Martens 1981, Harris et al. 1982, Durbin et al. 1983). Apparently, water depth per se rather than salinity tolerance is the decisive factor selecting between these copepod assemblages. Thus, as shallow basins are flushed more rapidly than deeper ones, _Acartia_ populations must exhibit behavioural features enabling retention within the former environments. Evidence for this type of behaviour has been reported by Show (1980) who showed that adjustment of the vertical position in relation to tidal cycle can indeed permit maintenance of the population within the estuary. Over longer time scales, _Acartia_ spp. maintain persistence during the non-growth season by means of benthic eggs (Grice and Marcus 1981). Other typical estuarine zooplankters such as tintinnids, rotifers and cladocerans are also known to have benthic resting stages (Hensen 1887). _Pseudocalanus_ type copepods on the other hand overwinter as pelagic adult or copepodite stages and hence require access to less turbulent, deeper water for this purpose.

The degree of gearing between phytoplankton and pelagic herbivores will depend on the life history types of the latter which in turn will determine the food supply of the benthos (Smetacek 1984). Thus, the factors triggering hatching of benthic stages or activation of overwintering populations will not necessarily coincide with those responsible for build-up of the food supply in the overlying water. Where large copepod stocks are present before significant growth of phytoplankton occurs, grazing pressure is heavy and the bloom prolonged. Under such conditions, residence time of nutrients in the surface layer increases as also the extent of their horizontal dispersal. Copepod grazing can also significantly contribute to deposition of river-borne clay particles by pelletization (Syvitski 1979). In contrast, where grazing pressure is low relative to growth of phytoplankton, nutrient uptake will be rapid and large blooms will develop resulting in subsequent heavy sedimentation within a fairly restricted region around the site of freshwater discharge. The impoverishment of the surface layer will result in low pelagic biomass levels. This will have far-reaching consequences for pelagic carnivores such as fish.

CONCLUDING REMARKS

We are now becoming increasingly aware of the importance of temporal gearing between phyto- and zooplankton in determining the patterns of utilization of organic matter (Longhurst 1983), although we know little about the underlying organism strategies. Parsons and Kessler (this volume) have demonstrated by means of a simulation model of an estuarine system how production and fate of organic matter in a freshwater plume can be significantly affected by small differences in initial factors such as extinction coefficient of the water or its organic load. They also showed that the initial size of the zooflagellate population can play an important role in determining the subsequent build-up of zooplankton biomass. It can be argued that similar small differences in the "seed populations" of phyto- or zooplankton species can be important in determining species dominance status and structure of the food web further downstream. Thus, the mechanisms by which the increasingly saline layer of surface water is successively colonized by organisms adapted to growing in such environments will also be important in determining production and fate of material.

We are yet far from achieving a comprehensive understanding of the factors influencing species composition of the marine plankton. Further, our appreciation of the role of species composition in influencing structure and functioning of pelagic systems is still in a rudimentary stage. Without this type of basic "natural history" knowledge we will not be able to make any environmental impact assessment as we will not be able to distinguish a response signal of a perturbed environment from natural fluctuation of the system under study (Hedgepeth 1979, Angel 1984a). For this purpose it is necessary to develop conceptual frameworks of the possible types of interactions between pelagic organisms and their environment that include both growth performance as well as survival strategy. The phytoplankton "life-forms" defined by Margalef (1978) represent a step in this direction but are based primarily on the smaller scales of the immediate growth environment. Extension of such a scheme to include the longer scales of survival and the mechanisms of departure from and reappearance in the environment of growth should now be attempted. An analysis of the descriptive literature available on species distribution patterns in relation to the environment can also help in the search for life history types whose presence or absence could enable assessment of a particular situation. Because of the limitations posed by the logistics of sampling of the marine environment, any information that helps in elucidating the past history of a given water mass will be of particular value. Knowledge of the biology of the species encountered could hence greatly aid in providing such information.

Acknowledgements: I thank J.-C. Therriault, U. Bathmann, B. v. Bodungen, J. Cloern, J. Lenz, M. Levasseur, T. Parsons, R. Peinert and S. Skreslet for their valuable comments on an earlier version of this paper. (Publ. Nr. 6 of the "Sonderforschungsbereich" 313 of Kiel University).

REFERENCES

Anderson, D.M. and D. Wall (1978). Potential importance of benthic cysts of Gonyaulax tamarensis and G. excavata in initiating toxic dinoflagellate blooms. J.Phycol. 14, 224-234.

Anderson, D.M., S.W. Chisholm and C.J. Watras (1983). Importance of life cycle events in the population dynamics of Gonyaulax tamarensis. Mar. Biol. 76, 179-189.

Angel, M.V. (1984). Deep-water biological processes in the northwest region of the Indian Ocean. Deep-Sea Res. 31, 935-950

Bamstedt, U. (1985). Spring-bloom dynamics in Kosterfjorden, Western Sweden: Variation in phytoplankton production and macrozooplankton characteristics. Sarsia 70, 69-82.

Bennekom, A.J. van and G.W. Berger (1984) Hydrography and silica budget of the Angola Basin. Neth. J. Sea Res. 17, 149-200.

Boynton, W.R., W.M. Kemp and C.W. Keefe (1982). A comparative analysis of nutrients and other factors influencing estuarine phytoplankton production. In: Estuarine comparisons, pp. 69-90, V.S. Kennedy (Ed.), Academic Press.

Burkill, P.H. (1982). Ciliates and other microplankton components of a nearshore food web: standing stocks and production processes. Ann. Inst. oceanogr. 58(S), 335-350.

Cadee, G.C., (1984). Particulate and dissolved organic carbon and chlorophyll a in the Zaire river, estuary and plume. Neth. J. Sea Res. 17: 426- 440.

Cloern, J.E., (1982). Does the benthos control phytoplankton biomass in south San Francisco Bay? Mar. Ecol Prog. Ser. 9: 191-202.

Cloern, J.E., A.E. Alpine, B.E. Cole, R.L.J. Wong, J.F. Arthur and M.D. Ball, (1983). River discharge controls phytoplankton dynamics in the northern San Francisco Bay estuary. Estuar. coast. shelf Sci. 16: 415-429.

Cohen, R.R.H. (in press). Physical processes and the ecology of a winter dinoflagellate bloom of Katodinium rotundatum. Mar. Ecol. Prog. Ser.

Cohen, R.R.H., P.V. Dresler, J.J. Phillips and L.R. Cory, (1984). The effect of the Asiatic clam, Corbicula fluminea on phytoplankton of the Potomac river, Maryland. Limnol. Oceanogr. 29: 170-180.

Colebrook, J.M. (1985). Continuous plankton records: overwintering and annual fluctuations in the abundance of zooplankton. Mar. Biol. 84: 261- 265.

Dahl, E. (1978). Effects of river discharge on the coastal plankton cycle. Mitt. Internat. Verein Limnol. 21: 330-341.

Dale, B. (1983). Dinoflagellate resting cysts: "Benthic plankton". In: Survival strategies of the algae, pp. 69-136, G.A. Fryxell (Ed.), Cambridge Univ. Press.

Dame, R., R. Zingmark and D. Nelson, (1980). Filter feeding coupling between the estuarine water column and benthic subsystems. In: Estuarine perspectives, pp. 525-526, V.S. Kennedy (Ed.), Academic Press.

Degens, E.T. and V. Ittekott (1984). A new look at clay-organic interactions. Mitt. Geol.-Paläont. Inst. Univ. Hamburg 56: 229-248.

Dowidar, N.M. (1984). Phytoplankton biomass and primary produc-

tivity of the south-eastern Mediterranean. Deep-Sea Res. 31, 983-1000.
Drebes, G. (1974). Marines Phytoplankton. Georg Thieme Verlag, 186 pp.
Durbin, E.G., A.G. Durbin, T.J. Smayda and P. Verity (1983). Food limitation of production by adult Acartia tonsa in Narragansett Bay, Rhode Island. Limnol. Oceanogr. 28, 1199-1213.
Edler, L. (1984). A mass development of Ceratium species on the Swedish west coast. Limnologica 15: 353-357.
Eilertsen, H.C., B. Schei and J.P. Taasen (1981). Investigations on the plankton community of Balsfjorden, northern Norway. The phytoplankton 1976-1978. Abundance, species composition and succession. Sarsia 66, 129-141.
Fosshagen, A. (1979). How the zooplankton community may vary within the a single fjord system. In: Fjord oceanography, pp. 399-405, H.J. Freeland, D.M. Farmer and C.D. Levings (Eds.), Plenum Press.
Føyn L. and F. Rey (1981). Nutrient distribtuion along the Norwegian Coastal Current. In: The Norwegian Coastal Current, pp. 629-639, R. Saetre and M. Mork (Eds.), Univ. of Bergen.
Fryxell, G.A. (Ed.) (1983). Survival strategies of the algae. Cambridge Univ. Press, 144 pp.
Garrison, D.L. (1981). Monterey Bay phytoplankton. II. Resting spore cycles in coastal diatom populations. J. Plankt. Res. 3: 137-156.
Graf, G., W. Bengtsson, U. Diesner, R. Schulz and H. Theede (1982). Benthic response to sedimentation of a spring phytoplankton bloom: process and budget. Mar. Biol. 67, 201-208.
Grice, G.D. and H.H. Marcus (1981). Dormant eggs of marine copepods. Oceanogr. mar. Biol. Ann. Rev. 19, 125-140.
Hargrave, B.T., G.C. Harding, K.F. Drinkwater, T.C. Lambert and W.G. Harri son (1985). Dynamics of the pelagic food web in St. George's Bay, southern Gulf of St. Lawrence. Mar. Ecol. Prog. Ser. 20, 221-240.
Harris, R.P., M.R. Reeve, G.D. Grice, G.T. Evans, V.R. Gibson, J.R. Beers and B.K. Sullivan (1982). Trophic interactions and production processes in natural zooplankton communities in enclosed water columns. In: Marine mesocosms, pp. 252-387, G.D. Grice and M.R. Reeve (Eds.), Springer Verlag.
Hedgpeth, J.W. (1978). As blind men see the elephant: the dilemma of marine ecosystem research. In: Estuarine interactions, pp. 3-15, M.L. Wiley (Ed.), Academic Press.
Hensen, V. (1887). Über die Bestimmung des Planktons oder des im Meere treibenden Materials an Pflanzen und Tieren. Ber. Komm. wiss. Unters. dt. Meere 5: 1-108.
Hernroth, L, and H. Ackefors (1977). The zooplankton in the Baltic proper. A long-term investigation of the fauna, its biology and ecology. Inst. Mar. Res. Lysekil Publ., 58 pp.
Hickel, W. (1982). Ceratium red tide in the German Bight in August, 1981: spatial distribution. ICES C.M. 1982/L:8.
Holland, A.F., R.G. Zingmark and J.M. Dean (1974). Quantitative evidence concerning the stabilization of sediments by marine benthic diatoms. Mar. Biol. 27: 191-196.
Karsten, G. (1905). Das Phytoplankton des antarktischen Meeres nach dem Material der deutschen Tiefsee-Expedition. Wiss. Ergebn. Tiefsee-Exped. 2: 1-136.

Kemp, W.M., W.R. Boynton, R.R. Twilley, J. C. Stevenson and L.G. Ward (1984). Influences of submersed vascular plants on ecological processes in upper Chesapeake Bay. In. The estuary as a filter, pp. 367-394, V.S. Kennedy (Ed.), Academic Press.

Kidd, R.and F. Sander (1979). Influence of Amazon river discharge on the marine production system off Barbados, West Indies. J. Mar. res. 37: 669- 681.

Lewis, M.R. and T.R. Platt (1982). Scales of variability in estuarine ecosystems. In: Estuarine comparisons, pp. 3-20, V.S. Kennedy (Ed.), Academic Press.

Longhurst, A.R. (1983). Summary - biological oceanography, Proc. Joint Oceanogr. Assem. 1982 - General Symposia pp. 143-146. CNC/SCOR Publ.

Mahoney, J.B. and F.W. Steimle (1979). A mass mortality of marine animals associated with a bloom of *Ceratium tripos* in the New York Bight. In: Toxic dinoflagellate blooms, pp. 225-230, D.L. Taylor and H.H. Seliger (Eds.), Elsevier, North Holland.

Malone, T.C. (1984). Anthropogenic nitrogen loading and assimilation capacity of the Hudson River estuarine system, USA. In: The estuary as a filter, pp.291-312, V.S. Kennedy (Ed.), Academic Press.

Malone, T.C. and M.B. Chervin (1979). The production and fate of phytoplankton size fractions in the plume of the Hudson River, New York Bight. Limnol. Oceanogr. 24, 683-696.

Margalef, R. (1978). Life-forms of phytoplankton as survival alternatives in an unstable environment. Oceanol. Acta 1: 493-509.

Martens, P. (1981). On the *Acartia* species of the northern Wadden Sea of Sylt. Kieler Meeresforsch., Sonderh. 5: 153-163.

Matthews, J.B.L. and B.R. Heimdal (1979). Pelagic productivity and food chains in fjord systems. In: Fjord Oceanography, pp. 377-398, H.J. Free land, D.M. Farmer and C.D. Levings (Eds.) Plenum Press.

Nichols, F. (in press). Increased benthic grazing: one explanation for low phytoplankton biomass in northern San Francisco Bay during the 1976-1977 drought. Estuar. coast. shelf Sci.

Noji, T., U. Passow and V. Smetacek. (in press). Interaction between pelagial and benthal during autumn in Kiel Bight. I. Development and sedimentation of phytoplankton blooms. Ophelia.

Officer, C.B., T.J. Smayda and R. Mann (1982). Benthic filter feeding: A natural eutrophic control. Mar. Ecol. Prog. Ser. 9: 203-210.

Parsons, T.R. and T.A. Kessler (this vol.). Computer model analysis of pelagic ecosystems in estuarine waters.

Peinert, R. (1985). Saisonale und regionale Aspekte der Produktion und Sedimentation von Partikeln im Meer. Ph.D thesis, Univ. Kiel, 105 pp.

Peinert, R. (this vol.). Production, grazing and sedimentation in the Norwegian coastal current.

Revelante, N. and M. Gilmartin (1976). The effect of Po river discharge on phytoplankton dynamics in the northern Adriatic Sea. Mar. Biol. 34: 259- 279.

Rey, F. (1981). The development of the spring phytoplankton outburst at selected sites off the Norwegian coast. In: The Norwegian Coastal Current, II, pp. 649-680, R. Saetre and M.

Mork (Eds.), University of Bergen.
Round, F.. (1981). The ecology of algae. Cambridge Univ. Press, 653 pp.
Sakshaug, E. (1976). Dynamics of phytoplankton blooms in Norwegian Fjords and coastal waters. In: Freshwater on the sea, pp. 139-143, S. Skreslet, R. Leinebo, J.B.L. Matthews and E. Sakshaug (Eds.), Assoc.Norwegian Oceanogr.
Schlieper, C. (1958). Physiologie des Brackwassers. Die Binnengewässer 22: 219-348.
Seliger, H.H., K.R. McKinley, W.H. Biggley, R.B. Rivkin and K.R.H. Aspden, (1981). Phytoplankton patchiness and frontal regions. Mar. Biol. 61: 119- 131.
Show, I.T. (1980) The movements of a marine copepod in a tidal lagoon. In: Estuarine and wetland processes, pp. 561-601, Hamilton, P. and K.B. MacDonald (Eds.) Plenum Press.
Sinclair, M., D.V. Subba Rao and R. Couture (1981). Phytoplankton temporal distributions in estuaries. Oceanol. Acta 4, 239-246.
Skreslet, S. (1976). Influence of freshwater outflow from Norway on recruitment to the stock of Arcto-Norwegian cod (Gadus morhua). In: Freshwater on the sea, pp. 233-237, S. Skreslet, R. Leinebo, J.B.L. Matthews and E. Sakshaug (Eds.), Assoc. of Norwegian Oceanogr.
Skreslet, S. (1981). Informations and opinions on how freshwater outflow to the Norwegian coastal current influences biological production and recruitment to fish stocks in adjacent seas. In: The Norwegian coastal current, II, pp. 712-748, R. Saetre and M. Mork (Eds.), University of Bergen.
Smayda, T.J. (1980). Phytoplankton species succession. In: The physiological ecology of phytoplankton, pp. 493-570, I. Morris (Ed.), Blackwell Scient. Publications.
Smayda, T.J. (1983). The phytoplankton of estuaries. In: Estuaries and enclosed seas, pp. 65-102, B.H. Ketchum (Ed.) Elsevier Scient. Publ. Comp.
Smetacek, V. (1980). Annual cycle of sedimentation in relation to plankton ecology in western Kiel Bight. Ophelia Suppl. 1, 65-76.
Smetacek, V. (1984). The supply of food to the benthos. In: Flows of energy and materials in marine ecosystems: Theory and practice, pp. 517-548, M.J.R. Fasham (Ed.), Plenum Press.
Smetacek, V. (1985a). The annual cycle of Kiel Bight plankton: A long-term analysis. Estuaries 8, 145-157.
Smetacek, V.S. (1985b). Role of sinking in diatom life-history cycles: ecological, evolutionary and geological significance. Mar. Biol. 84, 239- 251.
Smetacek, V., B. v. Bodungen, B. Knoppers, R. Peinert, F. Pollehne, P. Stegmann and B. Zeitzschel (1984). Seasonal stages characterising the annual cycle of an inshore pelagic system. Rapp. P.-v. Reun. Cons. int. Explor. Mer 183: 126-135.
Steidinger, K.A. and L.M. Walker (Eds.) (1984). Marine plankton life cycle strategies. CRC Press, 158 pp.
Svendsen, H. (this vol.). Mixing and exchange processes in estuaries, fjords and shelf waters.
Syvitski, J.P.M. (1979). Flocculation, agglomeration, and zooplankton pelletization of suspended sediment in a fjord receiving glacial meltwater. In: Fjord Oceanography, pp.

615-623, H.J. Freeland, D.M. Farmer and C.D. Levings (Eds.), Plenum Press.

Therriault, J.C., J. Painchaud and M. Levasseur (1985). Factors controlling the occurrence of <u>Protogonyaulax tamarensis</u> and shellfish toxicity in the St. Lawrence estuary: Freshwater runoff and the stability of the water column. In: Third international conference on toxic dinoflagellates. D.M. Anderson, A.W. White and D.G. Baden (Eds.), Elsevier (in press).

Therriault, J.C. and M. Levasseur (this vol.). Freshwater runoff control of the spatio-temporal distribution of phytoplankton in the lower St. Lawrence estuary.

Thordardottir, T. (this vol.). Timing and duration of spring blooming off S and SW Iceland related to hydrography and zooplankton.

Tyler, M.A. and H.H. Seliger (1981). Selection for a red tide organism: Physiological responses to the physical environment. Limnol. Oceanogr. 26: 310-324.

Tyler, M.A., D.W. Coats and D.M. Anderson (1982). Encystment in a dynamic environment: Deposition of dinoflagellate cysts by a frontal convergence. Mar. Ecol. Prog. Ser. 7, 163-178.

Walker, L.M. (1984). Life histories, dispersal and survival in marine, planktonic dinoflagellates. In: Marine plankton life cycle strategies, pp. 19-34, K.A. Steidinger and L.M. Walker (Eds.), CRC Press.

Walsh, J.J. (1983). Death in the sea: Enigmatic phytoplankton losses. Progr. Oceanogr. 12, 1-86.

Wassmann, P. (1983). Sedimentation of organic and inorganic particulate material in Lindaspollene, a stratified, landlocked fjord in western Norway. Mar. Ecol. Prog. Ser. 13: 237-248.

A FRAMEWORK FOR DISCUSSION OF MARINE ZOOPLANKTON PRODUCTION
IN RELATION TO FRESHWATER RUN-OFF

J.B.L.Matthews
Dunstaffnage Marine Research Laboratory
Oban, Scotland

ABSTRACT

In the absence of any comprehensive method for estimating zooplankton production, the article does not attempt a definitive review of the effect of freshwater outflow on coastal zooplankton, though reference is made to recent comprehensive articles which relate to the subject. It is suggested that a good framework for the discussion of research priorities is provided by simulation modelling from which key processes can be identified. The combined study of biological and physical processes in considered with emphasis on the choice of appropriate, but not necessarily the same, time and space scales. It is suggested that "The Ecology of Advection" may be a theme amenable to both theoretical and observational development.

INTRODUCTION

No comprehensive method of estimating secondary (herbivore) production in a pelagic system has yet been devised, so it is not possible to review this aspect in the context of coastal systems affected by freshwater run-off. Four recent papers, however, summarise relevant information and may be cited here to record the extent of present knowledge and to indicate current lines of thought in tackling the problem; Matthews & Heimdal (1980) attempted some standardisation of published information on levels of planktonic production in fjordic systems, tentatively identified planktonic communities associated with different degrees of water mass isolation, and suggested that massive water exchange ("flushing") is necessary for sustaining high levels of production. Miller (1983) considered the zooplankton of estuaries in a strict sense, i.e. where a river meets and mixes with the sea; he typified the communities in well studied North American and Australian estuarine systems, reviewed estimates of zooplankton production in the light of special features in the biology of organisms concerned (copepods), and considered the problems of population maintenace in relation to flushing. Denman & Powell (1984) included estuarine plume fronts briefly in their review of the effects of physical processes on planktonic systems. Burrell (in press) has comprehensively discussed biological interactions between fjords and coastal regions of the Gulf of Alaska, with extensive reference to investigations in other fjordic systems; the extent to which fjords possess individual features of water circulation and exchange becomes very apparent, making generalisations difficult at this, largely descriptive, stage of knowledge of fjord communities.

Against this background and in the context of a book intended to identify the scientific problems associated with the effects of freshwater run-off on coastal regions and to suggest ways of tackling them, I propose to summarise the biological aspects involved, list the physical processes that impinge on the zooplankton, and then combine these two aspects in the "Biophysics" of estuarine plume water masses. I shall follow the example of Longhurst (1984) and adopt the terminology of simulation modelling in defining the aspects which must be attended to and about which we have very disparate levels of understanding, but in so doing I do not want to restrict the reader's thinking about ecosystems to that particular methodology.

CONCEPTUAL BASE (FLOW DIAGRAM)

Although this provides the essential framework for the investigations and defines the form that they should take, it does not need extensive elaboration in this chapter. The structure of this book reflects a considerable degree of conceptualisation, starting with the physical matters of water movement, mixing, etc., and going on to look at the biological effects of these. I am only concerned that we should not stop there but keep an open mind on the possibility, if not of biological effects on the physical environment, at least of feedback from biological to physical oceanographers. It is, moreover, appropriate for concepts to be considered in greater detail in the reports of the working group discussions held at the meeting. I would emphasise Longhurst's point (loc.cit.) that, for successful integration, the conceptual base must be both adequate and comprehensible to all participants.

STATE VARIABLES

These are the biological units chosen to represent the zooplankton, be they species, dominants, size fractions, trophic groupings, feeding types or perhaps one lump total, all expressed as numbers or, more likely, as biomass.

RATE VARIABLES

Since state variables are expressed in terms of numbers or biomass, the relevant processes are those which effect changes in numbers and/or biomass, i.e. the standard processes of population dynamics - recruitment, growth and mortality. In any particular study two decisions need to be made: a) whether these processes need to be subdivided, e.g. distinguishing that part of mortality due to predation; b) whether underlying processes can and should be substituted for the more apparent processes already mentioned; for example, are needs satisfied by parameter values obtained from an empirical growth curve or need one go beyond this to parameters rooted in physiological truths?

The traditional measuring of stock size or biomass, because this can be dome by means of "snap-shot" sampling and because one thought that it was numbers or size that really mattered, has been partly superseded by the emphasis given in recent years to processes, on the argument that it is not how much but how fast that is important. A crucial lesson to be learnt from simulation modelling, however, is that knowledge of both, as state and rate variables respectively, is essential for a proper understanding of ecological systems.

FORCING FUNCTIONS AND BOUNDARY CONDITIONS

The forces that drive biological systems are well known (though not always well understood) and I do not intend to discuss them at length here; they are summarised below, under Environmental Processes.

Boundaries must be set firmly and conditions at those boundaries must be evaluated, but one must not ignore the often artificial and chance nature of those boundaries. There is no reason why political boundaries should coincide with the limits of scientific projects. Oceanography has until recently been able to benefit from free access to all but inshore waters, but that has now changed and any comparative studies of coastal systems must involve multinational collaboration. Fortunately, most of the systems under consideration in this book are under the jurisdiction of a single coastal state, so research proposals should not founder reasons of international politics.

Other limitations may be imposed for logistic reasons; one simply cannot disregard such factors as accessibility in choosing a site on topographic and hydrographic grounds; nor, having chosen the site, is one often free to sample or analyse it in as much detail or as frequently as one would wish.

If the boundaries in space or time hamper rather than help, they must be changed. It is the nature of the scientific problem which must determine what boundaries are acceptable.

BIOLOGICAL PROCESSES

I shall now summarise briefly the biological processes that seem important and, at the same time, indicate the time scales that operate and, in the vertical plane only, the space scale. As already mentioned, the processes are those of population dynamics:

1) Maturation and Reproduction - This involves the processes of change in the parent generation and, in addition to such obvious aspects as fecundity, can include, for example, changes in proximate biochemical composition associated with egg production or in behaviour associated with spawning. These processes operate generally in time scales of days to weeks.

2) Growth and Development - For present purposes, concerning the functioning of populations in a dynamic environment, I think it is sufficient to consider growth and development empirically, measuring performance in a changing environment and comparing such observations with theoretical or experimentally derived growth expectations in order to assess the effect of environment. These processes are generally measurable in days, weeks and months.

3) Mortality - This chapter is concerned only with zooplankton in the pelagic community, so I have intentionally termed this group of processes Mortality rather than Predation, though they may be considered as Predation (and Loss) in the synthesis of the whole system. The time scales operating in this category extend from days (the senile mortality of a whole generation that can occur after spawning) to months or more in the case of the mortality rate of a generation from hatching onwards.

4) Vertical Migration - This term is used here in a broad sense, to encompass any changes in vertical distribution. One is not concerned here with why these changes take place, but very much concerned with the extent of the phenomenon (both in terms of species and stages and in terms of distance covered), and with correlations that may exist with environmental conditions in the water column. If the

correlations are convincing enough causal relationships can be assumed for predictive purposes.

Diurnal migration operates (obviously!) on a daily time scale, while ontogenetic or seasonal migration operates on monthly to yearly scales. Vertical movements are of the order of tens to hundreds of metres, though the scale may be finer near the surface.

ENVIRONMENTAL PROCESSES

Planktonic populations are contained in, and to a large extent constrained by, water masses. The physical processes that affect populations are of two types, the static and the dynamic. The former, seasonal changes in temperature or light conditions, for example, can have an effect, physiological or otherwise, on the populations contained in the water mass, but they do not, per se, affect the integrity of the populations themselves. The processes of phytoplankton development belong in this category of processes (boundary conditions in the present context) since these profoundly affect the living conditions of zooplankton. The difference from physical processes is that in this case there can be reciprocal effects between the phytoplankton and the zooplankton, for example, net uptake of nutrients by phytoplankton and regeneration by zooplankton.

Dynamic processes, on the other hand, do affect the integrity of populations is so far as they involve transport and mixing and are certainly among the most important processes in determining and controlling near-shore pelagic systems, particularly those affected by freshwater run-off.

Steep gradients and a high frequency of change are features of both types of processes in the areas we are dealing with. Time scales range from hours to a year, with some year-to-year variations, and space scales from metres in the vertical plane to kilometres in the horizontal.

THE "BIOPHYSICS" OF PLUME WATER MASSES

A well-established requirement for a successful ecological investigation is correct identification of the operative time and space scales and execution of a sampling programme at one scale below them. This is undoubtedly true, but extension of the reasoning, to say that satisfactory interdisciplinary research depends on identifying common interests in particular scales, is not. Denman & Powell (1984) when discussing estuarine plume fronts point out that physical processes operating at one time scale can be crucial to the functioning of plankton at a larger scale. Such divergence of scale can also occur in the opposite direction, as in the case of local stocks of pelagic organisms, e.g. Meganyctiphanes norvegica in Korsfjorden (Matthews, 1973) being maintained from year to year, apparently as a result of a deep-water exchange which extends horizontally for hundreds of kilometres.

One must, therefore, be prepared to develop integrated ecological studies with the parts operating at different scales. It is not only because physical factors operate on biological ones rather than vice versa, but also because the scales may differ and the physical data are more precise that collaboration often breaks down - or never gets started. But perhaps the benefits can be mutual. Measurement of physical conditions gives an instantaneous picture of a situation. A succession of such measurements enables one to estimate rates of processes, and some processes, current flow for example, can be measured directly, but again instantaneously or for a limited period. A zooplankton organism, on the other hand, can contain a potted history of itself and the environments it has passed through - a "multiracer" if one can only interpret the signals. This, of course, is a function of the longevity of zooplankton organisms, relative to many physical processes. The operative time scale is often too long to relate to the speed of water movement in frontal areas, in marked contrast to the situation with phytoplankton. Fronts are, therefore, popular places for phytoplanktologists, but I suggest they should not necessarily be so for zooplanktologists, who should be looking in the water masses, where populations can live and grow. There are, of course, situations where patterns of water circulation and of plankton behaviour, or attraction to a space beyond a barrier, can concentrate zooplankton actually in fronts, but advection more usually concentrates swarms in water masses from which their behaviour or tolerance limits hinders their escape.

Advection is undoubtedly a key word in the range of coastal oceanographic processes under consideration in this book. I shall end, therefore, with a few remarks on what may be termed "The Ecology of Advection". This is the passive version of the Ecology of Migration. The opportunities may be more limited than in the case of migrating birds, for example, but the energy expenditure involved in exploiting passive means of dispersal is minimal compared with active migration. There may be few cases, on a large scale, where advection can provide a sure means of return, but perhaps the Selfish Gene (pace Dawkins 1976) has as small a role to play in the stage IV Calanus finmarchicus caught up in a one-way coastal current as it has in a worker ant.

The importance of retention of herring larvae near the spawning grounds has recently been discussed by Sinclair & Tremblay (1984) as a mechanism whereby discrete stocks are maintained. It might be supposed that planktonic organisms not destined to grow into the nekton would have an even greater need for retention.

On the other hand, evolutionary theory lays emphasis on the failure of almost any population to achieve potential recruitment levels. Positive consequences for genetic fitness, however, may not be the only result of excessive production: the high fecundity of many parasites seems to be assiciated with the low chance of any individual finding a suitable host. The ecology of plankton populations and the behaviour of plankton organisms may invoke similar strategies in strongly advective systems.

How should the Ecology of Advection be studies? Eulerian techniques can provide an assessment of the population or community transport, while Lagrangian techniques can be used to trace the fate of a chosen section of the population or community. On their own, however, these techniques are inadequate to tackle the added complexity introduced by biological action. In this respect the approach to diffusion studies adopted by Talbot (1974) is of great interest. This should be extended with the aim of developing what may be called a Theory of Biological Diffusion.

REFERENCES

Burrell, D.C. : Interaction between silled fjords and costal regions of the Gulf of Alaska. In: Hood, D.W., Zimmerman, S. (eds.). The Gulf of Alaska. Physical Environment and Biological Resources, Chap. 17 (in press).
Dawkins, R. : The Selfish Gene. Oxford: OUP 1976.
Denman, K.L., Powell, T.M. : Effects of physical processes on planktonic ecosystems in the coastal ocean. Oceanogr.Mar.Biol.Ann.Rev. $\underline{22}$, 125-168 (1984).
Longhurst, A.R. : The importance of measuring fluxes in marine ecosystems. In: Fasham, M.J.R. (ed.). Flows of Energy and Materials in Marine Ecosystems. NATO Conference Series IV, $\underline{13}$. New York: Plenum 1984, pp.3-22.
Matthews, J.B.L. : The succession of generations in a population of Meganyctiphanes norvegica, with an estimate of the energy flux. Int.Cl.Explor.Sea, C.M. 1973/L:19, 6pp.
Matthews, J.B.L., Heimdal, B.R. : Pelagic productivity and food chains in fjord systems. In: Freeland, H.J., Farmer, D.M., Levings, C.D. (eds.). Fjord Oceanography. NATO Conference Series, IV, $\underline{4}$. New York: Plenum 1980, pp.377-398.
Miller, C.B. : The zooplankton of estuaries. In: Ketchum, B.H. (ed.). Estuaries and Enclosed Seas. Elsevier, Amsterdam 1983, pp. 103-149.
Sinclair, M.E., Tremblay, M.J. : Timing of spawning of Atlantic herring (Clupea harengus harengus) populations and the match-mismatch theory. Can.J.Fish.Aquat.Sci. $\underline{41}$: pp. 1055-1065 (1984).
Talbot, J.W. : Diffusion studies in fisheries biology. In: Harden Jones (ed.). Sea Fisheries Research. London: Elek Science 1974, pp. 31-54.

THE DEPENDENCE OF FISH LARVAL SURVIVAL ON FOOD AND PREDATOR DENSITIES

William C. Leggett
Department of Biology
McGill University
1205 Avenue Dr. Penfield
Montréal. Canada. H3A 1B1

ABSTRACT

The results of field and laboratory investigations into the role of prey abundance and size, and of predators, as regulators of larval survival in fishes are reviewed. Laboratory evidence of the sensitivity of larval survival to prey available at the time of first feeding continues to mount. Estimates of minimum prey concentrations necessary to ensure survival have declined by approximately 1-2 orders of magnitude in recent years. Field evidence of a "critical period" at yolk sac absorbtion remains elusive. Knowledge of the impact of predation on larval survival also remains scanty. The use of negative correlations between predator and larval density as evidence of this effect has recently been effectively criticized. It is clear that predators are capable of exerting a major impact, but unclear whether this impact is realized in nature.

INTRODUCTION

Mortality in the larval stages of fishes is typically high (5-25% d^{-1}, Dahlberg 1979). It is widely believed that year-to-year variation in mortality rates during the larval period is the major cause of variability in year-class strength (Leiby 1984).

The dominant hypotheses relating to the regulation of larval survival in fishes are that 1) survival is a function of food availability, particularly at the onset of exogenous feeding; and 2) survival is a function of predation during the larval stage. This paper reviews the evidence for the dependence of larval survival on predators and prey. Space constraints prevent a consideration of evidence concerning the importance of the combined effects of predator prey dynamics on larval survival.

FOOD AND SURVIVAL - THE CRITICAL PERIOD CONCEPT

The concept that early larval survival in fishes is primarily food-related derives from Hjort (1914, 1926). His hypothesis, as re-stated by May (1974) proposes that "the strength of a year-class is determined by the availability of planktonic food shortly after the larval yolk supply has been exhausted" (the so called "critical period").

May (1974) reviewed the available evidence for the existence of a maxima in mortality rates at or near the onset of exogenous feeding and found it wanting... "The available data do not allow one to answer the question of whether or not mortality is concentrated at the end of the yolk-sac stage in natural populations." Dahlberg (1979) expanded May's data set by including mortality curves for freshwater, anadromous, and marine fishes. These data, too, fail to provide clear evidence of a major increase in mortality at or near the end of the yolk-sac stage. This is true even in cases where total mortality varied significantly either between years or between locations for a single species (plaice, _Hippoglossoides platessoides_; striped bass, _Roccus saxalilis_).

These negative findings do not, however, justify rejection of the idea that the major mortality experienced by larval fishes is highly concentrated at one or more critical development stages. Most of the larval survival curves developed from field data have been based on sampling programs which suffer several limitations: 1) time/space resolution may be insufficient to detect contracted periods of high mortality related to particular developmental stages or to detect and correct for the effect of patchiness. 2) Time averaging of age-abundance data for several cohorts produced in a single year may artificially smooth the survival curve and obscure detail actually present in the curves of synchronously developing cohorts; 3) averaging across more than one cohort may also cause a loss of information on meso-scale variations in survival (1-10 km, hours to days); 4) mortality estimates have typically not been corrected for losses due to advection or diffusion.

Fortier and Leggett (in press) attempted to circumvent these problems by repeatedly sampling a single cohort of larval capelin (Mallotus villosus) once each 2 h over a 46-h interval that spanned the transition from yolk-sac nutrition to exogenous feeding. The larvae were sampled during their seaward drift in the St. Lawrence estuary and were tracked by means of a Challenger-type drogue. The average daily survival rate over the entire period was 43.7% d^{-1}. This is substantially lower than survival rates averaged over the yolk-sac period (75% d^{-1} to 95% d^{-1}) reported by Dahlberg (1979). Advective losses were eliminated as a source of bias in the estimate, and similar loss rates were observed in other cohorts sampled in the same area. The authors concluded that the lower survival rates experienced by capelin in the 46 h period coincident with first feeding were real and differed from estimates for other species principally because the latter were averaged over longer time intervals thereby including both pre- and post-yolk sac as well as "transition" survival. A logistic model used to extrapolate the survival curve of larval capelin from emergence to final yolk sac exhaustion (Fig. 1) explained 57% of the variance in larval density over the sampling interval. The realized rate of mortality along the survival curve was maximal at t = -60 h when the population yolk-sac frequency was 50%. This result is consistent with Hjort's hypothesis which predicts a maximum mortality at the exhaustion of the yolk reserves. However the change in mortality rate was far from dramatic.

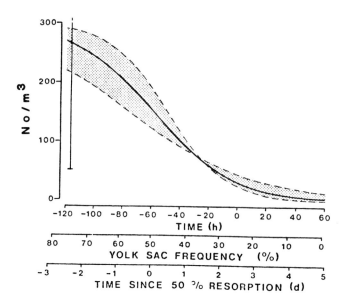

Figure 1 Mortality curve for larval capelin in the St. Lawrence estuary during transition to exogenous feeding. Shading represents the 95% confidence area for predicted abundance with time.

Further evidence of the sensitivity of larval survival to sub-optimal food levels at first feeding is provided by the results of laboratory studies. Lasker et al. (1970) reported no change in mortality rate when larval anchovy (Engraulis mordax) were provided with appropriate food within 0.5 d of yolk-sac absorption (2 days post-hatching). However, when the time of food presentation was delayed to 1.5 d after yolk-sac absorption a sharp increase in mortality rate was observed 1-2 d later. Mortality approached 100% in larvae offered food 2.5 d after yolk-sac absorption. Simultaneously, O'Connell and Raymond (1970) reported that anchovy larvae provided with food at concentrations of less than 0.1 nauplii/ml on day 2 post-hatching (0.5 d post yolk-sac absorption) experienced complete mortality. Lawrence (1974) reported similar resuls for larval haddock (Melanogrammus aeglefinus) reared under experimental conditions. Haddock began feeding two days after hatching and suffered high mortality if deprived of food beyond day 6, the approximate time of yolk-sac absorption. Even if fed, no larvae survived at food levels less than 0.1 plankters/ml but survival rates increased progressively at higher prey concentrations ($z = 0.06$ and 0.02 at 0.5 and 3.0 plankters/ml respectively). Houde (1974, 1975) reported survival in sea bream (<u>Archosargus rhomboidalis</u>), bay anchovy (<u>Anchoa mitchilli</u>) and lined sole (<u>Archirus lineatus</u>) to be marginal at prey concentrations less than 0.1 plankters/ml. At higher prey densities growth and survival were markedly affected if food was withheld more than 32 h post yolk-sac absorption.

These findings clearly demonstrate a sensitivity of larval fishes to low food levels at first feeding under laboratory conditions. They do not, however, contribute greatly to a resolution of the question of whether or not such a "critical" period is an important determinant of total larval survival under field conditions.

CRITICAL FOOD CONCENTRATIONS

It has long been recognized that the critical food levels identified in the laboratory studies cited above (approximately 0.1 particles/ml) were much higher than concentrations typically sampled at sea (O'Connell and Raymond 1970; Houde 1975; Vlymen 1977). This has been attributed to the patchy distribution of zooplankton in the sea, and the averaging effect created by tow net sampling. This led to the conclusion that larvae in the sea must become associated with high-density patches of zooplankton if they are to survive (Cushing 1972). Lasker (1975) subsequently extended his laboratory studies by subjecting larval anchovy that were maintained at sea in 8-L jars to food pumped directly from the surface and from the chlorophyll maximum layer. Feeding was poor in larvae maintained in water drawn from the surface but high in those maintained in water drawn from the chlorophyll maximum. Lasker hypothesized that the existence of an intact chlorophyll maximum was essential for successful first feeding of anchovy, and that storm events known to destroy the maximum cause variable recruitment in anchovy. Vlyman (1977) subsequently noted that the prey available in the chlorophyll maximum layer would be unlikely to support the growth of larvae to a length exceeding 12 mm, and suggested that the existence of other suitable prey aggregations would be essential to continued strong survival.

More recent studies involving larvae reared in tanks, meso- and macro-scale enclosures, indicate that larvae of cod (Gadus morhua), capelin, sea bream, bay anchovy, and lined sole can survive and grow at food concentrations ranging from 0.001 - 0.005 particles/ml (Houde 1978; Oiestad 1985; Frank and Leggett Fig. 4). These prey densities are commonly encountered in nature (Hunter 1981). Houde (1978) attributed the lower estimates of critical food densities he obtained to better culture techniques and better control of experimental food levels. The lower critical prey concentrations indicated from macro- and meso-scale enclosure experiments presumably also result from experimental conditions which more closely replicate those found naturally. These findings call for a critical reappraisal of the question of required prey concentrations in situ and of the importance of prey patchiness to larval survival.

FIELD EVIDENCE OF FOOD MEDIATED MORTALITY

Several field studies have recently provided indirect evidence of a link between in situ food levels and larval survival. Yoder (1983) analyzed the distribution of fish eggs and larvae on the continental shelf of the south eastern United States. During winter and spring the highest numbers of larvae occurred on the outer shelf in upwelling areas where plankton productivity was enhanced. Landward and seaward from this zone egg and larval densities were significantly reduced. Other investigators have reported similar relationships between larval survival and upwelling conditions in the sea (Walsh et al. 1980; Bakun and Nelson 1977; Parrish and MacCall 1978).

Nelson et al. (1977) analyzed year-class variation in Atlantic menhaden (Brevoortia tyrannus) in relation to the intensity of onshore Ekman transport which would provide a means by which larvae are transported from offshore spawning sites located over most of the continental shelf to food rich nursery areas located near shore. Nelson et al. (1977) reported that greater than 85% of the variation in year-class strength of menhaden (<u>Brevoortia tyrannus</u>) over the period 1955-1970 was explained by the estimated intensity of Ekman transport. Yoder (1977) criticized Nelson's use of monthly or seasonally averaged winds to estimate transport rates, noting that these did not conform to observations of wind events in the area. In reality, both onshore and offshore Ekman transport occurs in the area at characteristic periods of 2-8 d (Lee and Brooks 1979). It is thus more likely that the timing, strength, and duration of onshore transport events is the proximate regulator of year-class fluctuations in menhaden.

Bailey (1981) reported that poor year-classes of Pacific hake were positively correlated with years of higher than average offshore Eckman transport which displaces larvae from the food rich inshore nursery areas.

Crecco and Savoy (1983) related larval feeding incidence, zooplankton abundance, and juvenile year-class strength of American shad (<u>Alosa sapidissima</u>) in the Connecticut River during the years 1979-1982. The feeding incidence of larvae was highest and most persistent for the strong 1980 year class, intermediate for the moderately weak 1979 and 1981 year classes, and lowest for the 1982 year class which was the weakest recorded in 17 years.

Fortier and Leggett (1984) examined small-scale covariability between the abundance of herring (<u>Clupea harengus</u>) larvae and their prey in the St. Lawrence estuary. Spectral analysis of the resulting 196 h time series of samples revealed two distinct maxima in the coherence spectrum between larvae and microzooplankton (fig. 2). One maximum (period approximately 12 h) was related to in-phase semidiurnal vertical migrations of the two groups. The second sub-tidal maximum (period 3.6 h) resulted from horizontal advection past the sampling station of patches of herring larvae and microzooplankton approximately 7 km in cross-

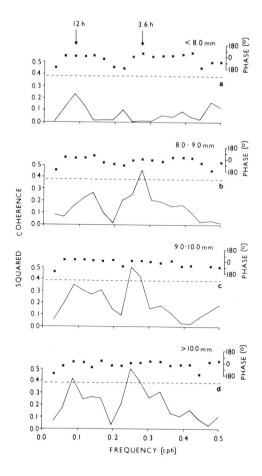

Figure 2. Coherence spectra for the relationship between increasing length categories of larval herring and micro-zooplankton abundance in the 20-40 m depth stratum of the St. Lawrence Estuary. Dotted line represents the 95% confidence limit for non-zero squared coherence.

sectional dimension. The swimming ability of larval herring was insufficient to directly explain this correlation between the observed distributions of larvae and microzooplankton. Significant coherence was absent in yolk-sac stage larvae (less than 8 mm) and appeared first in immediate post yolk-sac larvae (8-9 mm) (Fig. 2). Differential mortality of larvae in water masses containing rich and poor plankton densities could cause this development of coherence between the abundance of larvae and their zooplankton prey. Mean growth rates observed over the sampling period indicates that the time required to develop significant coherence after the transition to exogenous feeding was 4-7 d. This is approximately one-half the time to death observed in larval herring denied food under laboratory conditions (Blaxter and Hempel 1963). These findings are consistent with the hypothesis that the observed coherence was due to differential mortality at the time of onset of oxogeneous feeding.

THE IMPORTANCE OF PREY SIZE

It is well known that the size of prey organisms consumed by fishes increases as they grow. These changes relate both to the positive correlation between mouth width and larval length (Shirota 1970), and to the increased caloric demand of larger larvae (Vlymen 1977). Early larval stages typically consume prey in the 50-200 um size range with the smallest size fractions predominating in the guts of first-feeding larvae (Hunter 1981).

The effects of prey size on growth and survival have been well documented in the laboratory (see Howell 1973; Hunter 1977; Hunter and Kimbrell 1980). Direct field evidence of the importance of variability in the prey size spectrum available to larval fish on growth and survival is, however, rare. Lasker (1981) reported that anchovy year-class strength during the years 1962-1977 was unrelated to total plankton production but showed some relationship to the size of the dominant plankter.

Frank and Leggett (1982) documented rapid and dramatic changes in the mean size composition of zooplankton communities in coastal Newfoundland in response to wind-induced water mass exchange. They recently employed large (4 m^3) in situ enclosures to study the relevance of this wind-induced variation in zooplankton size structure to growth and survival in larval capelin (<u>Mallotus villosus</u>). Replicate enclosures were stocked with 1000 capelin in the late yolk-sac stage. The larvae were then provided with food collected from the nearshore waters during onshore or offshore winds. Analysis of the size structure of the zooplankton communities stocked into the enclosures verified that the community sampled during onshore wind has a significantly higher concentration of zooplankton in the lower end of the particle size distribution (Fig. 3).

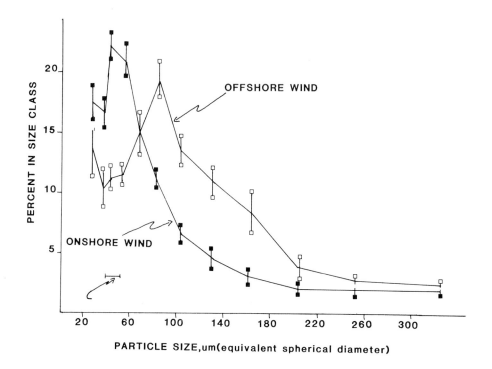

Figure 3. Plankton size spectra in coastal Newfoundland during onshore and offshore conditions.

The mortality rates of larvae reared in enclosures stocked with smaller zooplankton collected during onshore winds was dramatically lower than that of larvae reared in enclosures stocked with larger zooplankton collected during offshore winds (Fig. 4). This effect was particularly pronounced at low zooplankton densities and became non-significant at densities approaching 125 particles/L. This suggests that at these higher particle densities the absolute abundance of particles in the edible size fraction in the deep waters approached the minimum levels required to ensure high survival.

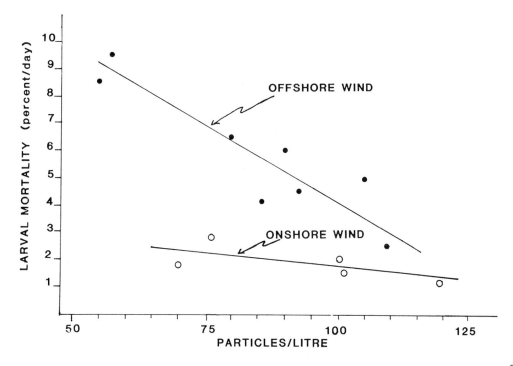

Figure 4. Survival curves for larval capelin reared in 4 m^3 enclosures moored at sea and stocked with zooplankton characteristic of communities dominating the nearshore during offshore and onshore wind conditions.

C. Taggart (McGill University) has used high frequency, small volume sampling to investigate the relationship between the survival of several cohorts of larval capelin in eastern Newfoundland. His preliminary data indicate that daily mortality is inversely related ($r = -0.71$, $n = 12$) to the average nearshore concentration of food in the 80 m (esd) size class.

The frequent changes in the size structure of the nearshore zooplankton community of eastern Newfoundland in response to changes in wind conditions is a temporal form of patchiness. The spatial distribution of zooplankton and larval fishes in the sea is also highly aggregated. The horizontal scale of these patches is known to vary from a few meters to hundreds of kilometers. The existence of temporal and spatial patchiness in the distribution of larvae and their food resources in the sea forms the basis of Cushing's (1972) match-mismatch hypothesis which assumes that variability in larval survival is a function of the degree to which the distribution of larval fishes overlaps high concentrations of suitable food organisms in time and space. The spawning site fidelity of many races and stocks of marine fishes are generally regarded as adaptations to maximize the probability of this coherence (for a general review of the subject see Norcross and Shaw, 1984).

BEHAVIORAL REGULATION OF FOOD AVAILABILITY

Until recently it was generally believed that the probability that larval fishes would become associated with environmental conditions favouring rapid growth and high survival was largely a function of spawning time and location. The results of our recent investigations of the early larval stages of several fish species have led us to question this assumption. Fortier and Leggett (1983) found that size and species-specific differences in patterns of vertical migration of larval capelin and herring both within and between the surface seaward current and the deep landward current in the estuary of the St. Lawrence River resulted in a complete separation of their ultimate distributions, and their concentration in two distinct high productivity zones in spite of the fact that they are spawned in the same area of the estuary at approximately the same time of year. Frank and Leggett (1981, 1982, 1983a,b) demonstrated that the episodic emergence of capelin larvae from beach incubation zones results from an active behavioural response to rapid temperature increases accompanying the appearance of surface waters in the nearshore region during onshore winds. This ensures that the larvae become immediately associated with a water mass rich in zooplankton of the appropriate size range. The episodic appearance of the larvae of winter flounder, radiated shanny, and sea snails also appears to result from active initiation of larval dirft during the brief occurrence of food-rich surface waters in the nearshore region (Frank and Leggett 1983b). These findings indicate that even the earliest larval stages of fishes may be capable of responding to signals in the environment in ways which maximize the probability of their becoming associated with a favourable food patch. If this is broadly true of fish larvae, the use of positive correlations between the abundance of larval fishes and prey as evidence for differential mortality in the presence/absence of suitable prey concentrations must be approached with considerable caution.

THE IMPACT OF PREDATORS ON LARVAL SURVIVAL

Numerous laboratory studies have demonstrated that macro-invertebrate predators can consume large numbers of larval fishes when predator and prey are together under confined conditions. Notwithstanding the container size effects which are known to influence the results of such studies, these findings and similar, frequently anecdotal, observations in the field have led to the widely held belief (see May 1974; Hunter 1976; Lawrence 1981) that predation by macroinvertebrates is a major cause of mortality among early larval fishes. The field evidence most frequently presented in support of this hypothesis is the existence of negative correlations between the abundance of macroinvertebrate predators and their prey (see Strasburg 1960; Pearcy 1962; Alvarino 1980; Möller 1979, 1980, 1984; Brewer et al. 1984) the assumption being that low larval numbers are the result of predation. Frank and Leggett (in press) reported a similar negative correlation between macrozooplankton and larval fish density in eastern Newfoundland (Fig. 5) which has no direct predatory basis. Rather, the correlations result from occupation by the larvae and predator complexes of two discrete water masses which alternatively dominate the nearshore areas in response to wind forcing. This finding led us to re-examine the previously reported negative correlations offered in support of the predation hypothesis with respect to available data on the dynamics of the physical environments in the areas studied (Frank and Leggett in press). We have concluded that most previously

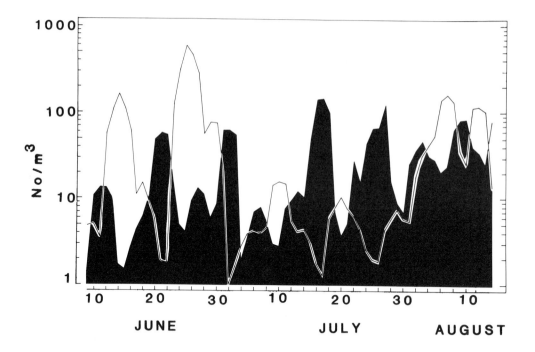

Figure 5. Reciprocal trends in ichthyoplankton (shaded) and invertebrate predators of Bryant's Cove, Newfoundland during July-August 1979. These patterns result not from predation, but from alternation of two discrete water masses in the nearshore in response to winds which are occupied independently by larval fishes and macroinvertebrates.

published reciprocal trends are not consistent with the predator-prey hypothesis advanced to explain them. Rather, the available evidence suggests the hypothesis that the combination of dynamic physical processes and the occupation of different water masses by the two separate components of the plankton is equally viable. Hence, until studies combining higher resolution sampling with a more detailed knowledge of the physical oceanography of the sampling areas are performed, negative correlations should not be accepted as meaningful evidence for the impact of macroinvertebrate predators on larval fishes.

A second line of evidence offered in support of the hypothesis that predation is a major source of mortality in larval fishes has been the results of diet analyses of invertebrate and vertebrate predators captured during plankton surveys (see Brewer et al. 1984 for a recent update of this literature). While these studies clearly demonstrate, as do laboratory studies, that macrozooplankton are capable of preying on larval fishes when confined in very small enclosures (net cod-ends, beakers, and small tanks) they do not provide reliable estimates of the magnitude of the predation. This fact has been dramatically demonstrated by Nicol (1984) who sampled swarms of Meganyctiphanes norvegica in the Bay of Fundy with dip nets and 10-min plankton tows. All samples were preserved immediately after capture. Only 0.23% of the specimens collected with dip nets contained food in their "food baskets" while 91% of those sampled with plankton nets contained food. Similar examples of net feeding by fishes have been reported by numerous authors (reviewed in Nicol 1984).

More reliable estimates of the potential impact of macro-zooplankton predation on larval fishes has recently been provided by Purcell (1981, 1984) who documented the diet of feeding intensity of hand-captured (dip nets and jars) specimens of Physalia physalis and Rhizophysa eysenhardti. Both species fed almost exclusively on larval fishes even though crustaceans of comparable size and activity were more abundant in the environment sampled. Purcell estimated that up to 60% of the larvae in the 0-5 m depth stratum could be consumed daily by P. physalis. The limited data available also suggest that the feeding rate of siphonophore predators is positively correlated to larval density. The global impact of siphonophore predators on larval fish populations is more difficult to assess. It clearly depends on yearly and seasonal variations in the abundance and distribution of predators and prey (Purcell 1984; Frank, in press).

Oiestad (1985) conducted macro (600,000 m^3) and meso (4,400 m^3) scale enclosure experiments on the interaction between larval fishes and potential predators. His results present a similar picture. In a series of experiments conducted between 1976 and 1983 the survival of larval herring, cod, plaice, and capelin was assessed in the absence of predators, and when exposed to predation from metamorphosed cod larvae, schooling herring larvae (25 mm) and hydromedusae. Average survival in the absence of predators, was 42% (sd 35, n = 9). When predators were present survival averaged 1% in spite of good feeding conditions and high specific growth rates of the larvae prior to their disappearance. These data indicate that marine fish larvae have high potential survival rates in the absence of predators when feeding conditions are good, and that predation, whether from metamorphosed fish larvae or macroinvertebrates, can have a significant impact on survival rates.

Y. Delafontaine (McGill University) has recently conducted a series of short-term (2-3 d) experiments in which a range of densities of larval capelin were exposed to a variety of jellyfish predators in 4 m^3 flowthrough enclosures moored at sea. Two important generalizations are apparent from the data: 1) larval mortality rate is proportional to total predator biomass (g/m^3) independent of the predator species employed and 2) larval mortality rates (at present uncorrected for the presence of alternate prey) on the order of 15% per day and higher are common.

CONCLUSIONS

May (1974) concluded his excellent review of the relationship between larval survival and food availability by noting that the data available from field and laboratory studies suggested starvation <u>may</u> be an important cause of larval mortality at the end of the yolk-sac stage, but that the relation between starvation-induced mortality and year-class strength remained unclear. He also noted that standard plankton sampling techniques were generally not precise or accurate enough to resolve this question. It is sobering to be forced to conclude, in spite of the considerable work that has been reported during the intervening 11 years, that the depth of our quantitative understanding of the causal relationships between food availability, larval survival, and year-class strength remains essentially unchanged. If anything the results of more recent experimental studies concerning the minimum food concentrations necessary to sustain growth and survival in larval fishes has served to weaken the hypothesis that larvae can survive only if they become associated with higher than average food concentrations (patches). Correlative studies of food levels and year class strength continue to provide indirect evidence for a positive relationship between the two, but detailed knowledge of the mechanism and timing remains elusive. On the positive side our improved knowledge of the importance of larval behavior as a regulator of larval drift and water mass association, of spawning site specificity, and of the relationship between retention zone stability and population size do provide sound evidence of a link between food resources and survival, at least on the evolutionary time scale.

Our quantitative knowledge of the importance of predators as regulators of larval survival has also advanced but here again perhaps too slowly for the effort expended. Recent studies do suggest, however, that this effect may be at least as important as that induced by food. Several lines of evidence do indicate, however, that larvae may have evolved effective behavioral and developmental mechanisms to minimize the extent to their co-occurrence with predators actually occurs.

The direct and indirect evidence continues to suggest that both food and predators can act to significantly alter survival rates in larval fishes. Still largely unresolved, however, is where, when, how, and by how much. The overriding conclusion of this review is that sufficient effort has been expended using conventional techniques and approaches without satisfactory resolution of these questions to clearly indicate that these approaches should be significantly altered. A resolution of these questions will almost certainly require _in situ_ approaches in which relevant parameters, both physical and biological, are sampled on time and space scales that are meaningful in terms of the biological processes in question. This idea is not new. Marr (1956) stated it explicitly almost 30 years ago. It's time we took it seriously. The results of the limited number of studies that have attempted this to date suggest that such approaches can be successful. They also indicate that in many instances the appropriate measurement scale may be on the order of hours rather than days or weeks, and of tens to hundreds rather than thousands of m^3.

REFERENCES

Alvarino, A. 1980. The relation between the distribution of zooplankton predators and anchovy larvae. Calif. Coop. Oceanic. Fish. Invest. Rep. 21: 150-160.

Bailey, H.M. 1981. Larval transport and recruitment of Pacific Lake *Merluccius productus*. Mar. Ecol. Prog. Ser. 6: 1-9.

Bakun, A. and C.S. Nelson. 1977. Climatology of upwelling related processes off Baja California. Calif. Coop. Oceanic Fish. Invest. Rep. 19: 107-127.

Brewer, G.D., G.S. Kleppel and M. Dempsey. 1984. Apparent predation on ichthyoplankton by zooplankton and fishes in nearshore waters of southern California. Mar. Biol. 80: 17-28.

Crecco, V.A. and T.F. Savoy. 1983. Effects of fluctuations in hydrographic conditions on year-class strength of American shad (*Alosa sapidissima*) in the Connecticut River. Can. J. Fish. Aquat. Sci. 41: 1216-1223.

Cushing, D.H. 1972. The production cycle and the numbers of marine fish. Symp. Zool. Soc. London 29: 213-232.

Dahlberg, M.D. 1979. A review of survival rates of fish eggs and larvae in relation to impact assessment. Mar. Fish. Rev. 41(3): 1-12.

Fortier, L. and W.C. Leggett. 1983. Vertical migrations and transport of larval fish in a partially mixed estuary. Can. J. Fish. Aquat. Sci. 40: 1543-1555.

Fortier, L. and W.C. Leggett. 1984. Small-scale covariability in the abundance of fish larvae and their prey. Can. J. Fish. Aquat. Sci. 41: 502-512.

Fortier, L. and W.C. Leggett (in review). A drift study of larval fish survival. Mar. Ecol. Prog. Ser.

Frank, K.T. (in press). Ecological significance of the ctenophore (*Pleurobrachia pileus*) off southeastern Nova Scotia. Can. J. Fish. Aquat. Sci.

Frank, K.T. and W.C. Leggett. 1981. Wind regulation of emergence times and early larval survival in capelin (*Mallotus villosus*). Can. J. Fish. Aquat. Sci. 38: 215-223.

Frank, K.T. and W.C. Leggett. 1982. Coastal water mass replacement: its effect on zooplankton dynamics and the predator-prey complex associated with larval capelin (*Mallotus villosus*) Can. J. Fish. Aquat. Sci. 39: 991-1003.

Frank, K.T. and W.C. Leggett. 1983a. Survival value of an opportunistic life stage transition in capelin (*Mallotus villosus*). Can. J. Fish. Aquat. Sci. 40: 1442-1448.

Frank, K.T. and W.C. Leggett. 1983b. Multispecies larval fish associations: Accident or adaptation. Can. J. Fish. Aquat. Sci. 40: 754-762.

Frank, K.T. and W.C. Leggett (in press). Reciprocal oscillations in densities of larval fish and potential predators: A reflection of present or past predation? Can. J. Fish. Aquat. Sci.

Frank, K.T. and W.C. Leggett (in review). The dynamics of trophic relationships between capelin (*Mellotus villosus*) larvae and their prey. Can. J. Fish. Aquat. Sci.

Hjort, J. 1914. Fluctuations in the great fisheries of northern Europe reviewed in the light of biological research. Rapp. P.-V. reun. Cons. Perm. Int. Explor. Mer. 20: 1-228.

Hjort, J. 1926. Fluctuations in the year classes of important food fishes. J. Cons. Perm. Int. Explor. Mer. 1: 5-38.

Houde, E.D. 1974. Effects of temperature and delayed feeding on growth and survival of larvae of three species of subtropical marine fishes. Mar. Biol. 26: 271-285.

Houde, E.D. 1975. Effects of stocking density and food density on survival, growth and yield of laboratory-reared larvae of sea bream *Archosargus rhomboidalis* (L.) (Sparidae). J. Fish. Biol. 7: 115-127.

Houde, E.D. 1978. Critical food concentratons for larvae of three species of tropical marine fishes. Bull. Mar. Sci. 28: 395-411.

Hunter, J.R. 1976. Report of a colloquium on larval fish mortality studies and their relation to fishery research. January 1975. NOAA Tech. Rept. NMFS CIRC-395, 6p.

Hunter, J.R. 1977. Behavior and survival of northern anchovy, *Engraulis mordax*, larvae. Calif. Coop. Oceanic. Fish. Invest. Rep. 19: 138-146.

Hunter, J.R. 1981. Feeding ecology and predation of marine fish larvae. pp. 34-77 in R. Laskei (ed.) Marine Fish Larvae, Morphology Ecology and Relation to Fisheries. Univ. Wash. Press. Seattle. 131 p.

Hunter, J.R. and C.A. Kimbrell. 1980. Early life history of pacific mackerel, *Scomber japonicus*. U.S. Fish. Bull. 78: 89-101.

Howell, B.R. 1973. Marine fish culture in Britain VIII. A marine rotifer, *Brachionus plicatilis* Muller, and the larvae of the mussel, *Mytilus edulis* L. as foods for larval flatfish. J. Cons. int. Explor. Mer. 35: 1-6.

Jacquaz, B.K.W. Able and and W.C. Leggett. 1977. Seasonal distribution, abundance and gorwth of larval capelin (*Mallotus villosus*) in the St. Lawrence estuary and northwestern Gulf of St. Lawrence. J. Fish. Res. Board Can. 34: 2015-2029.

Lasker, R. 1981. Factors contributing to variable recruitments of the northern anchovy (*Engraulis mordax*) in the California current. Contrasting years, 1975 through 1978. Rapp. P.-V. Reun. Cons. int. Explor. Mer. 178-375-388.

Lasker, R., H.M. Feder, G.H. Theilacker and R.C. May. 1970. Feeding, growth and survival of *Engraulis mordax* larvae reared in the laboratory. Mar. Biol. 5: 345-353.

Lawrence, G.C. 1974. Growth and survival of Haddock (*Melanogramnus aeglefinus*) larvae in relation to planctonic prey concentration. J. Fish. Res. Board Can. 31: 1415-1419.

Lawrence, G.C. 1981. Overview - modelling - an esoteric or potentially utilitarian approach to understanding larval fish dynamics? pp. 3-6 in Lasker, R. and K. Sherman (eds.). The early life history of fish: recent studies. Rapp. P.-V. Reun. Cons. int. Explor. Mer. 178. 607 p.

Lee, T.M. and D.A. Brooks. 1979. Initial observations of current, temperature and coastal sea level response to atmospheric and Gulf Stream forcing on the Georgia Shelf. Geophysical Research Letters 6: 321-324.

Leiby, M.M. 1984. Life history and ecology of pelagic fish eggs and larvae in K.A. Steidinger and L.M. Walker (eds.) Marine plankton life cycle strategies CRC Press, Boca Raton.

Marr, J.C. 1956. The "critical period" in the early life history of marine fishes. J. du Cons. 21: 160-170.

May, R.C. 1974. Larval mortality in marine fishes and the critical period concept, p. 3-19 in J.H.S. Blaxter (ed.). The early life history of fish. Springer-Verlag, N.Y.

Möller, H. 1979. Significance of coelenterates in relation to other plankton organisms. Meeresforchung. 27: 1-18.

Möller, H. 1980. Scyphomedusae as predators and food competitors of larval fish. Meeresforchung. 28: 90-100.

Möller, H. 1984. Reduction of a larval herring population by jellyfish predator. Science 224: 621-622.

Nelson, W.R., M.C. Ingham and W.E. Schaff. 1977. Larval transport and year-class strength of Atlantic menhaden Brevoortia tyrannus. Fishery Bull. 75(1) 23-41.

Nicol, S. 1984. Cod end feeding by the euphausid Meganyctyphanes norwegica. Mar. Biol. 80: 29-33.

Norcross, B.L. and R.F. Shaw. 1984. Oceanic and estuarine transport of fish eggs and larvae: A review. Trans. Am. Fish. Soc. 113: 153-165.

O'Connell, C.P. and L.P. Raymond. 1970. The effects of food density on survival and growth of early post yolk-sae larvae of the northern anchovy (Engraulis mordax Girard) in the laboratory. J. exp. Mar. Biol. Ecol. 5: 187-197.

Parrish, R.H. and A.D. MacCall. 1978. Climatic variation and exploration in the Pacific mackerel fishery. Calif. Dept. Fish and Game, Fish. Bull. 167, 110 pp.

Oiestad, V. 1985. Predation on fish larvae as a regulatory force, illustrated in mesocosm studies with large groups of larvae. NAFO Sci. Coun. Studies 8: 25-32.

Purcell, J.E. 1981. Feeding ecology of Rhizophipa eyoenhardti, a siphonopore predator of fish larvae. Limnol. Oceanogr. 26(3): 424-432.

Pearcy, W.G. 1962. Ecology of an estuarine population of winter flounder, Pseudopleuronectes americanus (Walbam). II distribution and dynamics of larvae. Bull. Bingham Oceanogr. Coll. 18: 16-37.

Purcell, J.E. 1984. Predation on fish larvae by Physalia physalis, the Portugnese man of war. Mar. Ecol. Prog. Ser. 19: 189-191.

Shirota, A. 1970. Studies on the mouth size of fish larvae (Jap. Engl. summary). Bull. Jap. Soc. Sci. Fish. 36: 353-368 (Trans. Fish. Res. Bd. Canada Transl. Ser. 1978).

Strasburg, D.W. 1960. Estimates of larval tuna abundance in the central Pacific. Fish. Bull. 60: 231-249.

Vlymen, W.J. 1977. A mathematical model of the relationship between larval anchovy (Engraulis mordax) growth, prey microdistribution and larval behavior. Env. Biol. Fishes. 2: 211-232.

Walsh, J.J., T.E. Whitledge, W.E. Esaias, R.L. Smith, S.A. Huntsman, H. Santander and B.R. de Mendida. 1980. The spawning habits of the Peruvian anchovy, Engraulis ringens. Deep-Sea Res. 27: 1-28.

Yoder, J.A. 1983. Statistical analysis of the distribution of fish eggs and larvae on the Southeastern U.S. continental shelf with comments on oceanographic processes that may affect larval survival. Estuarine, Coastal and Shelf Science 17: 637-650.

ASSESSMENT OF EFFECTS OF FRESHWATER RUNOFF VARIABILITY ON FISHERIES PRODUCTION IN COASTAL WATERS

M. Sinclair,[1] G.L. Bugden,[2] C.L. Tang,[2]
J.-C. Therriault,[3] and P.A. Yeats[2]

[1] Invertebrates and Marine Plants Division
Fisheries Research Branch
Halifax Fisheries Research Laboratory
Department of Fisheries and Oceans
Scotia-Fundy Region
P.O. Box 550
Halifax, Nova Scotia B3J 2S7
Canada

[2] Atlantic Oceanography Laboratory
Bedford Institute of Oceanography
Department of Fisheries and Oceans
P.O. Box 1006
Dartmouth, Nova Scotia B2Y 4A2
Canada

[3] Champlain Centre for Marine Science and Surveys
Department of Fisheries and Oceans
P.O. Box 15,500
Québec, Province of Québec G1K 7Y7
Canada

ABSTRACT

The effect of freshwater runoff on the fisheries production of the Gulf of St. Lawrence is reviewed. In addition, the impact of the runoff from the Gulf of St. Lawrence on the fisheries production downstream (the Scotian Shelf and the Gulf of Maine) is considered. The relative importance of local freshwater runoff and larger scale ocean circulation on continental shelf fisheries production is evaluated. Conclusions are drawn relating to the impact of freshwter runoff variability on fisheries production in the northwest Atlantic as well as on the possible mechanisms of impact. Within the Gulf of St. Lawrence there is persuasive evidence that inter-annual variability in runoff has an impact on fisheries production. The mechanisms by which the variability is induced, however, are not understood. Even though there is entrainment of nutrients into the surface layer of the Gulf of St. Lawrence due to freshwater runoff, and some evidence for relatively high primary production, there is little evidence that suggests that the zooplankton, benthos, and fisheries production are higher than on contiguous shelf areas. Regulation of the Gulf of St. Lawrence system of rivers does not seem to have had a measurable impact on fisheries production. Non-local ocean circulation may have a greater impact than freshwater runoff from the Gulf of St. Lawrence on at least the cod fisheries production outside the Gulf of St. Lawrence (on the Scotian Shelf and in the Gulf of Maine). It is inferred that there is limited support for the freshwater runoff-driven food-chain hypothesis of Sutcliffe for generating fisheries production variability. There is accumulating evidence that the physical oceanographic processes of advection and diffusion can have a direct impact (i.e. without links through primary production) on population abundance of zooplankton as well as on fish populations during their early life history stages.

INTRODUCTION

In the early 1970's it was recognized that inter-annual variability of freshwater runoff into the Gulf of St. Lawrence, as well as the artificial damping of the natural seasonal cycle in the runoff pattern by hydroelectrical developments, may have profound effects on the oceanography of the receiving waters. Neu (1970; 1975; 1976), using a simple two-layered physical model, evaluated the magnitude of entrainment of deeper, more saline water into the surface layer due to freshwater runoff in the St. Lawrence Estuary. He concluded that freshwater could move nearly 30 times its own volume. Given that the entrained water is nutrient rich, he further inferred that the damping of the seasonal runoff cycle may have a critical influence on the timing of biological processes in the Gulf and beyond. Sutcliffe (1972; 1973) found positive correlations between river runoff into the Gulf and landings of commercially important fish and invertebrate species when appropriate time lags (reflecting the time between spawning and capture) were used. He inferred that events progress from runoff → nutrient entrainment → increased primary production → effect on larval survival (the freshwater runoff-driven food-chain hypothesis). Sutcliffe et al. (1976; 1977) also inferred that freshwater runoff variability into the Gulf of St. Lawrence may influence the strength of the Nova Scotia current along the inner Scotian Shelf and that fish production as far as the Gulf of Maine could be affected. In sum, it seems that the annual variability in freshwater runoff can account for annual variability in fish production of the Gulf of St. Lawrence, and possibly of the continental shelf waters beyond the Gulf. This issue has been reviewed in detail by Bugden et al. (1982).

The aim of this paper is twofold: 1) to summarize the review of Bugden et al. (1982); and 2) to evaluate the relative importance to continental shelf fisheries production of freshwater runoff and non-local ocean circulation. Both physical features (runoff and ocean circulation) have strong low-frequency signals, and to some degree they may be coherent in that both are components of the global air-sea interaction system. The difficulty of separating out the relative importance of local versus large-scale phenomena on coastal fish and invertebrate production is represented schematically in Figure 1.

SUMMARY OF GULF OF ST. LAWRENCE STUDY

In this section the study by Bugden et al. (1982) is summarized. In their review coastal zone fluctuations in a variety of oceanographic and fisheries parameters were viewed predominantly from a river runoff perspective (i.e. the

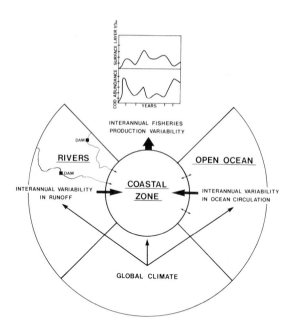

Figure 1. Schematic representation of the difficulty of interpreting the relative importance of local freshwater runoff and large-scale ocean circulation in generating inter-annual variability in fisheries production.

left-hand side of Figure 1). The Gulf of St. Lawrence receives a large input of freshwater relative to many other coastal areas. Of the total annual freshwater runoff into the north Atlantic from North America, 53% (1,245 km^3/yr) occurs between 45° and 55°N. The St. Lawrence River system (with tributaries) discharges 424 km^3/yr into the Gulf of St. Lawrence (Sutcliffe et al. 1976) or 34% of the freshwater input to this 10° zone of latitude. This volume may be compared to El-Sabh's (1977) estimate of annual outward transport of approximately 1.3×10^4 km^3/yr through Cabot Strait (giving a ratio of approximately 30 to 1 between St. Lawrence runoff and Gulf of St. Lawrence outflow). While discharge from the St. Lawrence River system itself dominates runoff into the Gulf of St. Lawrence, other rivers contribute substantially to the total annual flow of freshwater into the

Gulf as a whole. Annual discharge of the St. Lawrence system exceeds freshwater runoff of the entire eastern United States from the Gulf of Maine to Florida (Sutcliffe et al. 1976). In order to evaluate the impact of this magnitude of freshwater runoff on fisheries production it is important to first consider to what degree the Gulf of St. Lawrence can be considered an identifiable ecosystem within which variability in fisheries production is internally generated.

Reviews of the oceanography of the Gulf (Trites 1970; 1972; Trites and Walton 1975; Dickie and Trites 1983) strongly support the view that the Gulf forms an integral physical system. The features include a local origin of the intermediate cold layer, restricted exchange of water through the Belle Isle Strait, an estimate of residence time on the order of 1 yr, and an estuarine-type density-driven circulation which permits extensive vertical transport of salt and nutrients. The concentration of nutrients in the deeper layer of the Gulf are approximately three times higher than concentrations at similar depths in the north Atlantic outside the Gulf. Also, there is a characteristic oxygen minimum in the deeper water of the Laurentian Channel, with a marked decrease from the Cabot Strait toward the west and north ends of the Gulf. Primary production in the Gulf has been reported to be higher than observed on contiguous shelf areas, but there is some doubt about the accuracy of the estimates (Bugden et al. 1982, p. 7). Zooplankton species composition is less diverse than that observed on the continental shelves, but the overall zooplankton biomass estimates are comparable. Benthic macrofauna biomass in the Gulf is also not substantially higher than in adjacent waters. Phytoplankton production, zooplankton, and benthic biomass in the Gulf of St. Lawrence and on the Scotian Shelf are compared in Table 2 of Bugden et al. 1982.

There is relatively convincing evidence from fisheries research that a number of important fish populations complete their life cycle within the Gulf of St. Lawrence and the shelf areas immediately adjacent to the Cabot Strait. In addition, there are marked differences in relative year-class strengths and in some cases timing of spawning between the Gulf populations and populations of the same species in adjacent waters. Biomass trends, as well as fish species composition and their rank order, are also markedly different on the Scotian Shelf and within the southern Gulf of St. Lawrence. The average trawlable biomass, however, in the two areas surveyed are essentially identical (Sinclair et al. 1984). This suggests, as does the zooplankton and benthos biomass comparative estimates, that at trophic levels above the primary production level the Gulf is not more productive than contiguous shelf areas. Finally, analysis of lobster landings since the turn of the century suggests that the Gulf of St. Lawrence and Cape Breton areas have different temporal patterns in abundance than those observed along the Scotian Shelf and in the Gulf of Maine (Campbell and Mohn 1983; Harding et al. 1983). There are, however, also complex patterns in the landings trends within the Gulf itself, suggesting that the overall population is not responding to the environment and fishing in a coherent fashion.

In summary, even though biological productivity at higher trophic levels may not be different from the adjacent shelf areas, there is considerable evidence that the Gulf of St. Lawrence can be viewed as a distinct ecosystem. Thus, one might expect inter-annual differences in its features to be a function of internal events.

The effects of runoff regulation are superimposed on the natural, seasonal, and inter-annual variability. The data are adequate to document both the increase in the reservoir capacity since the first dam construction in 1908, as well as the effect of these developments on the natural runoff variability. At present the reservoir capacity in the St. Lawrence system alone (70 km^3) is approximately 15% of the estimated average annual discharge of the St. Lawrence River system (424 km^3). While storage itself does not alter total annual discharge, the important effect of runoff control is the displacement of discharge from the spring to the winter months. In relation to evaluating impacts it is to be noted that the volume of storage capacity approximately doubled in the 1970's.

It is estimated that during the 1960's, prior to the doubling of the reservoir capacity, the peak discharge during the spring was reduced by 15 to 35% with a corresponding increase in discharge during the winter months. The estimate of the inter-annual variability in runoff of the St. Lawrence River system up to and including the Saguenay [i.e. the RIVSUM of Sutcliffe (1973)], which is not influenced by regulation, is modest (the coefficient of variation from 1950 to 1979 being 10%). From these statistics both the magnitude of the natural inter-annual runoff variability, as well as the degree of damping of the seasonal cycle due to regulation, can be put in perspective. The relatively modest inter-annual variability is being significantly damped seasonally (by up to perhaps 50%). The freshwater runoff, however, can be amplified by up to 30 times within the Gulf by estuarine-type circulation.

The time series data on oceanographic and fisheries parameters within the Gulf of St. Lawrence are limited. There are sufficient temperature and salinity data available to generate a relatively complete surface-layer (top 10 m) monthly time series from 1944 to 1977. The data can usefully be aggregated into three regions: the Magdalen Shallows, the northeastern Gulf, and the northwestern Gulf. Preliminary analysis of these hydrographic data indicates that: 1) during the winter months there is coherence in the surface-layer features throughout the Gulf; and 2) during the spring and summer months the inter-annual estimates of surface-layer temperature do not fluctuate coherently. This simple analysis suggests that except for the winter months different areas of the Gulf respond differently to the various physical forcing functions. In addition to the surface-layer time series, several parameters characterizing the intermediate cold layer are available. Time series from 1951 to 1975 for depth of the temperature minimum, value of the temperature minimum, and the upper and lower depths of the 1.5°C isotherms (interpolated to mid July) were generated from the deep-station observations.

The above hydrographic data are essentially the only "hard" time series information on oceanographic properties of the Gulf (physical, chemical, and biological). The nutrient chemistry data are inadequate for detailed comparisons with runoff variability but do indicate marked inter-annual variability in nutrient profiles. The same conclusion is valid for phytoplankton biomass and production time series data. Except for the Lower St. Lawrence Estuary there are insufficient data available to analyse potential dependence on runoff variability. Marked differences between years have been observed in both the timing and the magnitude of the summer bloom. Therriault and Levasseur (this volume) deal with this aspect in detail. There are no relevant data on zooplankton and benthos available for the Gulf. There are, however, several useful time series on commercially important species. Lobster landings statistics since the turn of the century are potentially of value if the assumption of constant effort is appropriate. In addition, there are some relatively long (15 to 20 yr) time series for fish populations. These data have the advantage (in contrast to lobster landings) of including estimates of year-class strengths and biomass. The longest time series is for southern Gulf cod (1947 to 1977).

Two approaches can be taken to evaluate the impact of runoff variability on the oceanography and fisheries production of the Gulf: an evaluation of the available time series data, and a conceptual approach based on theoretical considerations.

Within the Estuary and the Gaspé Current it is to be expected that an increase in the runoff should increase the vertical entrainment of saline nutrient-rich water due to estuarine circulation processes. However, downstream the increased runoff is expected to suppress mixing, and vice versa.

The overall effect of the regulation of river discharge for the purpose of hydroelectric power production is to decrease the river discharge in the spring and to increase the winter discharge. The spring discharge decrease reduces the buoyancy flux to the surface layers of the ocean during the time when the surface-mixed layer is being set up, and thus may have a profound effect on the characteristics of this layer.

The surface salinity of the Gulf waters can be compared for high and low runoff years. Those years which have an average discharge which was more than one-half a standard deviation higher than the overall series mean were termed high runoff years. Those with one-half a standard deviation below the mean were defined as low runoff years. The 3 mo average surface salinities over the Magdalen Shallows are 1‰ less during the high runoff years.

The reduction in surface-layer salinity over the Magdalen Shallows during high runoff years supports the conclusion that the entrainment ratio (i.e. the ratio of the sum of the freshwater discharge and the entrained volume to the freshwater discharge) is not constant. If it were constant no change in surface salinity should be observed between high and low runoff years.

An attempt can be made to evaluate the impact of damping the spring runoff into the Gulf of St. Lawrence on the spring/summer surface temperatures. The approach is outlined by Bugden et al. (1982, p. 47 to 52). The results of the analysis indicate that regulation could have caused changes in temperature of the upper layer on the order of 0.5°C (i.e. damping of the spring runoff may have decreased surface temperatures due to the decrease in the buoyancy flux).

The increased winter runoff due to regulation may also have an impact on mixing processes. The intermediate cold layer is almost entirely formed by surface cooling during the winter in the Gulf of St. Lawrence (Banks 1966). This water forms the bulk of the water upwelled and entrained to the surface layer during the spring to autumn months, and thus the effect of winter runoff on its characteristics bears careful examination. The increased runoff in the winter due to regulation may inhibit convection through an increased buoyancy flux (i.e. enhanced winter salinity stratification). An indication that at least natural variations in fall and winter river discharge have had an effect on the depth of vertical mixing is suggested by a correlation of -0.73 between the depth of the lower 1.5°C isotherm [considered by Banks (1966) to be the lower bound of the intermediate water] and the winter freshwater discharge (i.e. the greater the winter runoff the shallower the winter mixed layer).

From correlation analyses between lobster landings and RIVSUM, Sutcliffe (1973) suggested that runoff variability controls recruitment to some adult lobster fishing grounds.

Sheldon et al. (1982) have more recently demonstrated that the
relationship (slightly modified in the second treatment)
between runoff and lobster landings has been predictive
(Figure 2).

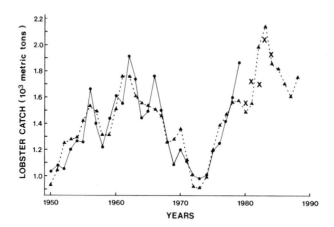

Figure 2. Comparison of observed Québec lobster catch in the
Gulf of St. Lawrence (predominantly Magdalen
Islands) and "predicted" landings from a regression
involving RIVSUM (from Sheldon et al. 1982). The
solid line represents the reported landings; the
broken line the "predicted" landings. The crosses
representing lobster catch from 1980 to 1984 have
been added to the original figure.

The southern Gulf of St. Lawrence cod recruitment time
series (1947 to 1976 year classes, as well as 1957 to 1976)
were analyzed in relation to RIVSUM and the Magdalen Shallows
summer surface layer (0-10 m) temperature (May to September)
and salinity (June) (Bugden et al. 1982, p. 56). The shorter
time series for cod recruitment was used as well as the
complete series due to less confidence in the earlier years
when sampling was scanty and the recording of catch by
geographic area poor. The Magdalen Shallows hydrographic data
were used in the statistical analysis since this is the major
area for the cod early life history distribution. Cod
recruitment was not statistically correlated with any of the
temperature or salinity time series. There was a significant

correlation with runoff the summer prior to spawning (r = 0.62 between the shorter time series and May and June runoff the previous summer). It is recognized, however, that the usual statistical confidence levels are not applicable when high correlations are being searched for. Nevertheless, the "suggestion" that events the year prior to spawning may be important is consistent with: 1) the findings that the number of oocytes are defined the year before spawning (K. Waiwood, Biological Station, St. Andrews, N.B. E0G 2X0, pers. comm.); and 2) the conclusion by Doubleday and Beacham (1982) that adult cod productivity in the southern Gulf is partially a function of RIVSUM. It must be concluded, however, that for the best finfish data set in the Gulf of St. Lawrence there is not a strong case (in a statistical sense) to be made that there is a major impact of freshwater variability on year-class strength.

In sum, both the natural inter-annual runoff variability into the Gulf of St. Lawrence, as well as the degree of damping in the seasonal cycle due to hydroelectrical developments, can be well described. Some effects on hydrographic characteristics of the surface waters of the Gulf can also be described. It is estimated that the regulation itself may reduce spring surface temperatures in the western Gulf by up to 0.5°C. Also, it is argued from the empirical data that increased winter runoff due to regulation decreases the depth of vertical mixing during winter convection. Since the marked increase in regulation during the 1970's, however, there has not been a dramatic impact on fisheries yield (i.e. cod year classes have been strong and lobster landings have not decreased dramatically during the late 1970's and early 1980's). To the degree that there is an effect higher up the food chain, it does not seem to have impaired the long-term productivity of the populations. The "predictive" nature of the runoff-lobster landings regression (Figure 2), however, continues to suggest that freshwater runoff is critical to recruitment variability of at least one commercially important species. It is not possible, given the available data within the Gulf, to evaluate the food-chain hypothesis of Sutcliffe (1973).

IMPACT OF NON-LOCAL OCEAN CIRCULATION ON CONTINENTAL SHELF FISHERIES PRODUCTION

In this section the right-hand side of Figure 1 is considered. In the above discussion the variability in salinity in the Gulf of St. Lawrence has been considered in relation to local river runoff. Since Sutcliffe et al. (1976; 1977) have suggested that Gulf of St. Lawrence runoff may have an impact on the Scotian Shelf and Gulf of Maine oceanography and fisheries, it is of interest both within the Gulf of St. Lawrence (perhaps particularly in the eastern part) and on the contiguous Shelf areas to evaluate the importance of large-scale ocean circulation on the "salinity signal." In this section we will briefly review the literature dealing with larger scale ocean circulation (those studies involving relatively low salinity fluctuations) on continental shelf biological productivity. From this broader literature some evaluation of the food-chain hypothesis is possible.

In the northwest Atlantic there are three papers of interest - Koslow 1984, Sutcliffe et al. 1977, and Sutcliffe et al. 1983. Sutcliffe et al. (1977) extended their analysis of statistical relationships between RIVSUM and fish landings from the Gulf of St. Lawrence to the Gulf of Maine. Both positive and negative correlations were found between RIVSUM and annual landings for a wide range of commercially important species, again using a time lag appropriate to the average age of capture. Dickie and Trites (1983) have recently reviewed the freshwater runoff studies by Sutcliffe and his collaborators. They conclude (p. 419):

> "There seems to be no serious doubt that environmental factors play a dominant direct role in the determination of relative year-class success of commercially important species throughout the entire area influenced by the Gulf of St. Lawrence. The mechanism of the effect is obscure and may involve factors related to distribution, direct physiological effects on survival, or influence through changes in food abundance or availability. The obvious and simple effects of river discharge

> on nutrients and on primary and secondary
> production, together with the evidence of density
> dependence of both growth and recruitment are
> incontrovertible evidence of the action of the
> physical system on biological production through
> the trophic chains."

The reviewers in sum seem to have found the evidence compelling that local runoff from the St. Lawrence River system has an impact on fisheries production as far downstream as the Gulf of Maine. In addition, they support to a certain degree the linear food-chain interpretation of Sutcliffe (1973) of the mechanism of impact of freshwater runoff variability on fish production.

Sutcliffe et al. (1983) show that year-class sizes of Labrador Atlantic cod are closely correlated with salinity changes at Station 27 off St. John's, Newfoundland. The year-class variability is interpreted to be a function of variation in nutrient entrainment within the Labrador current driven by variable freshwater runoff through the Hudson Strait. In this case high runoff is interpreted to suppress vertical entrainment of nutrients within the Labrador Current. The positive correlation observed between surface-layer salinity and year-class size (i.e. high surface-layer salinity generates large year classes) is again explained by a simple food-chain hypothesis (in this case increased Hudson Strait runoff → decreased nutrient entrainment in Laborador Current → decreased primary production → decreased cod larval survival).

The results of Koslow's (1984) study, particularly on Atlantic cod recruitment variability in the northwest Atlantic, seriously question the interpretation of Sutcliffe et al. (1977; 1983) and the above-cited endorsement by Dickie and Trites (1983). Two observations are important in our view. First, coherence in year-class strength variability was observed for all cod populations from "west Greenland cod" to "Gulf of Maine cod," except for the two populations within the Gulf of St. Lawrence which are themselves coherent. Second, the inter-annual variability in cod recruitment was "explained" in relation to large-scale north Atlantic circulation processes rather than by the more local effects of either Hudson Strait freshwater runoff on nutrient entrainment within the Labrador Current or Gulf of St. Lawrence runoff on the Scotian Shelf and Gulf of Maine area. The strength of the Labrador Current itself is driven by larger ocean-scale processes.

These results support the conclusions that: 1) Gulf of St. Lawrence runoff variability may not impact significantly the bank-spawning fish populations along contiguous continental shelf areas (i.e. Scotian Shelf and Gulf of Maine); 2) the Gulf of St. Lawrence fisheries production is regulated by processes within the Gulf (including freshwater runoff variability); and 3) continental shelf fisheries production is influenced significantly by large-scale ocean circulation variability (which in the northwest Atlantic has a strong surface salinity signal as observed at Station 27 off St. John's). An alternate interpretation of the observed positive correlation between Labrador cod year-class strength and salinity off St. John's, Newfoundland, does not involve food-chain events. High salinity reflects reduced strength of the Labrador Current which may permit enhanced retention of eggs and larvae within the appropriate distributional area for the population. In this interpretation year-class strength is defined directly by physical processes.

There is further support in the literature that large-scale physical oceanographic phenomena, rather than local biological processes, control much of the observed inter-annual variability in continental shelf plankton dynamics. The most detailed work has been done respectively in the northeast Atlantic by Colebrook and in the California Current by Bernal, Chelton, and McGowan. We will argue that these studies, in addition to indicating the critical importance of large-scale physical oceanographic processes on inter-annual variability in plankton dynamics, do not support the above-mentioned local freshwater runoff-driven food-chain hypothesis for fisheries production variability (Sutcliffe 1973; Sutcliffe et al. 1983).

Chelton et al. (1982) review essentially all the relevant material for the California Current. They conclude that "over inter-annual time scales, zooplankton abundance is primarily influenced by large-scale variations in the flow of the California Current" (high flow is associated with relatively low surface-layer salinity). Further, they interpret that "these large-scale inter-annual fluctuations are in many cases related to El Nino phenomena in the eastern tropical Pacific." The food-chain hypothesis holds up well at lower trophic levels (increased nutrient transport in California Current → increased primary production → increased zooplankton biomass). The links to fisheries production variability, however, do not appear to be related to trophic dynamics. Bernal and McGowan (1981) describe a tight spatial relationship within the California Current between surface-layer temperature, salinity, nutrients, and phytoplankton (their Figures 2, 3, 4, and 5). They note, however, that fish larval distributions (their Figures 6 and 7) do not follow the same pattern. Sinclair et al. (1985) have further shown that over time (1928 to 1965) Pacific mackerel egg-larval-post-larval survival is high during high-salinity years (as is the case for Labrador cod). However, in this case the detailed time series information on zooplankton biomass and advection (as estimated from sea-level data) led to the conclusion that a food-chain hypothesis was not supported. In their interpretation of the observations the direct role of advection (due to variable strength of the California Current) on retention of eggs, larvae, and post larvae within an appropriate distributional area is highlighted.

In the northeast Atlantic, a food-chain hypothesis does not appear to be substantiated even at the lower trophic levels. In an extensive ongoing analysis of zooplankton distributions (from the continuous plankton records) in the northeast Atlantic, Colebrook (1978; 1979; 1981; 1982a; 1982b; 1982c; 1984; 1985), Colebrook and Taylor (1984), and Garrod and Colebrook (1978) address the question of the causes of inter-annual variability in plankton. Radach (1984) has reviewed the Continuous Plankton Recorder Survey published results, including Colebrook's work, up to 1982. Several points are relevant to the freshwater runoff-driven food chain hypothesis of fisheries production variability of Sutcliffe (1973) and Sutcliffe et al. (1983). Even though there are still many uncertainties in the interpretation of this massive overall data set, several observations and/or conclusions of considerable significance have been made and are listed below.

1. At least half of the observed variability in the annual means of the plankton time series in the north Atlantic can be attributed to density-independent physical environmental processes (Colebrook 1978).
2. The response of the plankton populations to ocean climate variability is "presumably through the effects of advection in the Gulf Stream-north Atlantic drift system, and its consequential effect on more peripheral systems" (Garrod and Colebrook 1978).
3. "For the North Sea, correlations were calculated between sea-surface temperature, salinity, and indices of phytoplankton and zooplankton entities, indicating significant relations between salinity and phytoplankton as well as zooplankton in all but one area. There were not significant correlations with temperature" (Radach 1984 reviewing Garrod and Colebrook 1978).
4. The plankton ecosystem of the northeast Atlantic and the North Sea "consists of an assemblage of species responding individually to environmental influences as opposed to being an integrated community" (Colebrook 1985).
5. The observations that the long-term trend in zooplankton abundance has its origin in the degree of persistence of zooplankton overwintering populations in geographic space "suggests that there may be a relatively weak link betweeen average levels of primary and secondary production" (Colebrook 1985).

In sum the evidence suggests, as interpreted in particular by Colebrook, that (in north temperate latitudes at least) linear food-chain dynamics (nutrients → phytoplankton → zooplankton) do not control the observed inter-annual variability in plankton biomass. The conclusion by Huntley and Boyd (1984), that zooplankton in coastal regions are not limited by food availability, also undermines the food-chain hypothesis. Large-scale ocean circulation (processes with a salinity signal in the North Sea) may well have a direct impact on abundance through variability in advective losses of overwintering zooplankton from appropriate geographic areas within which persistence is possible (Colebrook 1985). The above selective review of the literature does not imply that on smaller spatial scales freshwater runoff has no impact. It does suggest, however, that the direct effect of the physical processes on fisheries production (i.e. the effect of runoff on the circulation itself in relation to population persistence) should be considered as an alternative hypothesis to Sutcliffe's food-chain hypothesis.

CONCLUSIONS ON IMPACT OF FRESHWATER RUNOFF VARIABILITY ON FISHERIES PRODUCTION

Conclusions are drawn relating to the impact of freshwater runoff variability on fisheries production in the northwest Atlantic (a, b, c, and d) as well as on the possible mechanisms of impact (e and f).

a. Within the Gulf of St. Lawrence there is persuasive evidence that inter-annual variability in runoff has an impact on fisheries production. The mechanisms by which the variability is induced, however, are not understood.

b. Even though there is entrainment of nutrients into the surface layer of the Gulf of St. Lawrence due to freshwater runoff and some evidence for relatively high primary production, there is little evidence that suggests that the zooplankton, benthos, and fisheries production are higher than on contiguous shelf areas. The average trawlable biomass in the southern Gulf is essentially identical to that on the Scotian Shelf.

c. Regulation of the Gulf of St. Lawrence system rivers does not seem to have had a measurable impact on fisheries production. If there has been an effect it has been masked by the natural inter-annual variability of runoff and the effects of fishing.

d. Non-local ocean circulation may have a greater impact than local freshwater runoff from the Gulf of St. Lawrence on at least cod fisheries production outside the Gulf of St. Lawrence (on the Scotian Shelf and in the Gulf of Maine).

e. From the broader literature on inter-annual variability on continental shelves it is inferred that there is limited support for a food-chain hypothesis generating fisheries production variability (Sutcliffe 1973; Sutcliffe et al. 1983). Even though we concur with Dickie and Trites (1983) that "environmental processes play a dominant direct role in the determination of relative year-class sizes," we are far from convinced that there is "incontrovertible evidence of the action of the physical system on biological production through the trophic chains."

f. There is accumulating evidence that the physical oceanographic processes of advection and diffusion can have a direct impact (i.e. <u>without</u> links through primary production) on population abundance of zooplankton as well as on fish populations during their early life history stages.

ACKNOWLEDGEMENTS

The reviews by B.T. Hargrave, M.J. Tremblay, and J.B.L. Matthews are gratefully acknowledged. We thank S. Dowell for editing and processing the manuscript, and G. Jeffrey for preparation of the figures.

REFERENCES

Akenhead, S.A. 1983. Mean temperatures and salinities from an ocean climate station by Newfoundland. Northwest Atlantic Fisheries Organization SCR Document 83/VI/30:28 p.

Banks, R.E. 1966. The cold layer in the Gulf of St. Lawrence. Journal of Geophysical Research 71:1603-1610.

Bernal, P.A. and J.A. McGowan. 1981. Advection and upwelling in the California Current. Pages 381-399 <u>in</u> F.A. Richards, editor. Coastal upwelling. American Geophysical Union, Washington, District of Columbia, U.S.A.

Bugden, G.L., B.T. Hargrave, M.M. Sinclair, C.L. Tang, J.-C. Therriault, and P.A. Yeats. 1982. Freshwater runoff effects in the marine environment: the Gulf of St. Lawrence example. Canadian Technical Report of Fisheries and Aquatic Sciences 1078:89 p.

Campbell, A. and R.K. Mohn. 1983. Definition of American lobster stocks for the Canadian Maritimes by analysis of fishery-landing trends. Transactions of American Fisheries Society 112:744-759.

Chelton, D.B., P.A. Bernal, and J.A. McGowan. 1982. Large-scale interannual physical and biological interaction in the California Current. Journal of Marine Research 40: 1095-1125.

Colebrook, J.M. 1978. Continuous plankton records: zooplankton and environment, north-east Atlantic and North Sea, 1948-1975. Oceanologica Acta 1:9-23.

Colebrook, J.M. 1979. Continuous plankton records: seasonal cycles of phytoplankton and copepods in the north Atlantic Ocean and the North Sea. Marine Biology 51:23-32.

Colebrook, J.M. 1981. Continuous plankton records: persistence in time-series of annual means of abundance of zooplankton. Marine Biology 61:143-149.

Colebrook, J.M. 1982a. Continuous plankton records: phytoplankton, zooplankton, and environment, north-east Atlantic and North Sea, 1958-1980. Oceanologica Acta 5: 473-480.

Colebrook, J.M. 1982b. Continuous plankton records: persistence in time-series and the population dynamics of *Pseudocalanus elongatus* and *Acartia clausi*. Marine Biology 66:289-294.

Colebrook, J.M. 1982c. Continuous plankton records: seasonal variations in the distribution of plankton in the north Atlantic and the North Sea. Journal of Plankton Research 4:435-462.

Colebrook, J.M. 1984. Continuous plankton records: relationships between species of phytoplankton and zooplankton in the seasonal cycle. Marine Biology 83: 313-323.

Colebrook, J.M. 1985. Continuous plankton records: overwintering and annual fluctuations in the abundance of zooplankton. Marine Biology 84:261-265.

Colebrook, J.M. and A.H. Taylor. 1984. Significant time scales of long-term variability in the plankton and the environment. Rapports et Procès-Verbaux des Réunions,

Conseil International pour l'Exploration de la Mer 183: 20-26.

Dickie, L.M. and R.W. Trites. 1983. The Gulf of St. Lawrence. Pages 403-425 in B.H. Ketchum, editor. Estuaries and enclosed seas. Elsevier, Amsterdam.

Doubleday, W.G. and T. Beacham. 1982. Southern Gulf of St. Lawrence cod: a review of multi-species models and management advice. In M.C. Mercer, editor. Multispecies approaches to fisheries management advice. Special Publication of the Canadian Journal of Fisheries and Aquatic Sciences 59.

El-Sabh, M.I. 1977. Oceanographic features, currents, and transport in Cabot Strait. Journal of the Fisheries Research Board of Canada 34:516-528.

Garrod, D.J. and J.M. Colebrook. 1978. Biological effects of variability in the north Atlantic Ocean. Rapports et Procès-Verbaux des Réunions, Conseil International pour l'Exploration de la Mer 173:128-144.

Harding, G.C., K.F. Drinkwater, and W.P. Vass. 1983. Factors influencing the size of American lobster (Homarus americanus) stocks along the Atlantic coast of Nova Scotia, Gulf of St. Lawrence, and the Gulf of Maine: a new synthesis. Canadian Journal of Fisheries and Aquatic Sciences 40:168-184.

Huntley, M. and C. Boyd. 1984. Food-limited growth of marine zooplankton. American Naturalist 124:455-478.

Koslow, J.A. 1984. Recruitment patterns in northwest Atlantic fish stocks. Canadian Journal of Fisheries and Aquatic Sciences 41:1722-1729.

Neu, H.J.A. 1970. A study on mixing and circulation in the St. Lawrence Estuary up to 1964. Bedford Institute of Oceanography, Atlantic Oceanography Laboratory Report 1970-9. Dartmouth, Nova Scotia, Canada:31 pp.

Neu, H.J.A. 1975. Runoff regulation for hydropower and its effects on the ocean environment. Canadian Journal of Civil Engineering 2:583-591.

Neu, H.J.A. 1976. Runoff regulation for hydro-power and its effect on the ocean environment. Hydrological Science 21: 433-444.

Radach, G. 1984. Variations in the plankton in relation to climate. Rapports et Procès-Verbaux des Réunions, Conseil International pour l'Exploration de la Mer 185:234-254.

Sheldon, R.W., W.H. Sutcliffe, and K. Drinkwater. 1982. Fish production in multispecies fisheries. Pages 28-34 in Multispecies approaches to fisheries management advice. Canadian Special Publication of Fisheries and Aquatic Sciences 59.

Sinclair, M., J.-J. Maguire, P. Koeller, and J.S. Scott. 1984. Trophic dynamic models in light of current resource inventory data and stock assessment results. Rapports et Procès-Verbaux des Réunions, Conseil International pour l'Exploration de la Mer 183:269-284.

Sinclair, M., M.J. Tremblay, and P. Bernal. 1985. El Nino events and variability in a Pacific mackerel (Scomber japonicus) survival index: support for Hjort's second hypothesis. Canadian Journal of Fisheries and Aquatic Sciences 42:602-608.

Sutcliffe, W.J. Jr. 1972. Some relations of land drainage, nutrients, particulate material, and fish catch in two eastern Canadian bays. Journal of the Fisheries Research Board of Canada 29:357-362.

Sutcliffe, W.H. Jr. 1973. Correlations between seasonal river discharge and local landings of American lobster (Homarus americanus) and Atlantic halibut (Hippoglossus hippoglossus) in the Gulf of St. Lawrence. Journal of the Fisheries Research Board of Canada 30:856-859.

Sutcliffe, W.H. Jr., K. Drinkwater, and B.S. Muir. 1977. Correlations of fish catch and environmental factors in the Gulf of Maine. Journal of the Fisheries Research Board of Canada 34:19-30.

Sutcliffe, W.H. Jr., R.H. Loucks, and K. Drinkwater. 1976. Coastal circulation and physical oceanography of the Scotian Shelf and the Gulf of Maine. Journal of the Fisheries Research Board of Canada 33:98-115.

Sutcliffe, W.H. Jr., R.H. Loucks, K. Drinkwater, and A.R. Coote. 1983. Nutrient flux onto the Labrador Shelf from Hudson Strait and its biological consequences.

Canadian Journal of Fisheries and Aquatic Sciences 40: 1692-1701.

Therriault, J.C. and M. Levasseur. 1985. Freshwater runoff control of the spatio-temporal distribution of phytoplankton in the lower St. Lawrence Estuary (Canada). (This volume.)

Trites, R.W. 1970. The Gulf as a physical oceanographic system. Pages 32-63 in Proceedings of the second Gulf of St. Lawrence workshop. Bedford Institute of Oceanography, Dartmouth, Nova Scotia, Canada (30 November to 3 December, 1970).

Trites, R.W. 1972. The Gulf of St. Lawrence from a pollution viewpoint. Pages 59-72 in M. Ruivo, editor. Marine pollution and sealife. Fishing News (Books) Limited, London, England.

Trites, R.W. and A. Walton. 1975. A Canadian coastal sea - the Gulf of St. Lawrence. Bedford Institute of Oceanography Report Series BI-R-75-15, Dartmouth, Nova Scotia, Canada: 29 p.

COMPUTER MODEL ANALYSIS OF PELAGIC ECOSYSTEMS

IN ESTUARINE WATERS

T.R. Parsons and T.A. Kessler
Department of Oceanography
University of British Columbia
Vancouver, B.C. V6T 1W5

ABSTRACT

A computer model of the principal biological components in a planktonic estuarine ecosystem have been described. The model in used to analyse the effect of organic additions, changes in extinction coefficient, initial nitrate concentrations and zooflagellates on the production of phytoplankton and zooplankton. The impact of these components on the model illustrate the effect of small phase shifts in the coupling between primary and secondary production. Collectively these minor changes perform a "phasing" function which greatly alters the effect of the main "forcing" functions (e.g. light, initial nutrient concentrations) on the planktonic ecosystems.

INTRODUCTION

The simulation of estuarine ecosystems may be carried out with an experimental model (e.g. Spies et al., 1983) or through a computer simulation (e.g. Petersen and Festa, 1984). In the latter reference the dynamics of estuarine phytoplankton productivity was considered primarily in terms of the physical dynamics of river flow, light attenuation and the sinking rate of particles. Little attention was given to the dynamics of an estuarine food web which may react in a dispensatory fashion to offset otherwise inhibitory effects on production. In contrast to the model described by Petersen and Festa (1984), the following Mixed Upper Layer Ecosystem Simulation (MULES) model will deal almost exclusively with the evolution of primary and secondary food web relationships within a homogeneously mixed surface layer.

The maximization of phytoplankton and zooplankton in the estuary of a major river usually occurs at some distance from the river mouth. An example is given in Parsons et al. (1969) for the Fraser River estuary where it is seen that the chlorophyll a maximum occurs at approximately 5 km from the delta, while the zooplankton biomass occurs at a distance of ca 10 km. The positioning of these plankton maxima is caused by a variety of effects such as the growth rate of the phytoplankton, the sediment load, the amount of nutrient entrainment in the mixed layer, the depth of the mixed layer and the characteristics of zooplankton grazing.

The simplest graphical representation of the phytoplankton and zooplankton maxima off the mouth of a river can be illustrated in Fig.1. Assuming a well mixed surface layer that is being advected horizontally, time-dependent changes in the depth averaged standing stock of phytoplankton (n) and zooplankton (H) can be used to evaluate the horizontal distribution of the plankton from the river mouth. The basic terms are growth rate (u), respiration rate (R) and all sinks (M) including grazing, sinking, etc. of the phytoplankton. Changes in zooplankton are similarly a function of the grazing intensity (g) on the available phytoplankton (P) minus respiration (R) and all zooplankton sinks (D) including all forms of mortality. It has been suggested however, that

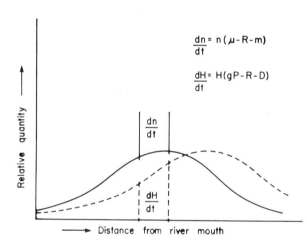

Fig.1 Summary application of the MULES model to an estuarine ecosystem.

this basic two-equation system is augmented by sundry food web relationships. For example, Parker et al. (1975) have shown the important role of heterotrophic bacteria in estuaries and their competition with autotrophic organisms for the limited supply of nitrate. Further, the role of zooflagellates in grazing the heterotrophic bacteria (Fenchel, 1982; Davis and Sieburth, 1984) and providing in turn an additional food supply for zooplankton has been discussed by Kopylov et al. (1981).

In the following discussion, an attempt has been made to incorporate the major driving functions for autotrophic plankton/herbivore production (e.g., light, nutrients) as well as alternate ecotrophic pathways in an analysis of estuarine food web dynamics.

METHODS

A detailed flow diagram of the Mixed Upper-Layer Ecosystem Simulation (MULES) model with respect to an estuarine ecosystem

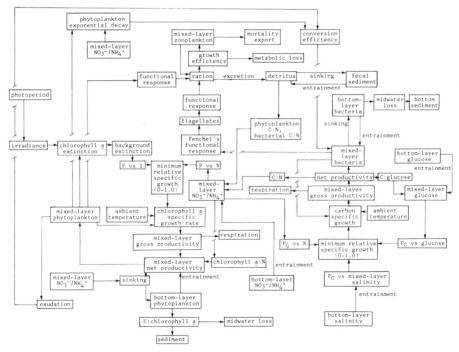

Fig.2 Flow diagram showing functioning components of the Mixed Upper Layer Ecosystem Simulation (MULES) model.

is shown in Fig.2. The actual input terms with some designated values are shown in Table 1 while a complete description of the model is given in Appendix I. A brief description of the use and origin of the terms in Table 1 is as follows: (Note: Words which are underlined reiterate descriptions presented in Table 1).

The simulation duration gives the number of hours the model was run. For the purposes of this assessment it was considered that the top 5m (depth of the mixed-layer) was a partially closed box (i.e. entrainment and sedimentation are the only points of import and export from the system) in which the state-variables were all recalculated at time steps of one hour. The initial amount of nitrate, ammonia, phytoplankton, glucose and zooplankton were entered as the starting mixed-layer values. Numerical values for the maximum surface irradiance (photosynthetically active) and the daylight fraction of 24hr day were chosen according to season and the background light extinction coefficient adjusted to approximate estuarine conditions. The entrainment velocity governed the input of nutrient-rich water from below the mixed layer. The chlorophyll specific extinction was held constant and determined the phytoplankton biomass-dependent feedback on the vertical light field. The reaction of bacteria to the uptake of glucose and inorganic nitrogen, and the reaction of phytoplankton to light-limited growth and inorganic nitrogen are expressed as Michaelis-Menten 1/2-saturation and maximum growth rate constants. Similarly, the grazing of zooplankton on phytoplankton and flagellates, and the grazing of heterotrophic zooflagellates on bacteria were characterized by maximum grazing response and 1/2-saturation constants in Michaelis-Menten functions. A grazing threshold was further introduced for both phytoplankton and flagellate standing stocks. Values for the respiration of phytoplankton and bacteria were expressed as a non-dimensional fraction of the maximum growth rate, PMAX and BMAX, respectively. Zooplankton ration excreted represents the fraction, faeces divided by ration, while zooplankton efficiency represents the fraction of the assimilated ration used for growth divided by growth plus respiration. The conversion of carbon to nitrogen and carbon to chlorophyll a for bacteria, flagellates and phytoplankton were necessary in order to convert lower levels of production into zooplankton production (e.g. C:N, C:CHLA etc.). Zooflagellate growth efficiency was taken from Fenchel (1982), as were other parameters for heterotrophic flagellate growth; these are discussed separately at the end of this section.

The ambient temperature controlled the maximum growth rate of phytoplankton and bacteria (Pmax and Bmax, respectively) according to temperature growth equations given by Eppley (1972) and Li and Dickie (in press); these are listed as Arrhenius regulated in Table 1. The nutrient-dependent sinking-rate of phytoplankton was controlled by a deplete/replete switch sensitive to in situ nitrate falling below the Michaelis-Menten 1/2-saturation constant for nitrate uptake by the particular phytoplankton being considered. Zooflagellate and zooplankton detritus sinking rates were entered as two independent parameters. Printouts of the model were entered for required time intervals (i.e. snapshots) and a total zooplankton mortality parameter accounted for all forms of zooplankton loss.

Depth of bottom layer was the total depth of the estuarine water column which was considered constant with distance from the river mouth. This parameter controlled the conversion of detrital POC according to Suess (1980). Constant phytoplankton exudation was the loss of photosynthetic products during growth. From a survey of the literature (Parsons et al., 1984) this value could range from 10 to 30% daily photosynthesis; a value of 10% was used in these simulations. Efficiency of bacterial conversion represents the efficiency of bacterial uptake of phytoplankton exudate and from phytoplankton organics released during decay following nitrate exhaustion as discussed below. The value for this efficiency has been discussed by Calow (1977) and may range from 0.1 to 0.8; a value of 0.5 was used in these simulations.

Typical values for most of the terms in Table 1 have been

discussed by Parsons et al. (1984); however, some special consideration had to be given to zooflagellate production, zooplankton grazing and the decay of phytoplankton following nutrient depletion.

The dynamics of zooflagellate production were taken from Fenchel (1982). The maximum consumption of bacteria by zooflagellates of 60 bacteria/flagellate/hr was converted to carbon assuming a bacterium volume of 0.25 μm^3 and a flagellate volume of 65 μm^3 to give 0.23 mgC/mgC/hr. Conversion factors for bacterial numbers and flagellate numbers to carbon were 2.5 x 10^5 bacteria/ml = 5 mgC/m^3 and 10^3 flagellates/ml = 5 mgC/m^3. Similarly, Fenchel's (1982) 1/2-saturation constant for zooflagellate grazing response on bacteria was 20 mgC/m^3 and zooflagellate growth efficiency was estimated to be 0.8 based on 3 x 10^{-3} flagellates/bacterium.

The maximum grazing of zooplankton on phytoplankton and zooflagellates was set at two different levels of 0.05 mgC/mgC/hr and 0.15 mgC/mgC/hr, respectively. The justification for this is the following: zooplankton in the model are assumed to include all herbivorous grazers from protozoa (e.g. tintinnids) to metazoans (e.g. copepods). Daily food consumption for this wide range of animals is given by Ikeda (1977) as decreasing with body size from > 500 %/day to > 100 %/day for animals from < 1 μg to > 1000 μg body weight. Assuming that feeding is also size dependent, the daily maximum food consumption of the relatively small flagellates by zooplankton was set at ca 300% per day (or 0.15 mgC/mgC/hr) while for the comparatively larger phytoplankton it was set at ca 100% per day (or 0.05 mgC/mgC/hr).

The decay of phytoplankton following nutrient depletion was taken from Antia et al. (1963) for a phytoplankton crop which was not allowed to sink and from which zooplankton grazers had been removed. The decay rate of this crop was $n_t = n_0 e^{-0.55t}$ where n_0 is the chlorophyll maximum, n_t is the chlorophyll after time t in days.

In the simulations reported here, four types of

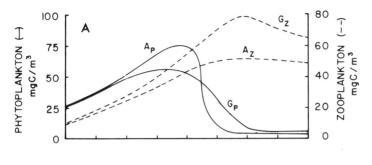

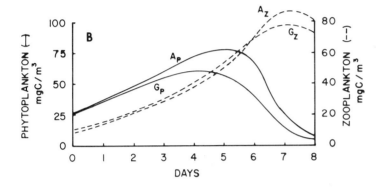

Fig. 3A Effect of an allochthonous substrate on summer production phytoplankton (——) and zooplankton (---) standing stock in the presence (G_p, G_z) and absence (A_p, A_z) of 5 mg/l glucose.

Fig. 3B Effect of an allochthonous substrate on spring production phytoplankton (——) and zooplankton (---) standing stock in the presence of (G_p, G_z) and absence (A_p, A_z) of 5 mg/l glucose.

experiments were considered (c.f. Table 2). These were effects of: 1. changes in the supply of an allochthonous metabolizable substrate (e.g., riverine glucose) during summer and spring growth conditions; 2. changes in the extinction coefficient of the water such as might be caused by the silt load of the river; 3. changes in the supply of nutrients with particular reference to the supply of particulate material to the benthos; and, 4. changes in the abundance of zooflagellates with reference to the production of herbivorous zooplankton. The purpose of the simulation experiments was to create scenarios of possible events rather than conduct a sensitivity analysis of the model.

RESULTS

Fig. 3A shows the effect of adding an allochthonous substrate, such as glucose, on primary and secondary production in an estuarine environment under summer conditions. In the absence of appreciable glucose, phytoplankton production (A_p) grows to utilize the available nitrate faster than it can be supplied by entrainment. The phytoplankton then decrease by sinking out and autolysis after nitrate exhaustion on day 4. Zooplankton production decreases after day 6 when the phytoplankton has virtually disappeared and when the bacterial/zooflagellate cycle, based on the decrease in phytoplankton, has become exhausted. In contrast, when 5 ppm of glucose is present at the beginning of the experiment, the heterotrophic cycle of bacteria/zooflagellates allows for more food to be available at the beginning of the simulation so that the early growth of herbivorous zooplankton is faster, depressing the initial standing stock of the phytoplankton by grazing, which in turn results in a slower utilization of the available nitrate. This has the effect of allowing the herbivorous zooplankton to better track the growth of the phytoplankton and, in so doing, more zooplankton is produced (ca 80 mgC/m^3 compared with ca 50 mgC/m^3 in the absence of glucose).

Fig. 3B shows the effect of glucose addition under spring (c.f. Table 2 parameters) phytoplankton growth conditions. Under these conditions, zooplankton grazers keep up with the growth of phytoplankton which is relatively slow, being controlled by light. The production of zooplankton reaches ca 80 mgC/m^3 in 7 days, which is about the same as is produced in Fig. 3A, under summer conditions in the presence of glucose. However, in Fig. 3B, the presence of glucose actually slightly depresses the total zooplankton production. Although the zooplankton initially grow faster in the presence of glucose, due to the bacteria/zooflagellate cycle, this cycle is actually less efficient than the purely autotrophic cycle which is well coupled with the zooplankton in a low light (spring) environment. This bacterial growth actually interferes with the close coupling of phytoplankton and zooplankton under spring conditions and consequently less zooplankton is produced, which is the opposite effect to the presence of glucose under summer conditions.

Fig. 4 shows the effect of three background extinction coefficients on the production of zooplankton. In very clear water (k = 0.2), phytoplankton grew rapidly, exhausted the nutrient supply and autolysed or sedimented to give only ca 50 mgC/m^3 of zooplankton, which was not able to keep up with the rapid growth of phytoplankton. When the extinction coefficient was increased 50% to 0.3 m^{-1}, phytoplankton growth was slowed down by the lack of light and zooplankton could graze the slower growing phytoplankton more efficiently, attaining maximum standing stock of ca 80 mgC/m^3. However, when the extinction coefficient was increased still further to 0.7 m^{-1}, phytoplankton growth was depressed to the point where zooplankton grazed more phytoplankton than was produced. The phytoplankton stock was then forced down to the grazing

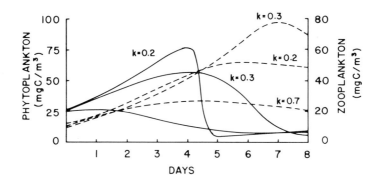

Fig.4   Production of phytoplankton (———) and zooplankton (- - -) different background extinction coefficients.

threshold level.

Figs. 5A and B show the effect of nutrient enrichment (nitrate) on the primary and secondary production and sedimentation of phyto- and zoodetritus. The response of the simulated ecosystem to high initial nitrate (20 μM) is to produce a large pulse of phytoplankton (Fig. 5B), much of which sediments out of the water column because the zooplankton are unable to keep up with the growth rate of the primary producers. The amount of phytodetritus greatly exceeds the amount of zoodetritus (Fig. A) under these conditions. At low initial nitrate concentration (2 µM), zooplankton grazing accounts for a larger fraction of the phytoplankton production and zoodetritus exceeds phytodetritus (Fig. 5A) - this is the opposite effect of a high nutrient regime.

Figs. 6A and B show the effect of having zooflagellates present or absent from an ecosystem. In the absence of zooflagellates (Fig. 6A), the zooplankton standing stock depended entirely on the phytoplankton and consequently only ca 10 mgC/m^3 of zooplankton grew before the phytoplankton disappeared due to a lack of nutrients. In the presence of zooflagellates, the zooplankton produced a standing stock of ca 50 mgC/m^3 by being sustained on a heterotrophic cycle after the autotrophic cycle had declined. Further, the higher the initial zooflagellate standing stock, the broader the pulse of zooplankton production. The underlying mechanisms is revealed by the time-series co-occurence of bacteria and zooflagellates (Fig. 6B) for the initial zooflagalle biomass conditions of 2 and 10 mgC/m^3. In the presence of 10 mgC/m^3, zooplankton grazed the zooflagellates to the grazing threshold of 5 mgC/m^3; subsequently the flagellates increased again in response to a pulse of bacteria from the decay of nutrient depleted phytoplankton after day 3. When zooflagellates were started at 2 mgC/m^3, they increased to just above the threshold level of 5 mgC/m^3, before being grazed, followed by an even larger zooflagellate pulse in response to an enhanced bloom of bacteria (from the delay of a larger standing stock of nutrient depleted phytoplankton). The zooflagellate grazing response to the enhanced bacterial bloom provided a larger available food supply to zooplankton which, on day 5 only, resulted in a greater zooplankton standing stock (Fig. 6A) in the 2 mgC/m^3 compared with the 10 mgC/m^3 initial zooflagellate simulations.

DISCUSSION

In this simulation analysis of an estuarine planktonic ecosystem using the MULES model, four conclusions can be

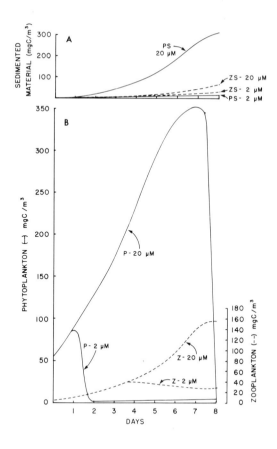

Fig. 5A Effect of nitrate enrichment on sedimented phytoplankton (PS———) and zooplankton (ZS---).

Fig. 5B Effect of nitrate enrichment on phytoplankton (P———) and zooplankton (Z---) production.

reached. The first is that an allochthonous substrate, under summer conditions of plankton growth, stimulates an initial increase in zooplankton production through heterotrophic production; this later results in more of the phytoplankton being grazed than was possible in the absence of glucose. A similar result of an increased zooplankton production in the presence of glucose has been obtained experimentally (Parsons et al., 1981), but the explanation for this increase was not clarified in terms of initally accelerated growth (i.e. a "phasing" function).

In the second experiment, it was shown that increasing the light extinction from 0.2 to 0.3 m^{-1} actually increased zooplankton production by allowing the zooplankton to keep pace better with the phytoplankton production. In this critical region of background extinction coefficients, one again observes a "phasing" function caused by a small change in a coefficient at the beginning of the experiment. An experimental result similar to this simulation has been obtained in a controlled ecosystem (Parsons et al., in press), when mine tailings were used to change the initial extinction coefficient of the water column; this was shown experimentally to result in an _increase_ in the standing stock of zooplankton under low light.

In the third experiment, it was shown that under conditions of high nitrate concentration, phytodetritus dominated over zoodetritus (i.e., fecal pellets), while the reverse was true for a low nitrate regime. Steele and Baird (1972) reported that, in the relatively nutient poor waters of Loch Ewe (< 5 µM nitrate), the predominant particles caught in sediment chambers were fecal pellets. In the study by Stephens et al. (1967) conducted in the nutrient rich waters of the Strait of Georgia (ca 20 µM nitrate), the predominant particles in sediment traps were from phytodetritus. While these two references may not be directly comparable to the model simulation due to differing nutrient flux regimes, the qualitative difference observed between the coastal Atlantic and Pacific seasonal cycles in zooplankton and phytoplankton could be explained on the basis of the inability of zooplankton growth to keep up with a large, rapid increase in phytoplankton abundance in nutrient rich waters (i.e. compare Fig. 5A and 5B).

Figs. 6A and B show the effect of having zooflagellates present or absent from an ecosystem. The important conclusion from this experiment is that the pulse of zooplankton production caused by autotrophic production is not sustained unless zooflagellates are present. This indicates an apparent major role for an alternate pathway in the ecosystem described in Fig. 1.

The concept of "phasing" functions has been introduced in the four experiments reported above. If a "forcing" function is defined for our purposes as a major driving component in the ecosystem (e.g. light, temperature or initial nutrient concentration) then a "phasing" function is one which can modify the effect of a "forcing" function but which can not have any effect in the absence of the "forcing" functions.

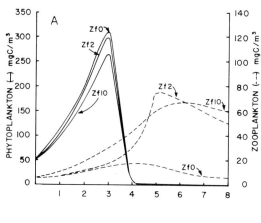

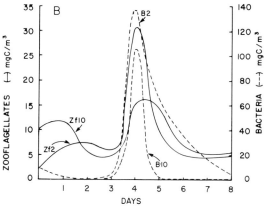

Fig.6A Changes in phytoplankton (——) and zooplankton (---) in the presence (Zf2 and 10) and absence (Zf0) of zooflagellates.

Fig.6B Changes in zooflagellates (Zf——) and bacteria (B---) for initial standing stocks of 2 and 10 mgC/m^3 of zooflagellates.

"Phasing" functions can be viewed as model elements (either parameters or initial conditions) which affect the time course of an ecosystem by causing non-linear dynamical effects (in the experiments cited, produced by small shifts in the producer growth - consumer growth phase relations). If our model assumptions are sufficiently close to the truth, small changes in the light extinction coefficient, initial soluble organic carbon concentration and the initial presence or absence of zooflagellates can be expected to affect secondary production (and presumably higher trophic levels) by several hundred percent. Thus estuarine production appears to be very sensitive to changes in these "phasing" functions.

ACKNOWLEDGEMENTS

The authors are grateful to the Natural Sciences and Engineering Research Council and the Max Bell Foundation for their support of this project.

REFERENCES

Antia, N.J., C.D. McAllister, T.R. Parsons, K. Stephens and J.D.H. Strickland. 1963. Further measurements of primary production using a large-volume plastic sphere. Limnol. Oceanogr. 8: 166-183.

Calow, P. 1977. Conversion efficiencies in heterotrophic organisms. Biol. Rev. 52: 385-409.

Davis, P.G. and J. McN. Seiburth. 1984. Estuarine and microflagellate predation of actively growing bacteria: estimation by frequency of dividing-divided bacteria. Mar. Ecol. Prog. Ser. 19: 237-246.

Eppley, R.W. 1972. Temperature and phytoplankton growth in the sea. Fish. Bull. 70: 1063-1085.

Fenchel, T. 1982. Ecology of heterotrophic microflagellates. II. Bioenergetics and growth. Mar. Ecol. Prog. Ser. 9: 35-42.

Ikeda, T. 1977. Feeding rates of planktonic copepods from a tropical sea. J. Exp. Mar. Biol. Ecol. 29: 263-277.

Kopylov, A.I., A.F. Pasternak and Ye. V. Moiseyev. 1981. Consumption of zooflagellates by planktonic organisms. Oceanology 21: 269-271.

Li, W.K.W. and P.M. Dickie. (in press). Rapid enhancement of heterotrophic but not photosynthetic activites in Arctic microbial plankton at mesobiotic temperatures. Polar Biol.

Parker, R.R., J. Sibert and T.J. Brown. 1972. Inhibition of primary productivity through heterotrophic competition for nitrate in a stratified estuary. J. Fish. Res. Bd. Canada 32: 72-77,

Parsons, T.R., M. Takahashi and B. Hargrave. 1984. Biological Oceanography Processes. Publ. Pergamon Press (Oxford) 330 pp.

Parsons, T.R., C.M. Lalli, P. Thompson, Wu Yong, Hou Shumin and Xu Huai-Shu. (in press). The effect of mine tailings on the production of plankton. Acta Ocean. Sinica.

Parsons, T.R., L.J. Albright, F. Whitney, C.S. Wong and P.J. LeB. Williams. 1981. The effect of glucose on the productivity of seawater: An experimental approach using controlled aquatic ecosystms. Mar. Environ. Res. 4: 229-242.

Parsons, T.R., K. Stephens and R.J. LeBrasseur. 1969. Production studies in the Strait of Georgia. Part I: Primary production under the Fraser River plume, February to May, 1967. J. Exp. Mar. Biol. Ecol. 3: 27-38.

Peterson, D.H. and J.F. Festa. 1984. Numerical simulation of phytoplankton productivity in partially mixed estuaries. Est. Coast Shelf Sci. 19: 563-589.

Spies, A., E.D. Nutbrown and T.R. Parsons. 1983. An experimental approach to estuarine microplankton ecology. Est. Coast Shelf Sci. 17: 97-105.

Steele, J.H. and I.E. Baird. 1972. Sedimentation of organic matter in a Scottish sea loch. Mem. Inst. Ital. Idrobiol. 29: (Suppl.) 73-88.

Stephens, K., R.H. Sheldon and T.R. Parsons. 1967. Seasonal variations in the availability of food for benthos in a coastal environment. Ecology 48: 852-855.

Suess, E. 1980. Particulate organic carbon flux in the oceans: surface productivity and oxygen utilization. Nature 288: 260-263.

Winter, D.F., K. Banse and G.C. Anderson. 1975. The dynamics of phytoplankton blooms in Puget Sound, a fjord in the Northwestern United States. Mar. Biol. 29: 139-176.

Wroblewski, J.S. 1977. A model of phytoplankton plume formation during variable Oregon upwelling. J. Mar. Res. 35: 357-394.

Table 1. Tabulated parameter values for the MULES model.

Numerical Value	Parameter
192.0	simulation duration (hrs)
5.0	depth of mixed layer (m)
1.0	time step (hrs)
2.0	starting mixed-layer nitrate (μM)
0.0	starting mixed-layer ammonia (μM)
1.0	starting mixed-layer phytoplankton (mgChla/m^3)
10.0	starting mixed-layer bacteria (mgC/m^3)
10.0	starting mixed-layer zooflagellates (mgC/m^3)
10.0	starting mixed-layer salt concentration (ppt)
1.0	starting mixed-layer glucose (mgGlucose/m^3)
10.0	starting mixed-layer zooplankton (mgC/m^3)
2000.0	maximum surface irradiance ($\mu E/m^2/sec$)
0.6	daylight fraction of 24 hr day (n.d.)
0.2	background light extinction coefficient (/m)
0.1	entrainment velocity (m/day)
0.025	chlorophyll specific extinction (m^2/mgChla)

Table 1. continued

Numerical Value	Parameter
1.0	1/2-saturation for glucose-limited growth (mgGlucose/m^3)
0.1	bacterial 1/2-saturation for N-limited growth (μM)
0.6	phytoplankton 1/2-saturation for N-limited growth (μM)
200.0	1/2-saturation for L-limited growth ($\mu E/m^2/sec$)
15.0	1/2-saturation for grazing on phytoplankton function (mgC/m^3)
20.0	Fenchel's 1/2-saturation bacterial grazing function (mgC/m^3)
15.0	1/2-saturation for grazing on zooflagellates function (mgC/m^3)
0.05	maximum zooplankton grazing response on phytoplankton (mgC/mgC/hr)
0.23	maximum zooflagellate grazing response on bacteria (mgC/mgC/hr)
0.15	maximum zooplankton grazing response on zooflagellates (mgC/mgC/hr)
5.0	grazing threshold on phytoplankton (mgCm^3)
5.0	grazing threshold on zooflagellates (mgC/m^3)
0.1	phytoplankton respiration as a fraction of Pmax (n.d.)
0.1	bacterial respiration as a fraction of Bmax (n.d.)
0.2	zooplankton ratio excreted (1 - A) (n.d.)
0.1	bottom-layer phytoplankton concentration (mgChla/m^3)
0.1	bottom-layer bacteria concentration (mgC/m^3)
6.0	phytoplankton C:N (n.d.)
10.0	bacterial C:N (n.d.)
6.0	zooflagellate C:N (n.d.)
25.0	phytoplankton C:Chla (n.d.)
13.0	ambient temperature (°C)
0.6	zooplankton (G/T + G) efficiency (n.d.)
0.8	zooflagellate growth efficiency (n.d.)
0.1	replete phytoplankton sinking rate (m/day)
2.0	depleted phytoplankton sinking rate (m/day)
0.5	deplete/replete switch (μM)
0.1	zooflagellate sinking rate (m/day)
0.5	zooplankton detritus sinking rate (m/day)
20.0	NO_3 concentration below surface mixed-layer (μM)
0.0	NH_4 concentration below surface mixed-layer (μM)
1.0	glucose concentration below surface layer (mgGlucose/m^3)
0.0	zooplankton detritus below mixed-layer (mgC/m^3)
0.1	zooflagellate concentration below mixed-layer (mgC/m^3)
0.0275	slope for phytoplankton Arrhenius regulation (/day·°C)
-0.7	Y-intercept for phytoplankton Arrhenius regulation (/day)

Table 1. continued

Numerical Value	Parameter
0.1	slope for bacterial Arrhenius regulation (/day·°C)
-1.5	Y-intercept for bacterial Arrhenius regulation (/day)
24.0	hours between simulation snapshots
0.15	zooplankton mortality rate (fraction/day)
20.0	depth of bottom layer (m)
0.1	constant phytoplankton exudation (fraction/day)
0.5	efficiency of bacterial conversion of phytoplankton exudate (n.d.)

Table 2. Changes in Table 1 parameters used to simulate boundary conditions for data in Figs. 3, 4, 5 and 6.

Fig. 3A - <u>Response to glucose, summer conditions</u>

    Glucose, mixed-layer: 0.001 or 5 mgGlucose/l
    Maximum surface radiation: 2000 $\mu E/m^2/sec$
    1/2-saturation constant for light growth: 200 $\mu E/m^2/sec$
    1/2-saturation constant for nutrient growth: 0.6 μM
    Ambient temperature: 18°C

Fig. 3B - <u>Response to glucose, spring conditions</u>

    Glucose, mixed-layer: 0.001 or 5 mgGlucose/l
    Maximum surface radiation: 500 $\mu E/m^2/sec$
    Daylight fraction of 24 hrs: 0.33
    1/2-saturation constant for light growth: 50 $\mu E/m^2/sec$
    1/2-saturation constant for nutrient growth: 0.1 μM
    Ambient temperature: 15°C

Fig. 4 - <u>Response to changes in the extinction coefficient</u>

    Extinction coefficients: 0.2, 0.3 and 0.7 m^{-1}
    Glucose, mixed-layer: 0.001 mgGlucose/l
    Maximum surface radiation: 2000 $\mu E/m^2/sec$
    Daylight fraction of 24 hrs: 0.6
    1/2-saturation constant for light growth: 200 $\mu E/m^2/sec$
    1/2-saturation constant for nutrient growth: 0.6 μM
    Ambient temperature: 18°C

Fig. 5 - <u>Response to changes in initial nitrate concentration</u>

    Starting mixed-layer nitrate: 2 or 20 μM
    Starting mixed-layer phytoplankton: 2 $mgChla/m^3$
    Starting mixed-layer zooplankton: 5 mgC/m^3
    Ambient temperature: 18°C

Table 2. continued

Fig. 6 - Response to changes in initial zooflagellate population

Starting mixed-layer zooflagellates: 0, 2 or 10 mgC/m³
Starting mixed-layer nitrate: 5 µM
All other conditions as in Fig. 5 and Table 1

Appendix I
A Description of the MULES Model

Under the assumption of a water column consisting of two homogeneous layers, the surface layer food web can be represented by a coupled system of ordinary differential equations describing the rate of change of depth-averaged concentrations in phytoplankton (P), bacteria (B), zooflagellates (F), zooplankton (H), detritus (D), nitrate and ammonia (N) and glucose (G).

$$\frac{d\overline{P}}{dt} = \frac{\alpha_P \overline{P}}{\nu_P} - \beta_P \overline{P} + (P' - \overline{P})(E - W_P)$$

$$\frac{d\overline{B}}{dt} = \alpha_B \overline{B} - \beta_B \overline{B} + (B' - \overline{B})(E - W_B)$$

$$\frac{d\overline{H}}{dt} = \gamma_P \overline{H} + \gamma_F \overline{H} - \beta_H \overline{H} + (H' - \overline{H})(E - W_H)$$

$$\frac{d\overline{F}}{dt} = \gamma_B \overline{F} - \beta_F \overline{F} + (F' - \overline{F})(E - W_F)$$

$$\frac{d\overline{D}}{dt} = \varepsilon_P \overline{H} + \varepsilon_F \overline{H} + (D' - \overline{D})(E - W_D)$$

$$\frac{d\overline{N}}{dt} = \frac{\varepsilon_P \overline{H}}{\phi_P} + \frac{\varepsilon_F \overline{H}}{\phi_F} - \frac{\alpha_P \overline{P}}{\eta_P} - \frac{\alpha_B \overline{B}}{\phi_B} + E(N' - \overline{N})$$

$$\frac{d\overline{G}}{dt} = -\alpha_B \overline{B} + E(G' - \overline{G})$$

where overbars denote depth-averaged concentrations, the primes denote bottom-layer concentrations, E the entrainment velocity, and W_P, W_B, W_H, W_F and W_D sinking velocities for phytoplankton, bacteria, zooplankton, zooflagellates and zoodetritus respectively. $\emptyset_P$, $\emptyset_B$ and $\emptyset_F$ are specified carbon:nitrogen ratios for phytoplankton, bacteria and zooflagellates; η_P and ν_P are the specified Chla:N and C:Chla ratios respectively. α_P, α_B, γ_P, γ_B, γ_F, ε_P, and ε_F are source coefficients defined as follows:

$$\alpha_P = (1 - \phi_P)10^{(\tau_P t + \lambda_P)} \cdot \min \begin{cases} \overline{N}/(n_P + \overline{N}) \\ \overline{L}/(1_P + \overline{L}) \end{cases}$$

where ϕ_P is the fraction of gross productivity lost to respiration, τ_P and λ_P are Arrhenius coefficients, n_P and l_P are the Michaelis-Menten 1/2-saturation constants for nitrogen and light limited growth respectively. The depth-averaged light level ($\overline{L}$) is defined as:

$$\overline{L} = 1/n \sum_{i=1}^{n} L_i$$

where L_i is calculated according to Wroblewski (1977).

$$\alpha_B = (1 - \phi_B)10^{(\tau_B t + \lambda_B)} \cdot \min \begin{cases} \overline{N}/(n_B + \overline{N}) \\ \overline{G}/(g + \overline{G}) \\ 1 - \overline{S}/(s + \overline{S}) \end{cases}$$

ϕ_B is the bacterial loss to respiration, τ_B and λ_B Arrhenius coefficients, and n_B, g and s Michaelis-Menten constants for nitrogen, glucose and salt limited growth respectively. The salt limited growth is based on Spies et al. (1983), who demonstrated bacterial growth rate dependence on ambient salt concentration.

$$\gamma_P = \frac{(R_P \cdot \overline{P} \cdot v_P)}{(r_P + \overline{P} \cdot v_P)} (1 - \pi) \cdot A_H$$

$$\gamma_F = \frac{(R_F \cdot \overline{F})}{(r_F + \overline{F})} (1 - \pi) \cdot A_H$$

$$\varepsilon_P = \frac{(R_P \cdot \overline{P} \cdot v_P) \pi}{(r_P + \overline{P} \cdot v_P)}$$

$$\varepsilon_F = \frac{(R_F \cdot \overline{F})}{(r_F + \overline{F})} \pi$$

R_P and R_F are maximum grazing responses by zooplankton on phytoplankton and zooflagellates respectively, r_P and r_F their respective Michaelis-Menten coefficients, π the ration fraction excreted and A_H the assimilation efficiency.

$$\gamma_B = \frac{R_B \cdot \overline{B}}{r_B + \overline{B}} \cdot A_F$$

R_B is the maximum grazing response on bacteria, r_B is the associated Michaelis-Menten coefficient and A_F the assimilation efficiency. Unlike zooplankton, all zooflagellate ration is assumed to be metabolized.

β_P, β_B and β_F are sink coefficients defined as:

$$\beta_P = \frac{(R_P \cdot \overline{P} \cdot v_P)}{(r_P + \overline{P} \cdot v_P)} \cdot \overline{H} + S\overline{P} \cdot e^{-\xi t} + \overline{P} \cdot \upsilon$$

P, H, R_P and r_P are defined as before, while ξ and υ are a time dependent and constant, phytoplankton to bacteria conversion factor, respectively. S is defined as:

$$S = \begin{cases} 1 & N < n_P \\ 0 & N \geqslant n_P \end{cases}$$

and therefore acts to limit the time dependent conversion of phytoplankton biomass to bacteria at times when ambient nitrogen is less than the phytoplankton Michaelis-Menten parameter for n-limited growth. ξ and υ are based on Antia et al. (1963) and Winter (1975) respectively where no account was taken of differences in chemical ratios. Hence this is also ignored here.

$$\beta_B = \frac{(R_B \cdot \bar{B})}{(r_B + \bar{B})} \cdot \bar{F} - S\bar{P} \cdot e^{-\xi t} + \bar{P} \cdot \upsilon$$

$$\beta_F = \frac{(R_F + \bar{F})}{(r_F + \bar{F})} \cdot \bar{H}$$

(All terms as defined above).

The zooplankton loss coefficient (β_H) is defined as a constant and interpreted as an export component of the food web.

THE CAPACITY OF LAKE MELVILLE FJORD TO STORE RUNOFF

S. A. Akenhead and J. Bobbitt[1]
Fisheries Research Branch
Department of Fisheries and Oceans
P.O. Box 5667
St. John's, Newfoundland Canada A1C 5X1

ABSTRACT

The massive hydroelectric development (1967-1971) on the Churchill River altered runoff, but did this affect the local ocean? Comparison of data from the 1950s and 1972-81 shows there has been a threefold increase in winter runoff and a one-third reduction in summer peak runoff. However, mixing in Lake Melville, especially near the outlet, and the lake's capacity to absorb a freshwater volume equivalent to the March to July runoff, effectively buffers the runoff changes. This fresh water passes out of Lake Melville during the winter replacement of deep water. We therefore observed no difference in the summer water properties of Groswater Bay before and after the Churchill development.

---

[1]Oceans Ltd., P. O. Box 13398, St. John's, Newfoundland.

INTRODUCTION

Construction of the hydro-electric development on the Churchill River began in 1967 and was completed in December 1971. No large dams were constructed but rather a series of dykes covering more than 65 km to link up countless existing lakes and muskeg to create the 3000 km^2 Smallwood Reservoir. The initial drainage area was increased by 11,400 km^2 through the construction of dykes at Orma and Sail Lakes (Coté 1972). This study examines the effects of controlled hydro discharge from the Churchill River upon the water properties of Lake Melville and Groswater Bay. Unpublished measurements made by the **BLUE DOLPHIN** expedition in the early 1950s (Nutt 1951) are compared with recent data collected in 1981; i.e. before and after the Churchill River development of 1967-71. Additional temperature and salinity data collected by the Bedford Institute of Oceanography during summer 1979 have been included. The results have been detailed in a technical report (Bobbitt and Akenhead 1982).

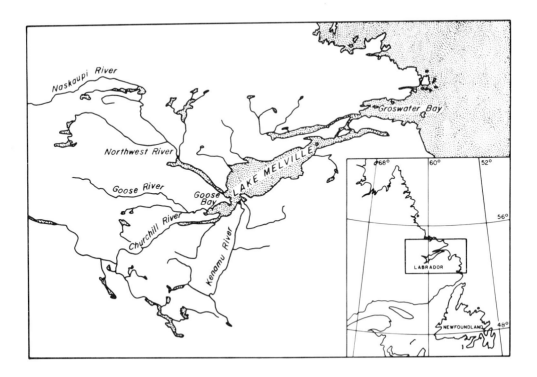

Fig. 1. Study area.

STUDY AREA

Groswater Bay, Lake Melville, and Goose Bay together make up Hamilton Inlet (Fig. 1), the largest inlet located along the Labrador coast. Groswater Bay extends west for approximately 50 km, ending at a narrow and shallow area about 22 km long, known as the Narrows. The most shallow cross section of the Narrows, 2 km south of the community of Rigolet, is approximately 30 m deep and 2.8 km wide. The shallow constriction at the Narrows restricts the exchange of water between Lake Melville and Groswater Bay. Lake Melville is approximately 180 km long, 35 km wide at its western (widest) end, and is over 200 m deep. It is classified as a landlocked fjord from the dimensions of the constriction in relation to the size and depth of Lake Melville.

Fresh water is discharged into Lake Melville at the western end by four major rivers (Fig. 1). In order of size, these are the Churchill River (which accounts for over 60% of the inflow), the Northwest, the Kenamu, and the Goose. Discharge rates for the four rivers, as determined by survey parties during the **BLUE DOLPHIN** expeditions, are shown in Table 1, for

Table 1. Discharge rates in m^3/sec for the four major rivers flowing into Lake Melville, as measured by the **BLUE DOLPHIN** expeditions.

Date	Churchill	Northwest	Kenamu	Goose
July, 1950	2,727	1,262	179	135
July, 1952	2,774	625	161	43
August, 1952	1,934	669	692	303
March, 1953	~226	366	7	5
May, 1953	10,919	1,800	290	532

years 1950-1953 (Coachman 1953). During spring runoff in May 1953 the discharge of the Churchill River was exceptionally high with a flow rate of approximately 10,900 m$^3 \cdot$s^{-1}. The maximum flow, as measured by the Water Resources Branch of Inland Waters Directorate since 1954, was usually in June with a mean value of 5,100 m$^3 \cdot$s^{-1} for years 1954-1966. The minimum discharge was usually in April with a mean flow rate of 400 m$^3 \cdot$s^{-1}. The low value is because the surface water of the Labrador plateau is mainly in the form of ice and snow during the winter months.

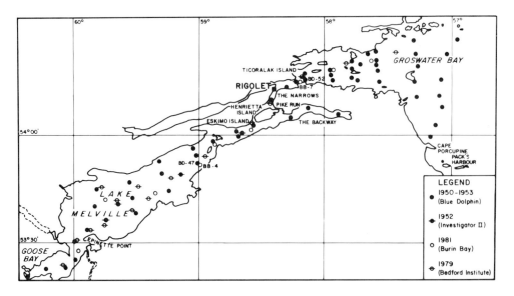

Fig. 2. Stations sampled in Lake Melville and Groswater Bay.

METHODS

In order to compare the salinity differences in Lake Melville, six stations inside the sill were sampled in August 1981 (Fig. 2), and the measurements compared to those taken in the early 1950's by the **BLUE DOLPHIN** expedition and **INVESTIGATOR II**. Since there is less fresh water currently flowing into Lake Melville during the summer months, a reduced volume of fresh water in the fjord was expected, and significant differences in the salinity structure.

The salinity profiles from the stations in August 1979 and August 1981 were compared with the profiles from the early 1950s at nearby locations.

The fresh water volume was determined by dividing Lake Melville into 4 to 18 segments depending on the number of stations sampled during each oceanographic survey. The area of each segment was extracted from bathymetric charts, and the concentration of fresh water calculated from the salinity profiles. The concentration C of fresh water was determined for 5 m depth intervals by the formula: $C = (S_0-S)/S_0$; where S was the mean observed salinity in the 5m depth interval, and S_0 the salinity of seawater. The salinity of seawater was taken as 30°/oo, the salinity between 15 and 30 m at the station sampled outside the Narrows. Only the fresh water volumes in the upper 30 m were considered because most of the river water remains near the surface with the halocline occurring between 5 and 20 m (Fig. 4). In addition, the sill is 30 m deep, so that water below this depth cannot easily leave the fjord.

The residence time is the time required for the rivers to replace the freshwater volume without mixing and can be expressed by t = (volume of fresh water)/(river discharge). The residence time can also be interpreted as the time for replacement of 1/e of the old freshwater in the lake by new runoff, under conditions of complete and instant mixing.

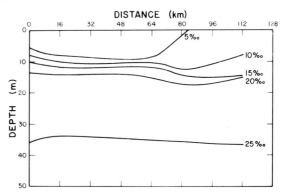

Fig. 3. Longitudinal salinity distribution in Lake Melville.

RESULTS

The monthly mean runoff values before and after the Churchill Falls development are shown in Fig. 2, indicating that the monthly flow rates have changed considerably since the development. The greatest differences are during the winter months. From December to April, the flow rates have tripled over those measured before hydro development. During June and July, the flow rates have decreased by about 30%. The overall quantity of freshwater has increased by about 10% (300 $m^3 \cdot s^{-1}$) due to the diversion of water which originally flowed into the Kanairiktok River.

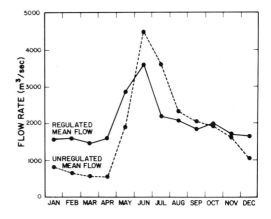

Fig. 4. Mean monthly flows before and after the Churchill hydroelectric development.

Profiles of the near surface water throughout Lake Melville showed that the salinity structure of the near-surface waters are remarkably similar, even though the fresh water inflow from the rivers has changed considerably.

The slight differences observed are in the range one would expect when sampling on different days during the same season due to tides, internal waves and wind mixing. At specific depths, the salinity of the water did not noticeably increase from head to mouth as normally found in most fjords. This indicates that horizontal advection and turbulent mixing are more important than vertical mixing by entrainment.

The calculated volumes of fresh water and the corresponding residence times are summarized in Table 2. The freshwater volume and residence time in Lake Melville was found to be comparable in August 1979 and 1981 to that during the months of July and August in the early 1950's. The long residence times, together with a decrease in freshwater volume during the winter months, indicate that Lake Melville acts as an effective filter between variations in river flow and changes in Groswater Bay. The freshwater volume in March 1953 was approximately 20 km³ or one-half that during the summer months. The river discharge from March to July into Lake Melville was about 25 km³. This means that most of the spring runoff is trapped by being mixed into Lake Melville.

Table 2. Freshwater volumes and their respective flushing times.

Observation date	Fresh water volume (km^3)	River discharge (m$^3 \cdot$s^{-1})	Flushing time (months)
March 1953	20.00	600	12.9
July 1950	40.27	4,300	3.6
July 1952	44.20	3,600	4.7
August 1951	34.52	~2,880	~4.6
August 1952	35.13	3,600	3.8
October 1952	33.15	3,090	4.1
August 1979	38.83	3,520	4.2
August 1981	41.05	2,740	5.8

In order to get vertical exchange between fjord deep water and adjacent coastal water, the density of coastal water at sill depth and above has to be greater than that of the deep water in the fjord. This happens in summer in Lake Melville. The density of the water between 12 and 30 m at station BB-7, located just outside the Narrows, is greater than that found anywhere in Lake Melville. A similar situation existed in 1952 except that the density of the adjacent coastal water was slightly lower in that year. The high density of the water outside the sill suggests that the deep water is readily being flushed. However, the water below sill depth in Lake Melville is approximately 1-3 degrees colder than that of the adjacent coastal water, and has negative temperature below 100 m. T-S curves (Fig. 5) of the water masses on opposite sides of the sill show that the water properties are considerably different. If the water outside the sill, at sill depth and above, is mixed with that found at the same depth inside the sill, the resulting mixture will not have the properties of the water found below 30 m in the lake. The properties of the deep water can only be obtained through cooling. Even though the water properties indicate no exchange taking place during summer, the deep water is not stagnant. The oxygen values of the water below sill depth were found to vary between 7.2 and 7.8 ml/L at station BD-47 in August 1952. Similarly high oxygen values were found throughout the lake.

Because the incoming sea water has a salinity greater than 24.7‰, its density will increase with cooling. Figure 6 compares March and August T-S characteristics with respect to bottom water formation in Lake Melville. In the summer, incoming Groswater Bay water would have to be mixed with colder, fresher water to get the properties found below sill depth in Lake Melville. By comparison, the potential for bottom water formation does exist during autumn and winter, particularly in the region of the Narrows. The T-S curves in Fig. 6 indicate that, by March, mixing of Lake Melville and Groswater Bay waters at sill depth will result in water with the properties of the deep water in Lake Melville. Moreover, the high oxygen values measured in 1952 suggest that complete exchange may occur over the winter period, even though there is a shallow constriction at the Narrows. Oxygens were not measured in 1981, but Vilks et al. (1982) report high oxygen values in the bottom waters of Lake Melville in August 1979.

In August 1979 and 1981, the deep water had lower salinities than during August 1950 and 1951. At a depth of 100 m the average salinity was

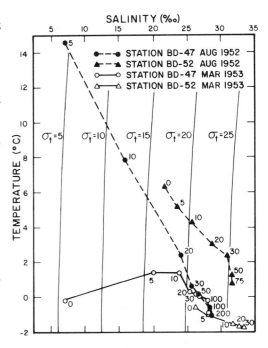

Fig. 5. T-S curves of water masses in Lake Melville and outside the sill during August 1952 and 1981.

lower by 0.7°/∘∘. This lower salinity is probably due to different mixing conditions caused by the larger volume of river water discharged during winter since the Churchill development.

The Narrows, located between Lake Melville and Groswater Bay, is 30 m deep, 2.8 km wide, and 22 km long. This constriction reduces the rise and fall of the tide from 1.3-2 m in Groswater Bay to 0.2- 0.5 m in Lake Melville (Nutt 1963). This gives a choking coefficient of approximately 0.2 where the choking coefficient is defined as the ratio of the inside tidal height to the outside tidal height. On July 24, 1950, ebb current in the Narrows was found to have a speed of 5 knots and flood current had a speed of 4 knots. On July 18, 1951, an ebb current of 6.6 knots and a flood current of 6.0 knots was reported. The constriction at the Narrows can convert an enormous amount of tidal potential energy to kinetic energy for turbulent mixing on both sides of the constriction. McClimans (1978) found that for frictionless tidal flow, the maximum energy will be transferred to the fjord when the tidal range within the fjord is 0.7 times as large as the external tidal range. This value is expected to be

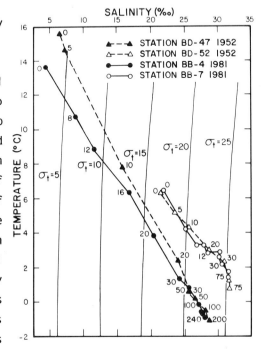

Fig. 6. T-S curves of water masses in Lake Melville and outside the sill during August 1952 and 1981.

different in the case of Lake Melville because of the Narrows are too long to neglect frictional effects. However, the choking coefficient of approximately 0.2 indicates that tidal kinetic energy is unlikely to be very effective in driving the circulation or in producing mixing inside the fjord.

The transport capacity of the constriction will increase during the times the barotropic tidal currents are strong enough to give a one way directed flow. The barotropic flow becomes important when its velocity is greater than $\frac{1}{2}(g'H)^{\frac{1}{2}}$ or approximately 2 knots in the case of the Narrows. For this condition, the flow is alternating between barotropic and baroclinic, or just barotropic (Stigebrandt 1977) during each tidal cycle.

The condition for supercritical flow is that the current be greater than $(g'H)^{\frac{1}{2}}$, or greater than approximately 3.4 knots in the case of the Narrows. For this situation, internal waves are not generated at the sill and the flow is similar to that expected when stratification is absent (Stigebrandt 1980). In supercritical flow, the flow expands on the lee side of the constriction and a turbulent jet develops. Since currents of 6 knots have been measured in the Narrows, it is probable that the flow goes from subcritical (less than wave velocity) to supercritical over each tidal cycle.

The jet on the lee side of a constriction can be important for turbulent vertical mixing, both in Lake Melville and in Groswater Bay, depending on the stage of the tide. Turbulent mixing in the vicinity of the Narrows is probably a large factor in why the salinity stratification of the upper layer of the fjord is practically constant. Therefore, in addition to restricting the flow below sill depth, the Narrows play a major role in the conversion of tidal potential energy to kinetic energy for turbulent mixing and circulation on both sides of the sill. The circulation and mixing mechanisms taking place in the Narrows need further study in order to understand the circulation and mixing processes within Lake Melville.

The water properties at the head of Groswater Bay were compared with stations sampled before and after the Churchill development. Since these stations were located immediately outside the Narrows, any changes in the water structure of Groswater Bay due to the Churchill development would be reflected at this location. The stations were sampled during July 1950, August 1952, October 1952, August 1979, September 1979, and August 1981. The temperature and salinity profiles are shown in Fig. 7. These profiles show that the water properties in 1979 and 1981 are within the natural variations found before the Churchill development.

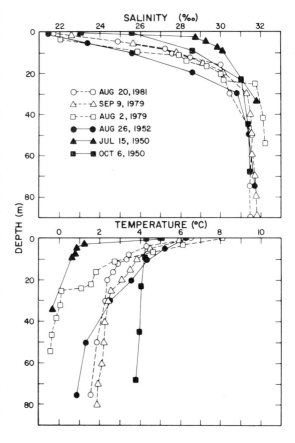

Fig. 7. Salinity and Temperature profiles outside the Narrows.

DISCUSSION

A comparison of the water properties in Groswater Bay shows that, in the region outside the Narrows, the water structure was similar in August 1981 to that in the early 1950's. If there had been changes in Groswater Bay due to the regulation of the Churchill River, these changes would have been most noticeable at this station. The similarity of 30 m and deeper salinities from Fig. 7 is emphasized by the TS plot of Fig. 5, which shows the water mass of August 1952 is identical to that of August 1981.

Energy from the wind is available only during summer, in contrast with the constant tidal energy expended on mixing in the Narrows. At the beginning of November, the ice cover eliminates wind energy input, and vertical mixing is reduced. This allows a stronger gradient in the surface waters of the lake. Approximately at this time the density of the Labrador Current water outside the sill increases and is the source of bottom water replacement throughout the winter. This new bottom water lifts the summer-formed brackish water above sill depth, and results in the lake losing freshwater content. Since winter represents the period of lowest freshwater runoff, all these effects occur together and result in the lake's freshwater content reaching a minimum just before ice breakup. The breakup of the ice cover allows wind energy for mixing to enter the lake at the start of the spring runoff maximum. These natural controls over freshwater volume in Lake Melville provide mechanisms whereby the alterations of freshwater runoff due to hydroelectric development can be absorbed by Lake Melville and not transmitted to Groswater Bay.

Our study describes the nature of freshwater volume changes and retention times, the effects of the Narrows on exchange with the coastal water, and compares Lake Melville and Groswater Bay during summer in the period before and after hydroelectric development. Although the available data is limited, there is apparently no effect of development either within the Lake or in the Groswater Bay area in the important summer months. More convincingly, we have identified mechanisms whereby no effect would be expected, despite the scale of the runoff alterations. We conclude that fjord dynamics, such as exhibited in Lake Melville, can ameliorate the effect of hydroelectric development on the discharge of fresh water onto the ocean. This is not to say there is no effect at other times of the year for which we have no data. The increased winter runoff, added to the winter discharge of accumulated fresh water in the lake, probably lowers the Groswater Bay salinity during winter.

REFERENCES

Bobbitt, J., and S. Akenhead. 1982. Influence of controlled discharge from the Churchill River on the oceanography of Groswater Bay, Labrador. Can. Tech. Rep. Fish. Aquat. Sci. 1097: 43 p.

Coachman, L. K. 1953. River flow and winter hydrographic structure of the Hamilton Inlet-Lake Melville estuary of Labrador. Blue Dolphin Labrador Expedition, unpublished manuscript. 19 p.

Coté, L. 1972. Heritage of power. Report prepared for Churchill Falls (Labrador) Corporation Limited by Cobana Ségium & Associates Inc., 75 p.

Farmer, D., and J. D. Smith. 1978. Nonlinear internal waves in a fjord, p. 465-494. In J. Nihoul [ed.] Hydrodynamics of Estuaries and Fjords. Elsevier Oceanography Series, Vol. 23.

Glenne, B., and T. Simensen. 1963. Tidal current choking in the landlocked fjord of Nordasnatnet. Sarsia 11: 43-73.

McClimans, T. A. 1978. On the energetics of tidal inlets to landlocked fjords. Mar. Sc. Commun. 4(2): 121-137.

Nutt, D. C. 1951. The Blue Dolphin Labrador Expeditions, 1949 and 1950. Arctic 4(1): 3-11.

1963. Fjords and marine basins of Labrador. Polar Notes 5: 9-24.

Stigebrant, A. 1977. On the effects of barotropic current fluctuations on the two-layer transport capacity of a constriction. J. Phys. Ocean. 7: 118-122.

1980. Some aspects of tidal interaction with fjord constrictions. Estuarine and Coastal Marine Science 11: 151-166.

Vilks, A., B. Deonarine, F. J. Wagner, and G. V. Winters. 1982. Foraminifera and mollusca in surface sediments of the southeastern Labrador Shelf and Lake Melville, Canada. Geol. Soc. Amer. Bull. 93: 225-238.

INORGANIC NUTRIENT REGENERATION IN LOCH ETIVE BOTTOM WATER

Anton Edwards and Brian E. Grantham
Scottish Marine Biological Association
P.O.Box 3, Oban, Argyll, Scotland

ABSTRACT

Loch Etive is a three basin Scottish fjord. Water above sill level in the deepest basin usually has low salinity because of locally high freshwater runoff. When runoff lessens, salinity and density rise and the loch bottom water is renewed by density current inflow. When runoff rises again, the new water is isolated by its high density and it stagnates for many months. Rates of oxygen consumption and inorganic nutrient regeneration were measured in stagnant water during 1980/81. Stoichiometric comparison shows the possibility of some denitrification and of a strong silicate source.

INTRODUCTION

Concentrations of nonconservative properties in the ocean change through biogeochemical processes, and physical processes of diffusion and advection. Biogeochemical rates can be measured only by experiments designed to measure or eliminate the physical effects. For example: sediment incubation work in the laboratory or in the field with bell jars (Balzer, 1980; Rowe et al., 1975) aims to remove them completely. Studies of the changing budgets of material in coastal water (Sen Gupta and Koroleff, 1973) account for physical exchange with neighbouring water before quoting mobilisation rates. However, drawbacks exist: laboratory and bell jar incubations are open to the criticism of sediment disturbance; bell jars are best for short term work in shallow water; in budget surveys it is difficult to quantify advection and diffusion from adjacent water and where these are comparable to or greater than the local fluxes, errors are particularly large. In this paper we use naturally trapped water at the bottom of Loch Etive as a natural bell jar, free from advection, whose diffusion is estimated by reference to conservative properties. Other fjordic studies use a similar approach (Almgren et al., 1975; Dyrssen, 1980). The advantages are natural conditions, no sediment disturbance, little advection and measured diffusion. The disadvantage is that, although the water is oxic, it and the sediment are not typical of shelf seas.

LOCH ETIVE

Loch Etive is a fjord with three basins (60, 60, 150 m deep) separated by sills about 13m deep (Figure 1). Salinity near sill level is much lower than shelf salinity because of freshwater runoff. Low runoff raises salinity and sill water becomes sufficiently saline and dense to flow as a renewing density current into the basin bottoms.

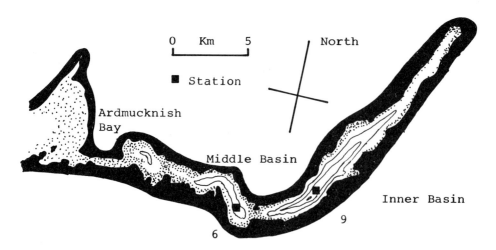

Figure 1. Loch Etive Bathymetry with sampling stations 6 and 9. Contours at 50, 100 and 150 metres. Shaded areas are less than 15 m. deep. More open shelf water starts at Ardmucknish.

After renewal the bottom water stagnates in the deepest basin for months or years (Edwards and Edelsten, 1977). Figure 2 shows typical conditions during a stagnation. Estuarine circulation develops in the upper thirty or so metres with a primary pycnocline separating outgoing brackish water from incoming sea water. The bottom water is separated from the estuarine circulation by a secondary pycnocline which inhibits mixing and turbulent transfer. The temperature and salinity change slowly, the density falls slowly - priming the system for the next renewal - and changes in the concentration of inorganic solutes are dominated by the nonconservative regenerative processes at work in the water and in the sediment. Oxygen concentration falls and major and minor (Edwards and Truesdale, 1980) inorganic nutrient concentrations rise. Figure 3 is schematic and shows the main processes.

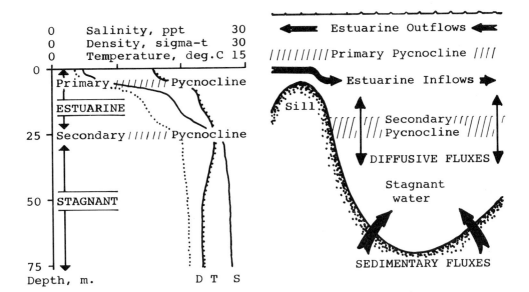

Figure 2. Stagnation, station 9 in Loch Etive, 11 Nov. 1971. A secondary pycnocline stops estuarine water from mixing with the deep stagnant water. Salinity controls the density.

Figure 3. Deep water, isolated by its high density, stagnates below the advective estuarine system where sill water flows down to its own density level. Fluxes change deep properties.

MEASUREMENTS

Measurements are summarised in figure 4. In 1980 to 1982, temperature, salinity and oxygen were measured at ten metre intervals at stations 6 and 9. Previous surveys have shown axial isotropy during stagnations, so these stations were taken to represent their basins. In June 1980, sharp rises in salinity, temperature and oxygen marked deep renewal and similarly spaced measurements of silicate, phosphate and nitrate (including nitrite) were started. The deep water remained stagnant for two years, although partial renewals in 1981 penetrated the upper layers between sill level and about 70 metres deep. The top of the stagnant water - the deepest limit of the advective estuarine circulation - had to be determined before making budgets of stagnant water properties.

Water flowing over the sill from station 6 can sink no deeper at station 9 than the depth where station 9 water has the same density as the inflow. Below this the water is stagnant. Figure 4 shows the stagnant boundary in 1980-82, superimposed on the measurements. At station 9, incursion of new oxygen-saturated water with low nutrient concentrations in June 1980 is clearly shown when sill water density exceeded any density in the inner basin and could therefore sink freely to the bottom. Fairly low summer rainfall in 1981 produced a sill density high enough to sink a partial renewal to about 70 metres, and stagnation ended completely in June 1982 by the incursion of high salinity sill water formed in the dry summer.

DIFFUSIVE AND BIOGEOCHEMICAL CHANGES IN THE BOTTOM WATER

Changes in oxygen and nutrients owe both to biogeochemical processes and to diffusive exchange with the estuarine circulation above. At the start of stagnation, biogeochemical effects dominated because vertical differences were small so that diffusion was negligible. As differences between the lower

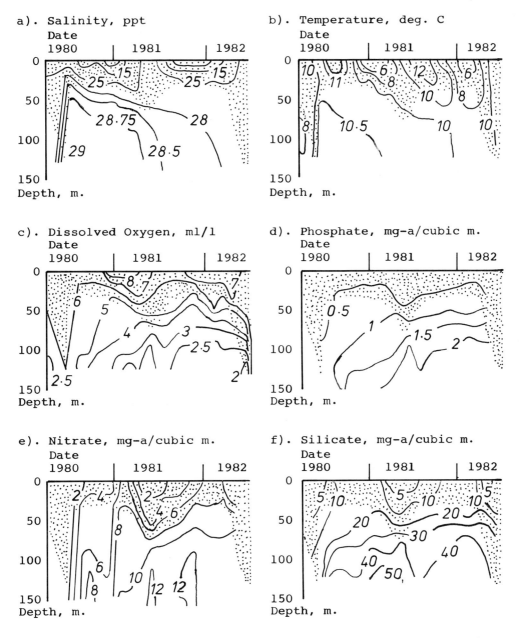

Figure 4. Measurements of conservative and nonconservative properties, station 9, Loch Etive, 1980 to 1982. The estuarine advective circulation is restricted to the shaded region. Below this the water stagnates with a general increase in inorganic nutrients and a decrease in dissolved oxygen concentration. Renewal is complete in 1980 and 1982 but only partial in 1981.

estuarine water and the stagnant water increased, so did the diffusive fluxes. Ultimately, diffusive and biogeochemical fluxes became roughly equal and opposite and the stagnant water reached a steady state with little observable change in late 1981 or 1982.

In this paper we restrict ourselves to the first part of the stagnation from July 1980 to April 1981, before partial renewals of 1981 complicated the picture and before steady state was attained. In this period, the top of the stagnant water was about 55 metres deep but eventually fell to about 85 metres as the partial renewals started.

To quantify the diffusion, assume that the rate of diffusive change of the mean concentration in the stagnant water is proportional to the difference between the mean in the stagnant water and the concentration in the lower layer of the estuarine circulation just above the top of the stagnant water. Using salinity as a conservative tracer similar to the nonconservative properties in its diffusive behaviour, the constant of proportionality was determined as about 0.25 per year, close to 0.2 per year found by Edwards and Edelsten (1977). Applied to the average differences in estuarine and stagnant concentrations, it was used to estimate the size of the diffusive changes in the nonconservative properties. From July 1980 to April 1981, the observed increases in nutrients and decrease in oxygen were about 20% less than those of biogeochemical origin that would have occurred in the absence of diffusion. From July to October 1980, the correction was only about 5% and during July itself the diffusion could be neglected.

The oxygen budget below any depth was calculated by vertical integration of the product of horizontal basin area and oxygen concentration. Three periods were considered: July 1980; July - Oct 1980; and the whole period July 1980 - April 1981. The mean consumption rate per unit area of seabed was then calculated from the rate of change of the total oxygen budget, after applying the small diffusive corrections above. It has already been suggested (Edwards and Truesdale, 1980) that fluxes from the sediments rather than the water dominate the regeneration processes in this fjord, so all fluxes are here related to the area of sediment in contact with the stagnant water rather than to water volume. On the large scale, the mean slope of the fjord bottom is only about 10 degrees, so that the sediment area was as a first approximation taken to be the same as its mapped area. Table 1 shows the results. Estimates from deeper than 55m may be compared with those deeper than 85m so as to assess the influence of possible undetected partial renewals in the 55 to 85m zone.

Oxygen consumption rates were highest at the start but soon diminished to a typical value of about 30 mg-atom/sq.m./day in winter. This rate falls in the range for many fjords reviewed by Gade and Edwards (1980). There is a small difference between results according to the depth range chosen. It might be expected that undetected partial renewals in the upper parts of

Table 1: Stagnant Water of Loch Etive, July 1980 to April 1981

Flux rates after corrections for diffusion. mg-atoms per square metre per day. Parenthetic values are relative to phosphate.

Period	Depth,m	Oxygen	Nitrate	Silicate	Phosphate
July 1 1980	55-145	-62(-365)	3.5(21)	5.8(34)	0.17(1)
- Aug 7 1980	85-145	-49(-490)	2.3(23)	5.0(50)	0.1 (1)
July 1 1980	55-145	-43(-331)	1.3(10)	3.4(26)	0.13(1)
- Dec 18 1980	85-145	-34(-378)	1.0(11)	2.8(31)	0.09(1)
July 1 1980	55-145	-32(-291)	1.3(12)	3.5(32)	0.11(1)
- Apr 9 1981	85-145	-25(-278)	0.8(9)	2.7(30)	0.09(1)

the stagnant water (55-85 m) would make estimates spuriously low: on the contrary, inclusion of the 55-85 metre water leads to larger rates. This rather confirms that the partial renewals did not affect the water below 55m until after the beginning of April 1981 and even suggests a tendency for a higher consumption rate in the upper part of the stagnant water than deeper.

Fluxes calculated from budgets of nitrate, silicate and phosphate are shown in table 1. As with the oxygen there was a tendency for the fluxes to be biggest in the first month of stagnation, falling later to typical rates of 1, 3.1, 0.11 mg-atom/sq.m./day respectively. Again, fluxes below 55m exceed those below 85 metres, suggesting slightly higher production rates in the 55-85m range than deeper.

The size of these fluxes is similar to those from other sites in the world. Hartwig (1974) used bell jars in the La Jolla Bight. He estimated an inorganic nitrogen flux of order 1 mg-atom/sq.m./day and a phosphate flux about 0.1 mg-atom/sq.m./day. Rowe and others (1975) measured a total inorganic nitrogen flux of 1.65 mg-atom/sq.m./day in Buzzards Bay, Massachusetts. Balzer (1980) reported a phosphate flux of 0.055 mg-atom/sq.m./day in oxic conditions in the Baltic, a little less than Dyrssen's (1980) Byfjord estimate of 0.083 mg-atom/sq.m./day. From this viewpoint, Loch Etive is perhaps more typical of oxic coastal waters than its peculiar hydrography at first suggests.

STOICHIOMETRIC RELATIONS

It is interesting to examine the relative proportions in which the various properties are produced or consumed. Table 1 shows the flux rates normalised to that of phosphate. After the first month, in which the phosphate flux seems particularly low in relation to the other components, and in which the system was probably recovering from the effects of sediment brought into suspension by the inflowing density currents (Stanley et al, 1981; Gade and Edwards, 1980), the relative flux rates were typically:

$O:N:Si:P = 320:10:31:1$

These ratios are similar to those found in the decomposition of marine organic matter, but the high silicon flux relative to the others (Grill and Richards, 1964) suggests that the regeneration in Loch Etive owes to a silicate rich source. The rather low nitrate ratio compared to oceanic values (Redfield et al., 1963) might indicate that some nitrogen was entering the water as undetected ammonia: however, with the long stagnation, this may be expected to have oxidised to nitrate. In this case observed nitrate represents total inorganic nitrogen and its low ratio may indicate some denitrification in the system. On the other hand, low N:P ratios are commonly found in coastal waters and may also be attributed to terrestrial influence or a slow nitrogenous regeneration (Sen Gupta and Koroleff, 1973).

SUMMARY

The deep stagnant water of Loch Etive has been used to measure the rates of oxygen consumption and dissolved inorganic nutrient production per unit area of sea bed. Rates are similar to those in some other coastal areas and their stoichiometric ratio $O:N:Si:P$ is about $320:10:31:1$. The rates are greatest in the initial stages of stagnation, in the upper stagnant water, and appear to owe to the decomposition of silicon rich marine organic material.

ACKNOWLEDGEMENTS

We thank Kirsty Petre, Neil Pascoe and David Ansell for all their help. The SMBA is grant-aided by the Natural Environment Research Council.

REFERENCES

Almgren, T., L.G.Danielsson, D.Dyrssen, T.Johansson & G.Nyquist (1975). Release of inorganic matter from sediments in a stagnant basin. Thalassia Jugoslavica 11(1/2), 19-29.

Balzer, W. (1980). Redox dependent processes in the transition from oxic to anoxic conditions. In "Fjord Oceanography", ed. H.J.Freeland, D.M.Farmer & C.D.Levings. Plenum. 659-665.

Dyrssen, D. (1980). Sediment surface reactions in fjord basins. In "Fjord Oceanography", ed. H.J.Freeland, D.M.Farmer & C.D.Levings. Plenum. 645-657.

Edwards, A. & D.J.Edelsten (1977). Deep water renewal of Loch Etive: a three basin Scottish fjord. Estuar. Coast. Mar. Sci. 5, 575-595.

Edwards, A. & V.W.Truesdale (1980). The speciation of iodine in Loch Etive. Marine Physics Group Rept 6. SMBA. Oban. 13pp.

Gade, H.G. & A.Edwards (1980). Deep water renewal in fjords. in "Fjord Oceanography", ed. H.J.Freeland, D.M.Farmer & C.D.Levings. Plenum. 453-489.

Grill, E.V. & F.A.Richards (1964). Nutrient regeneration from phytoplankton decomposing in seawater. J. Mar. Res. 22, 51-69.

Hartwig, E.O. (1974). Physical, chemical and biological aspects of nutrient exchange between the marine benthos and the overlying water. University of California Institute of Marine Resources, 74-14, Sea grant Publication No. 40.

Redfield, A.C., B.H. Ketchum & F.A. Richards (1963). The influence of organisms on the composition of sea-water. in "The Sea", 2, ed. M.N. Hill, Wiley Interscience.

Rowe, G.T., C.H.Clifford, K.L.Smith & P.L.Hamilton (1975). Benthic nutrient regeneration and its coupling to primary productivity in coastal waters. Nature 255, 215-217.

Sen Gupta, R. & F.Koroleff (1973). A quantitative study of nutrient fractions and a stoichiometric model of the Baltic. Estuar. Coast. Mar. Sci. 1, 335-360.

Stanley, S.O., J.W.Leftley, A.Lightfoot, N.Robertson, I.M.Stanley & I.Vance (1981). The Loch Eil Project: sediment chemistry, sedimentation and the chemistry of the overlying water in Loch Eil. J. Exp. Mar. Biol. Ecol. 55, 299-313.

PHYSICAL EXCHANGE AND THE DYNAMICS OF PHYTOPLANKTON IN SCOTTISH SEA-LOCHS

Paul Tett

Scottish Marine Biological Association
Dunstaffnage Marine Research Laboratory
PO Box 3, Oban, Argyll, Scotland

ABSTRACT

Freshwater inflows and salinity distributions were used with a single-box model to estimate the effects of exchange on the dynamics of phytoplankton populations in 5 Scottish fjordic sea-lochs. Results suggest that although growth can substantially exceed losses within a loch, these processes are in most cases in rough balance, with little net export of primary production.

INTRODUCTION

The rate of change of phytoplankton biomass X in an estuarine segment of cross-section A and length dy is

$$dX/dt = \mu \cdot X - l \cdot X + (d\phi/dy) \cdot (1/A) \quad (1)$$

where μ and l refer to specific rates of phytoplankton growth and loss by grazing or sedimentation (and are individually equal to $(dX/dt)(1/X)$ in the absence of other changes), and ϕ refers to down-estuary biomass flux. All values are means over the water column and a tidal cycle.

Officer (1979) derives an equation for the flux of a non-conservative substance in an estuary. Applied to phytoplankton biomass, it is

$$\phi = F(X - S \cdot (dX/dS)) \quad (2)$$

where S refers to salinity and F to freshwater inflow, assumed to contain no viable phytoplankton. It is further assumed that, averaged over depth and a tidal cycle, there is no net flux of salt in or out of the segment. The simplest solution to equations (1) and (2) is for the case of a single well-mixed box, volume V:

$$dX/dt = (\mu - l)X - (F/(A \cdot \Delta y)) \cdot (X - S(\Delta X/\Delta S)) \quad (3)$$

Using the approximations $V = A \cdot \Delta y$, $\Delta X = X - X_0$ and $\Delta S = S - S_0$, where X_0 and S_0 are mean biomasses and salinities in the segment supplying the estuarine box; introducing exchange rate:

$$E = (F/V).(S/(S_o - S)) \qquad (4)$$

and rearranging so that all main terms have dimensions of time^{-1}, gives:

$$(dX/dt).(1/X) = (\mu - 1) - F/V - E((X - X_o)/X) \qquad (5)$$

showing that the net rate of change of biomass is given by the biological terms (intrinsic growth less grazing and sedimentation) less dilution by river flow and marine exchange. E includes estuarine circulation as well as tidal diffusion effects (see Officer, 1976).

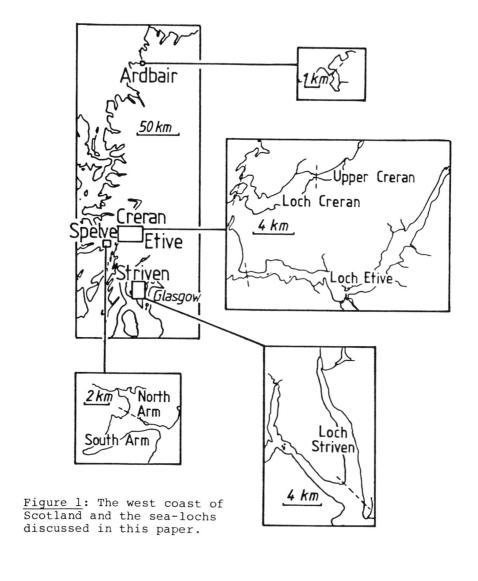

Figure 1: The west coast of Scotland and the sea-lochs discussed in this paper.

The aim here is to apply this theory to several fjordic sea-lochs on the west coast of Scotland (figure 1). Relevant accounts of loch Ardbhair have been published by Gowen et al. (1983), and of lochs Creran and Etive by Edwards & Edelsten (1966a, 1977), Landless & Edwards (1976), Tett & Wallis (1978) and Wood et al (1973). A thesis by Tyler (1983) provides further information about Loch Creran.

METHODS

Loch volumes were estimated (to c.10%) from Admiralty charts, a loch's boundary being taken as the shallowest part of the entrance sill. Catchment areas were estimated from Ordnance Survey maps, and include the loch surface. Volumes, areas and depths refer to mid-tide, except in the case of relative tidal exchange (mean tidal prism/volume at MHW). Relative catchment is catchment divided by loch area.

Runoffs into lochs Creran and Etive were mostly estimated by scaling continuous guaging of the main rivers to the extents of nonguaged catchment. In other cases mean freshwater inflow was estimated as 0.84 of the catchment mean annual rainfall after a comparison of rainfall and measured flow in the river Creran in 1977 and 1978

These measured flows also suggested that maximum weekly flow could be taken as four times mean weekly flow. Single measurements by well-calibrated guage, and mean annual flows however determined, are likely to be corrrect to within 30%; but flows estimated from a single month's rainfall may be more erroneous; the proportion of monthly precipitation appearing as guaged flow in the River Creran varied from 1/4 during dry summer months through 1 for rainy autumn months to as much as 2 during snow melt in spring.

Salinities were generally measured with simple electronic probes calibrated against precision salinometer measurements of watebottle samples. Basin mean salinities were computed from measurements made at 1 to 5 stations and weighted according to the proportion of basin volume found in a horizontal segment at each depth measured. Error analysis of probe calibrations suggests that absolute salinities (i.e. values of S) might be in error by up to 0.2 p.p.th (95% confidence limits, roughly equal to 2 s.d.), and that salinity differences measured with the same instrument (i.e. $(S_0 - S)$) might have a standard deviation of about 0.1 p.p.th. Other difficulties in estimating S are discussed below. Phytoplankton biomass was measured as chlorophyll a, after retention on a glass-fibre filter. Pigments were extracted in 90% acetone, and, in most cases, measured fluormetrically before and after acidification (Tett & Wallis 1978). Some of the Striven extracts were determined spectrophotometrically. Loch mean chlorophyll X was determined by interpolating logarithmically between sampled depths and weighting as for mean salinity. Measurement errors and those due to small scale patchiness result in a standard deviation of about 30% for a single chlorophyll measurement (Tett & Grantham 1980); single estimates of the ratio $(X-X_o)/X$ derive from 2 to 24 samples for X_o and 4 to 30 samples for X, and may be in error by up to 40% (95% confidence limits).

Table 1 : Main Physical Parameters of the Lochs

	ARD-BHAIR	CRERAN UPPER BASIN	SPELVE NORTH ARM	CRERAN	STRIVEN	ETIVE	
Volume	2.4	18.6	40	183	567	935	Mc.m
Sill depth	3	3	–	7	50	10	m
Mean depth	5	11	15	13	37	33	m
Greatest depth	18	37	35	50	70	160	m
Catchment	9	87	61	179	98	1400	sq.km
Rel.catchment	27	51	21	13	6	50	
Mean rainfall	1.5	2.1	2.2	1.9	1.8	2.2	m/yr
Runoff from rain	0.031	0.42	0.31	0.78	0.41	7.1	Mt/d
Runoff by guage :							
mean	–	0.44	0.39	0.81	–	7.0	Mt/d
max.weekly mean	–	1.8	4.2	3.4	–	38	Mt/d
Freshwater flushing :							
mean F/V	0.01	0.02	0.01	<0.01	<<0.01	0.01	/d
max. F/V	(0.05)	0.10	0.10	0.02	<0.01	0.04	/d
Relative mean tidal exchange :	1.06	0.34	0.27	0.30	0.13	0.08	/d
'Central' exchange rate, from Tables 2 and 3 :	0.67	0.67	0.84	0.16	>0.02	0.07	/d

DIMENSIONS AND RIVER FLOWS

The sea-lochs in Table 1 show a range of sizes and relative freshwater inflows. Ardbhair, the smallest, has a volume of just over 2×10^6 m³ mid-tide, about one fourhundredth of that of Loch Etive. With increasing size goes increasing depth : Ardbhair does not exceed 20 m, whereas depths up to 160 m can be found in Etive. All the lochs are fjords; Ardbhair, Creran and Spelve might also be called 'polls' in the terminology of Matthews & Heimdal (1980) since they are small and have shallow sills at roughly the depth of the main pynocline. Landless & Edwards (1976) thought that "in its dimensions, run-off and tides, Creran is close to the unrealized typical Scottish fjord". Striven has a deep sill, whereas the two-layered North Arm of Spelve, which receives most of the loch's freshwater inflow, exchanges with the generally well-mixed South Arm without any restriction. Edwards & Edelsten (1977) point out that Etive has a rainwater catchment larger than any other Scottish fjord. Its relative freshwater inflow may however be exceeded by smaller lochs, such as the upper basin of Loch Creran, which receives nearly 60% of the loch's drainage into only 10% of its volume.

Freshwater runoff is not only difficult to measure but also greatly variable in the mountainous terrain surrounding the lochs. Flows in the River Creran might increase a hundredfold in 24 hours after heavy rain; the range of weekly means during May - October 1978 was more than fortyfold. A larger catchment, including a lake and a hydro-electric scheme, somewhat buffers the inflow to Loch Etive, the maximum weekly mean flow being about 16 times the minimum (Edwards & Edelsten, 1977). The buffering power of the smaller catchment of Loch Spelve appears less: during a two month period of guaging the main river in August and Spetember 1981, the maximum weekly mean flow was about 70 times the minimum.

The ratios F/V in Table 1 represent the (relative) rates at which water (and phytoplankton) are displaced from these lochs as a direct result of the volume added by runoff. For Loch Striven the rates are less than 0.01 day^{-1} under all conditions, and the term can be neglected so far as removal of phytoplankton is concerned. Indeed, unless the flow estimates in Table 1 are seriously in error, normal freshwater inflow can have little direct impact on the phytoplankton since the greatest value of F/V for mean runoff is 0.02 day^{-1} (for upper Loch Creran). Given the Creran runoff pattern, weekly-averaged flows more than three times the mean are however likely to obtain during 4 weeks of the year. The resulting volume displacement rates of 0.04 to 0.10 day^{-1} may remove a significant proportion of biomass from those lochs (Creran, Etive, and Striven) where the net rate of population growth is small (see below and Table 4).

EXCHANGE

Equation (4) assumes a steady state with resect to salinity. Runoff variability preludes short-term equilibrium, and

Table 2 : Estimates of Exchange Rate using Equation (4).

LOCH and period	no. visits	stat- istic	S_o-S p.p.th.	$(S_o-S)/S$	F Mt/d	E /d	Note
ARDBHAIR							
1/81-7/82	8	min.	-0.24	-	-	-	(1)
		mean	0.58	57	0.028	0.67	
		max.	1.58				
UPPER CRERAN							
12/77-12/78	20	min.	0.25	132	0.06	0.44	(2)
12/77-4/78	9	median	0.76	41	0.52	1.18	(3)
		up90%ile	1.16	26	1.30	1.88	
5/78-10/78	11	median	0.50	65	0.14	0.51	(3)
		up90%ile	1.76	16	0.93	0.83	
SPELVE NORTH ARM							
8/82-9/82	4	min.	0.23	-	-	-	(4)
		mean	0.39	84	0.4	0.84	
		max.	0.68	-	-	-	
CRERAN							
12/77-12/78	20	min.	0.28	118	0.12	0.08	(2)
12/77-4/78	9	median	0.86	37	0.96	0.19	(3)
		up90%ile	0.92	34	2.40	0.45	
5/78-10/78	11	median	0.24	135	0.26	0.19	(3)
		up90%ile	1.41	21	1.72	0.20	
8/83-10/83	5	min.	0.37	-	-	-	(5)
		mean	0.50	62	0.60	0.20	
		max.	0.63	-	-	-	
10/83-12/83	5	min.	0.58	-	-	-	(5)
		mean	1.77	18	1.23	0.12	
		max.	2.59	-	-	-	
STRIVEN							
7/79-4/80	4	min.	-0.67	-	-	-	(6)
		mean	-0.28	-119	0.31	-	
		max.	0.13	-	-	-	

Notes and references to sources of data. (1) Gowen et al.,1983. (2) Tyler, 1983. (3) Tyler et al.,1984. (4) Tett, unpublished. (5) Raine & Tett, 1984. (6) Clyde River Purification Board, unpublished. Some of the Loch Creran data was subject to preliminary statistical analysis by Tyler, 1983, and Tyler et al. 1984, who present tables of the maximum, median and lower 90%ile salinity at each depth at each station for the relevant period. The maximum salinities were used to calculate minimum salinity differences, and the lower 90%ile salinities to calculate the upper 90%ile salinity differences. Corresponding minimum, median and upper 90%ile runoffs were extracted from weekly mean river flow data for the relevant period.

observed mean salinities can in one week change by an amount equal to $S-S_0$. It is thus necessary to average values of F, $(S_0-S)/S$, and $(X-X_0)/X$ to obtain valid estimates of exchange rate and its effect of phytoplankton dynamics. Such averaging of salt and chlorophyll differences also helps to satisfy the requirement for tidal averaging in equations (1) to (5).

Resulting estimates of exchange rate are summarized in Table 2. Those in Table 3 were calculated by alternative methods,

Table 3 : Additional Estimates of Flushing Rate.

LOCH	D /day	Method	Ref.
CRERAN	0.16	Correlation of nearsurface salinities and convoluted runoff (from rainfall) 1971-72; n=50.	(1)
	0.07	Correlation of nearsurface salinities and convoluted flow in R.Creran, 2 to 4/79 and 11 to 12/79; n=15.	(2)
	0.20	Correlation etc; 5/79-10/79; n=27.	(2)
	0.12	Equation (6) with S_2 = mean salinity below 20m, and S_1 = mean salinity 0 - 6 m, 12/77-12/78; n=23.	(3)
STRIVEN	0.024	Equation (6) with S_2 = salinity at 45 m in entrance region and S_1 = mean loch salinity; 7/79-4/80; n=4.	(4)
ETIVE	0.07	Equation (6) with S_2 = flood salinity of 30 p.p.th. and S_1 = ebb salinity of 27 p.p.th.	(5)
	0.13	Correlation of nearsurface salinities with convoluted flow in R.Awe, 1970-71; n=50.	(6)
	0.04	Correlation etc for 14m, 1970-71; n=50.	(7)

References : (1) Landless & Edwards, 1976. (2) Tett et al., 1981. (3) Tyler, 1983. (4) Clyde River Purification Board, unpublished. (5) Data in Wood et al., 1973. (6) Landless & Edwards, 1976. (7) Edwards & Edelsten, 1976b.

both giving flushing or dilution rate D, equivalent to the sum of F/V and E. In the first (from Bowden, 1967 and Saelen, 1967):

$$D = (F/V) \cdot (S_2 / S_2 - S_1) \qquad (6)$$

where either (a) S_2 is the mean flood, and S_1 the mean ebb, salinity at the loch mouth, or (b) S_2 is the mean salinity of the deep layer and S_1 that of the surface layer of a loch. The second method is that of Landless & Edwards (1976):

the flushing time $1/D$ is that giving the best correlation between point salinities in a loch and 'convoluted' timevarying river flows (i.e. flows summed with a weighting function reflecting assumed flushing time).

Conditions in Loch Striven could not be analysed using equation (4) since outside salinities were on average less than those inside. This loch, with the smallest relative catchment of the 6 basins considered in this paper, is probably not well described by equation (5).

Some idea of the errors attached to estimates of exchange and flushing rates can be gained from the tabulated range of 0.07 to 0.45 day^{-1} for Loch Creran. Some of this variation may have arisen from errors in estimating $S_o - S$ due to nonuniformity in the loch; and some from errors in estimating inflow over, or in deciding on a suitable averaging period for, the whole catchment. Single estimates of E in Table 2 may be up to 50% in error, and thus probably none of the Creran values can be considered significantly different from the central estimate of 0.16 day^{-1}. Despite such variablility there seems little doubt that the three small basins (Ardbhair, Upper Creran and the North Arm of Loch Spelve) exchange considerably faster than the larger basins, at more than 0.6 day^{-1} compared to 0.16 day^{-1} for Creran and less than 0.1 day^{-1} for Striven and Etive. These differences may be partly explained by the relatively larger tidal prism of the smaller basins (Table 1) and the likelihood that their tidal exchange with the sea or external basin is more efficient. In contrast, Loch Etive has a relatively small tidal prism and only about half its ebb volume is returned on the next flood (Wood et al., 1973).

PHYTOPLANKTON DYNAMICS

Table 4 presents mean values of the biomass ratio $\Delta X/X$ $(= (X-X_0)/X)$ based as far as possible on measurements made at the same time, and analysed numerically in the same way, as the salinity data. These ratios were combined with relevant estimates of exchange rates to estimate the biological term $(\mu-1)$ in equation (5). In the case of some data sets from lochs Creran and Spelve, each covering no more than 9 weeks in late summer or autumn, good estimates of $(dX/dt).(1/X)$ could be made from $\Delta \ln X / \Delta t$, and the means of these are given in the table. It was assumed that mean $(dX/dt).(1/X)$ tended to zero as time series grew longer. In the case of fast-exchanging lochs the major errors in such estimates of $(\mu-1)$ derive mainly from errors of estimation of E. $\Delta X/X$. With slow flushing lochs the main errors derive from $(dX/dt).(1/X)$ and are probable no worse than ± 0.05 day^{-1}.

These rough estimates of errors can be used to divide values of $(\mu-1)$ into three classes.

(a) Lochs Creran, Etive and Striven had mean net phytoplankton growth rates of less than 0.1 day^{-1}. The results suggest that a low exchange rate allows a small excess of growth over internal loss to maintain a phytoplankton biomass about twice that outside these lochs : each day a few percent of new biomass is exported by physical exchange.

(b) Two of the three small fast-flushing basins seem to have had a low rate of net population increase : indeed in Upper Creran consumption or sedimentation of algae might have exceeded their production. The data for Ardbhair are more variable than those for Creran, but suggest that although growth did exceed internal loss, rapid flushing nevertheless prevented any substantial enhancement of biomass in Ardbhair.

(c) The results from Loch Spelve's North Arm demonstrate that a substantial imbalance between growth and internal loss is possible, and that a relatively fast net rate of

Table 4 : Phytoplankton Dynamics.

LOCH and period for X	no. visits	$\frac{dX.1}{dt \ X}$	$(\mu -1)$	$-F/V$	$-E.\frac{\Delta x}{X}$	E	$\frac{\Delta x}{X}$ (s.d.)	Ref
ARDBHAIR								
1/81-7/82	4	(0)	(+0.12)	-0.01	-0.11	0.67	0.16 (0.19)	(1)
UPPER CRERAN								
5/78-10/78	13	(0)	(-0.02)	-0.01	+0.03	0.51	-0.05	(2)
SPELVE NORTH ARM								
8/78-9/78	4	+0.07	(+0.45)	-0.01	-0.37	0.84	0.44 (0.15)	(3)
CRERAN								
5/78-10/78	13	(0)	(+0.05)	0.00	-0.05	0.19	0.24	(2)
8/83-10/83	5	-0.05	(+0.07)	0.00	-0.12	0.20	0.61 (0.08)	(4)
10/83-12/83	5	-0.05	(+0.02)	-0.01	-0.06	0.12	0.51 (0.13)	(4)
						Using D (eqn. 6) and relevant X_o.		
STRIVEN								
7/79-4/80	4	(0)	(+0.01)	0.00	-0.01	0.024	0.24 (0.77)	(5)
ETIVE								
7/70-12/71	21	(0)	(+0.05)	-0.01	-0.04	0.07	0.55 (0.18)	(6)

References and notes : (1) Gowen et al., 1983. (2) Unpublished data, SMBA. Preliminary statistical analysis gave median (water-column mean) chlorophyll for each station. (3) Unpublished data, SMBA. (4) Raine & Tett, 1984. (5) Unpublished data, CRPB. X_o based on chlorophyll at 45 m in entrance region (corresponding to S_o in Table 3). (6) Wood et al., 1973. X value based on mean of 0-10 m, and X_o on mean of 10-50 m, chlorophyll concentrations.

growth can maintain a substantial excess of biomass despite rapid flushing. Perhaps a third of new production is exported each day.

DISCUSSION

Equation (5) is based on a single-box model of a fjordic estuary in which exchange with the sea is described by one-dimensional mixing (Figure 2). The model employs salt as

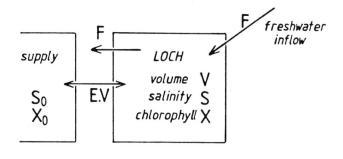

$$\frac{dX}{dt}\cdot\frac{1}{X} = (\mu - g) - \frac{F}{V} - \frac{E\cdot(X - X_0)}{X}$$

| rate of change of chlorophyll | growth minus loss | freshwater flushing | exchange |

$$E = \frac{F}{V}\cdot\frac{(S_0 - S)}{S}$$

Figure 2: The box model of sea-loch exchange.

a tracer of water movements, and was inapplicable to a loch (Striven) receiving relatively little freshwater input. It is likely to work best in lochs in which diffusion rather than advection dominates the transport of salt and phytoplankton : that is, those falling into the 'partially mixed category' of Hansen & Rattray (1966). Calculations based on data in Edwards & Edelsten (1977), Tyler et al (1984) and Wood et al (1973) suggest that the 'almost typical Scottish loch' Creran falls into this category, whereas Loch Etive, with a larger relative catchment and reduced tidal mixing, belongs to the stratified category.

The model seems most relevant to the smaller basins (ranging from Ardbhair at 2.4 to Creran at 183×10^6 m^3) and most applicable during periods of normal run-off; during high runoff it is likely that some of these lochs develops a marked two-layer flow. The three smaller basins (Ardbhair, Upper Creran, and the North Arm of Spelve) appear to exchange rapidly with the sea or adjacent basin, perhaps as a result of efficient tidal mixing. Enhanced chlorophyll concentrations in one of these small lochs (Spelve) suggests a substantial excess of phytoplankton growth over losses within the loch. Data from Loch Ardbhair suggests a somewhat smaller excess of growth over internal loss. In other cases growth and grazing plus sedimentation appear to be roughly in balance: generalizing these results suggests that many fjordic sea-lochs export relatively little of their planktonic primary production.

Further investigation of these conclusions seems desirable; in particular comparison of direct measurements of primary production in the three smaller basins might help to explain the substantial differences in the growth/grazing balance estimated by the present indirect method.

ACKNOWLEDGMENTS

SMBA observations described in this paper were funded by Ardvar Fish Farmers, Golden Sea Produce Ltd., the Highlands and Islands Development Board, the Manpower Services Commisiion Job Creation Programme, the Ministry of Agriculture, Fisheries and Food, and the National Environment Research Council, and were carried out in collaboration with the Clyde River Purification Board. I am grateful to for facilities in the Marine Science Laboratories University College of North Wales, and to the Royal Society Browne Research Fund and Marshall & Orr Bequest for financial aid, during the writing of this paper.

REFERENCES

Bowden, K.F. 1967.
    Circulation and diffusion. In Lauff, G.H., Ed.,
    Estuaries, American Association for the Advancement
    of Science, Washington, D.C., 83, 15-36.

Edwards, A. & Edelsten, D.J. 1976b.
    Marine fish cages - the physical environment.
    Proc. R. Soc. Edinb., sect. B. 75, 207-221.

Edwards, A. & Edelsten, D.J. 1976b.
 Control of fjordic deep water renewal by runoff modification. Hydrol. Sci. Bull., 21, 445-450.

Edwards, A & Edelsten, D.J. 1977.
 Deep water renewal of Loch Etive : a three basin Scottish fjord. Estuarine Coastal Mar. Sci., 5, 575-595.

Gowen, R.J., Tett, P & Jones, K.J. 1983.
 The hydrography and phytoplankton ecology of Loch Ardbhair : a small sea-loch on the west coast of Scotland. J. Exp. Mar. Biol. Ecol., 71, 1-16.

Hansen, D.V. & Rattray, M. 1966.
 New dimensions in estuary classification. Limnol. Oceanogr., 11, 319-326.

Landless, P.J. & Edwards, A 1976.
 Economical ways of assessing hydrography for fish farms. Aquaculture, 8, 29-43.

Matthews, J.B.L. & Heimdal, B.R. 1980.
 Pelagic productivity and food chains in fjord systems. In Freeland, H.J. Farmer, D.M. & Levings, C.D., Eds., Fjord Oceanography, Plenum Press, New York, 377-398.

Officer, C.B. 1976.
 Physical Oceanography of Estuaries John Wiley & Sons, New York, 465 pp.

Officer, C.B. 1979.
 Discussion of the behaviour of nonconservative dissolved constituents in estuaries. Estuarine Coastal Mar. Sci., 9, 91-94.

Raine, R. & Tett, P. 1984.
 The autumn decrease of phytoplankton in Loch Creran, Sept-Dec 1983. Scott. Mar. Biol. Assoc. Internal Rep., 106, 84 pp.

Saelen, O.D. 1967.
 Some features of the hydrography of Norwegian fjords. In Lauff, G.H., Ed., Estuaries American Association for the advancement of Science, Washington, D.C. 83, 15-36.

Tett, P., Drysdale, M. & Shaw, J. 1981.
 Phytoplankton in Loch Creran during 1979 and its effect on the rearing of oyster larvae. Scott. Mar. Biol. Assoc. Internal Rep., 52, 77 pp.

Tett, P. & Grantham, B. 1980.
 Variability in sea-loch phytoplankton. In Freeland, H.J., Farmer, D.M. & Levings, C.D., Eds., Fjord Oceanography., Plenum Pess, New York, 435-438.

Tett, P. & Wallis, A. 1978.
 The general annual cycle of chlorophyll standing crop in Loch Creran. <u>J. Ecol</u>., 66, 227-239.

Tyler, I.D. 1983.
 <u>A carbon budget for Creran, a Scottish sea loch.</u> PhD thesis, University of Strathclyde, Scotland, 202 pp.

Tyler, I.D., Grantham, B., MacNaughton, E. & Tett, P. 1984.
 Salinity and temperature profiles in Loch Creran, 1978. <u>Scott. Mar. Biol. Assoc. Internal Rep</u>., 84, 133 pp.

Wood, B.J.B., Tett, P.B. & Edwards, A. 1973.
 An introduction to the phytoplankton, primary production and relevant hydrography of Loch Etive. <u>J. Ecol</u>., 61,

FRESHWATER OUTFLOW EFFECTS IN A COASTAL, MACROTIDAL ECOSYSTEM AS REVEALED BY HYDROLOGICAL, CHEMICAL AND BIOLOGICAL VARIABILITIES (BAY OF BREST, WESTERN EUROPE)

B. Quéguiner[*] and P. Tréguer[**]
Laboratoire de Physiologie Végétale[*]
Laboratoire de Chimie des Ecosystèmes Marins[**]
Université de Bretagne Occidentale
Faculté des Sciences
6 Avenue Le Gorgeu 29287 Brest Cédex France

ABSTRACT

High nutrient loading originating from low freshwater outflows characterize the semi-enclosed macrotidal Bay of Brest. This ecosystem appears as an interactive system including the bay itself, estuaries and adjacent shelf waters. In these ecosystems, phytoplankton are ecologically and physiologically adapted to the mode of fertilization. Tidal action plays the main role in the water renewal thus preventing eutrophication processes.

INTRODUCTION

Coastal ecosystem productivity is primarily related to fertilization from atmospheric precipitation, urban and industrial sewage outfalls or riverine discharges (Leach, 1971 ; Goldman, 1976 ; Topping, 1976 ; Solorzano and Ehrlich, 1977 ; Correll and Ford, 1982 ; Jaworski, 1982). This is also the case in the coastal Bay of Brest (fig. 1) which receives riverine inputs from two main rivers : Elorn in the Northwest and Aulne in the Southwest. These waters are also exchanged with the adjacent ecosystem : Iroise. From 1979 to date studies have been carried out in order to assess the seasonal pattern and regulation mechanisms of nutrient discharges originating from freshwater outflows.

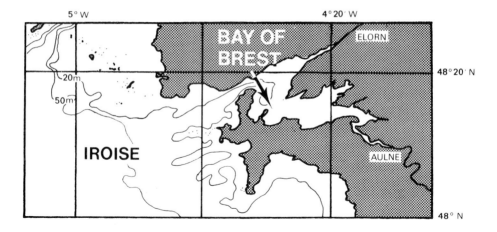

Figure 1 : Geographical location of the Bay of Brest showing main rivers (Aulne and Elorn) and adjacent ecosystem Iroise.

METHODS

Physical (salinity), chemical (organic and inorganic nutrients) and biological (chlorophyll a, primary production, counting and identification of phytoplankton) were measured according to methods referenced in Delmas et al. (1983).

RESULTS

FRESHWATER OUTFLOWS AND TIDAL MIXING

Due to its geographical location in the temperate zone of the Northern hemisphere, the Bay of Brest and its river drainage basins receive heavy and variable rainfalls (range : 800-1000 mm per year). Accordingly, river flows exhibit great variations in both seasonal and annual trends. The two main rivers, Aulne and Elorn, responsible for 80 % of the total freshwater inputs in the bay, show a well-characterized seasonal pattern with high winter flows (up to 250 m^3/s in Aulne) and low summer flows (1 m^3/s) (fig. 2).

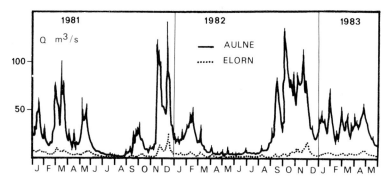

Figure 2 : Daily flows variability of Aulne and Elorn.

As compared with systems influenced by major rivers of Western Europe (Seine, Thames, Rhine), mean annual freshwater inputs are relatively low and the Bay of Brest remains over the year as a typically marine system in which salinity never decreases down to 32 °/oo in winter (Delmas et al., 1983 ; Quéguiner and Tréguer, 1984). This results from the action of tides (semi-diurnal and fortnightly periodicities) on the renewal and mixing of the water masses within the bay and the connected estuaries. Then, during spring tides, the tidal variation reaches 8 m which represents an oscillating volume of 40 % (about 10^9 m^3) of the bay volume at high tide and induces a renewal rate of 1/30 at each tide (Delmas and Tréguer, 1983).

NUTRIENTS INPUTS INTO THE BAY OF BREST

Intensive land use for agriculture and cattle rearing on river drainage bassins is responsible for high nitrogen loading of the ecosystem, while silicate inputs principally result from metamorphic rocks dissolution (Delmas and Tréguer, 1983). Mixing diagrams (fig. 3) summarize the seasonal evolution of these nutrients. During winter, freshwater nutrients reach particularly high levels : 500 µM-N for nitrate which is the principal source of dissolved inorganic nitrogen (DIN) and 150 µM-Si for silicate ; as compared to Loire this is two fold higher. At this time of the year nitrate and silicate evolution is conservative as there is weak phytoplankton production. During the productive period nitrate concentrations decline to 100 µM-N in freshwaters and < 1 µM-N in the bay, while noticeable digressions from straight dilution are observed, resulting from decreasing river flows and phytoplankton uptake. Silicate exhibits the same seasonal pattern showing a spring decrease related to phytoplankton outburst and river flow

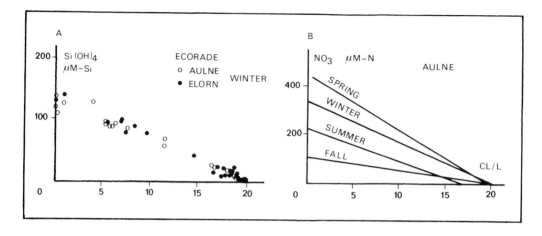

Figure 3 : A. Mixing diagram silicate concentrations-chlorosity during winter.
B. Seasonal evolution of mixing diagram nitrate concentrations-chlorosity (redrawn from Delmas and Tréguer, 1983).

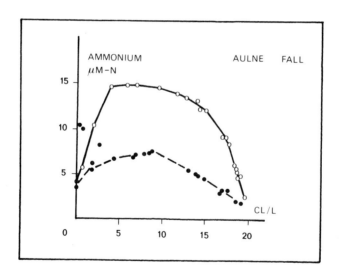

Figure 4 : Fall mixing diagram ammonium-chlorosity in Aulne estuary.

decline. Ammonium, the second important source of DIN, reaches 7.5 % of the total DIN stock in Elorn ; this reflects urban outfalls in the North basin of the bay. Ammonium seasonal evolution varies from nitrate and silicate, by showing maximum values at the end of summer and in the fall due to regeneration processes occurring in estuaries (fig. 4). Nitrite is of small importance in the nitrogen cycle as it represents 0.4-0.8 % of DIN ; its evolution is identical to ammonium. Phosphate levels are greater in Elorn (0.7-6 µM-P) than in Aulne (0.7-3 µM-P), reflecting urban disturbances in the North basin. They are regulated through adsorption-desorption reactions at the water-suspended matter interface (Delmas and Tréguer, 1983). Dissolved organic matter is regulated through the same mechanism and concentrations are in the range 5-50 µM-N and 0.5-1 µM-P showing seasonal variations due to uptake, excretion and regeneration processes (Le Jehan and Tréguer, 1984).

ECOLOGICAL AND PHYSIOLOGICAL ADAPTATION OF COASTAL PHYTOPLANKTON

Owing to the fast seawater renewal of the bay, induced by tidal action, freswater outflows never induce stratification of the water column, their main role consisting in discharging nutrients in the system. The temporal variability of river inputs affects the phytoplankton species succession and primary production on both a seasonal and annual scale (Quéguiner and Tréguer, 1983 ; Hafsaoui *et al.*, 1985). Phytoplankton respond to nutrient pulses by successive outbursts during the production period (fig. 5), from March to September and, accordingly, communities exhibit a lack of maturity which is also related to the short residence time within the ecosystem (Quéguiner and Tréguer, 1983). During this period, despite ammonium concentrations exceeding 1 µM-N, nitrate uptake by phytoplankton is not prevented. Laboratory experiments made on natural populations (fig. 6) clearly demonstrate that nitrate uptake occurs at ammonium levels as high as 8 µM-N suggesting a physiological adaptation of coastal phytoplankton, as already observed on oysterpond microalgae by Maestrini *et al.* (1982).

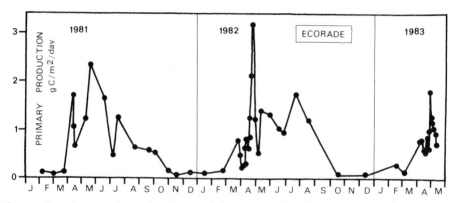

Figure 5 : Seasonal variations of integrated primary production in the Bay of Brest.

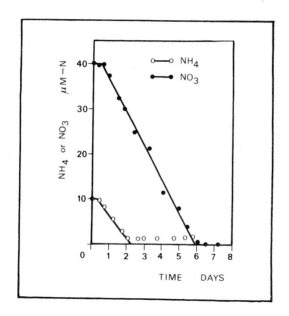

Figure 6 : Simultaneous uptake of ammonium and nitrate by natural phytoplankton populations under laboratory experiments.

EXCHANGE BETWEEN ECOSYSTEMS

Renewal of the waters of the bay is performed through exchange with the shelf waters of Iroise. In winter (fig. 7 A) most of the nutrients entering the Bay of Brest are then ejected outside by means of tidal action. In spring (fig. 7 B), primary production takes place within the coastal ecosystem and phytoplankton populations are exported and mix together with Iroise phytoplankton, inducing increasing structure complexity from the bay to shelf waters, as evidenced by rank-frequency curves. At this time of the year, short term evolution of chemical parameters in the bay show unpredictable increases of silicate content in near-bottom waters (Quéguiner, 1982) which would reflect an inversion of flux for this nutrient as already observed in comparable ecosystem (Peterson et al., 1975). As phytoplankton of the bay are potentially limited by silicate levels (Hafsaoui, 1984), this mechanism appears to enable the continuation of primary production during low freshwater outflows in summer.

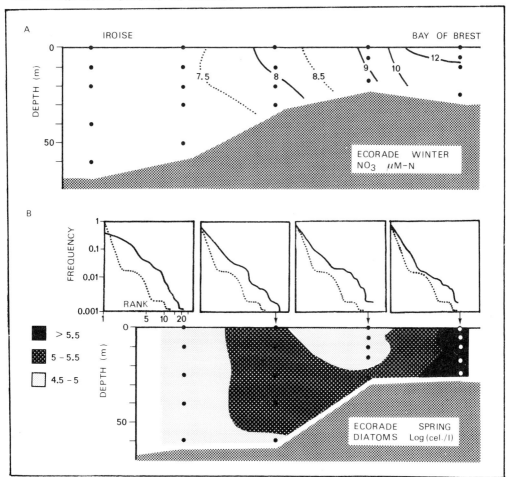

Figure 7 : A. Spatial variations in nitrate concentrations at the mouth of the Bay of Brest in winter.
B. Spatial variations of diatom number in spring. Rank frequency diagram is presented for each station (solid lines) superimposed on rank-frequency diagram within the bay (dotted line).

DISCUSSION

Despite heavy winter fertilization by means of freshwater outflows, no ecological disequilibrium is apparent in the coastal, macrotidal ecosystem studied, because of fast seawater renewal induced by tidal action (Delmas and Tréguer, 1983) and physiological and ecological adaptation of phytoplankton populations (Quéguiner and Tréguer, 1984 ; Hafsaoui, 1984). Accordingly, annual primary productivity (255-280 g C/yr) reaches values in good agreement with data obtained in other coastal ecosystems (Table I). This kind of macrotidal ecosystem commonly found on the West European coast line, is however very sensitive to nitrogen inputs as revealed by an increasing accumulation of transitory forms (ammonium and nitrite) in estuaries (fig. 4), as already observed in other eutrophic ecosystems (Mc Carthy et al., 1984 ; Seitzinger et al., 1984).

Table I : Pelagic primary production in coastal ecosystems.

Geographical location	$gC/m^2/a$	Reference
Puget Sound, U.S.A.	465	Winter et al., 1975
Great South Bay, U.S.A.	450	Lively et al., 1983
Bay of Morlaix, France	314	Wafar, 1981
"Etang de Berre", France	250-290	Kim, 1983
Bay of Brest, France	255-280	present study
Narragansett Bay, U.S.A.	220	Smayda, 1973
Long Island Sound, U.S.A.	205	Smayda, 1973
Bristol Channel, U.K.	165	Joint and Pomeroy, 1981
Gulf of San José, Argentina	161	Charpy-Roubaud et al.,1982
Kiel Bight, Germany	158	Smetacek et al., 1982
Korsfjorden, Norway	74	Erga and Heimdal, 1984
Beaufort estuaries, U.S.A.	68	Thayer, 1971
Kattegat, Danemark	51-82	Steemann Nielsen, 1964
Biscayne Bay, U.S.A.	13-46	Roman et al., 1983

REFERENCES

Charpy-Roubaud, C.J., Charpy, L.J. and Maestrini, S.Y. Fertilité des eaux côtières nord-patagoniques : facteurs limitant la production du phytoplancton et potentialités d'exploitation mytilicole. *Oceanol. Acta* 5, 179-188 (1982).

Correll, D.L. and Ford, D. Comparison of precipitation and land runoff as sources of estuarine nitrogen. *Est. Coast. Shelf Sci.*, 15, 45-56 (1982).

Delmas, R., Hafsaoui, M., Le Jehan, S., Quéguiner, B. and Tréguer, P. Impact de fertilisations à forte variabilité saisonnière et annuelle sur le phytoplancton d'un écosystème eutrophe. *Oceanol. Acta.* Actes 17ème Symposium Européen de Biologie Marine, Brest, 17 septembre-1er octobre 1982, 81-85 (1983).

Delmas, R. and Tréguer, P. Evolution saisonnière des nutriments dans un écosystème eutrophe d'Europe Occidentale (la Rade de Brest). Interactions marines et terrestres. *Oceanol. Acta*, 6, 345-355 (1983).

Erga, S.R. and Heimdal, B.R. Ecological studies on the phytoplankton of Korsfjorden, western Norway. The dynamics of a spring bloom seen in relation to hydrographical conditions and light regime. *J. Plank. Res.* 6, 67-90 (1984).

Goldman, J.C. Identification of nitrogen as a growth-limiting nutrient in wastewaters and coastal marine waters through continuous culture algal assays. *Water Res.* 10, 97-104 (1976).

Hafsaoui, M. Fertilisation d'un système eutrophe à forte variabilité saisonnière et annuelle (Rade de Brest). Mise en évidence des facteurs limitants de la production phytoplanctonique. Assimilation simultanée des différentes formes d'azote inorganique et organique. *Thèse de spécialité*, U.B.O., Brest, 167 p (1984).

Hafsaoui, M., Quéguiner, B. and Tréguer, P. Production primaire et facteurs limitant la croissance du phytoplancton en rade de Brest (1981-1983). 11, 181-195 (1985).

Jaworski, N.A. Sources of nutrients and the scale of eutrophication problems in estuaries. In : Estuaries and Nutrients. B.J. Neilson and L.E. Cronin ed., Humana Press, Clifton, New Jersey, 83-110 (1981).

Joint, I.R. and Pomroy, A.J. Primary production in a turbid estuary. *Est. Coast. Shelf Sci.* 13, 303-316 (1981).

Leach, J.H. Hydrology of the Ythan estuary with reference to distribution of major nutrients and detritus. *J. Mar. Biol. Ass. U.K.* 51, 137-157 (1971).

Le Jehan, S. and Tréguer, P. Evolution saisonnière de composés organiques dissous dans un écosystème eutrophe d'Europe Occidentale (rade de Brest). *Oceanol. Acta* 7, 181-190 (1984).

Lively, J.S., Kaufman, Z. and Carpenter, E.J. Phytoplankton ecology of a barrier island estuary : Great South Bay, New York. *Est. Coast. Shelf Sci.* 16, 51-68 (1983).

Mc Carthy, J.J., Kaplan, W. and Nevins, J.L. Chesapeake Bay nutrients and plankton dynamics. 2. Sources and sinks of nitrite. *Limnol. Oceanogr.* 29, 84-98 (1984).

Maestrini, S.Y., Robert, J.M. and Truquet, I. Simultaneous uptake of ammonium and nitrate by oyster-pond algae. *Mar. Biol. Lett.* 3, 143-153 (1982).

Peterson, D.H., Conomos, T.J., Broenkow, W.W. and Scrivani, E.P. Processes controlling the dissolved silica distribution in San Francisco Bay. In : Estuarine Research, Vol. 1, L.E. Cronin ed., Academic Press, New York, 153-187 (1975).

Quéguiner, B. Variations qualitatives et quantitatives du phytoplancton dans un écosystème eutrophe fortement soumis aux effets des marées : la rade de Brest. *Thèse de spécialité*, U.B.O., Brest, 123 p (1982).

Quéguiner, B. and Tréguer, P. Studies on the phytoplankton in the Bay of Brest (Western Europe). Seasonal variations in composition, biomass and production in relation to hydrological and chemical features (1981-1982). *Bot. Mar.* 27, 449-459 (1984).

Roman, M.R., Reeve, M.R. and Froggatt, J.L. Carbon production and export from Biscayne Bay, Florida. I. Temporal patterns in primary production, seston and zooplankton. *Est. Coast. Shelf Sci.* 17, 45-59 (1983).

Seitzinger, S.P., Nixon, S.W. and Pilson, M.E.Q. Denitrification and nitrous oxide production in a coastal marine ecosystem. *Limnol. Oceanogr.* 29, 73-83 (1984).

Smayda, T.J. A survey of phytoplankton dynamics in the coastal waters from Cape Hatteras to Nantucket. In : Coastal and offshore environmental inventory. Cape Hatteras to Nantucket Shoals, pp. 3-1 à 3-100, Marine Publication, series N° 2, Univ. Rhode Island (1973).

Smetacek, V., Bodungen, B.V., Bröckel, K.V., Knoppers, B., Peinert, R., Pollehne, F., Stegmann, P. and Zeitzschel, B. Phytoplankton primary production and species succession in relation to the environment in Kiel Bight. Symposium I.C.E.S., Kiel, mars 1982, communication N° 23 (1982).

Solorzano, L. and Ehrlich, B. Chemical investigations of Lock Etive, Scotland. 1. Inorganic nutrients and pigments. *J. Exp. Mar. Biol. Ecol.* 29, 45-64 (1977).

Steemann Nielsen, E. Investigations of the rate of primary production at two danish light ships in the transition area between the North Sea and the Baltic. *Meddelelser fra Danmarks Fiskeri-og Havundersøgelser* 4 (3), 31-77 (1964).

Thayer, G.W. Phytoplankton production and the distribution of nutrients in a shallow unstratified estuarine system near Beaufort, N.C. *Chesapeake Sc.* 12, 240-253 (1971).

Topping, G. Sewage and the sea. In : Marine Pollution. R. Johnston ed., Academic Press, New York, 303-351 (1976).

Wafar, M. Nutrients, primary production, and dissolved and particulate organic matter in well-mixed temperate coastal waters (Bay of Morlaix, Western English Channel). *Thèse de spécialité*, Univ. Paris VI, 226 p (1981).

Winter, D.F., Banse, K. and Anderson, G.C. The dynamics of phytoplankton blooms in Puget Sound, a fjord in the North-western United States. *Mar. Biol.* 29, 2, 139-176 (1975).

RIVER INPUT OF NUTRIENTS INTO THE GERMAN BIGHT

U. H. Brockmann and K. Eberlein
Institut für Biochemie und Lebensmittelchemie
der Universität Hamburg
Martin-Luther-King-Platz 6
D-2000 Hamburg 13

ABSTRACT

In the present paper we discuss the relationship between nutrient discharge and eutrophication conditions in the German Bight during summer. Nutrient inputs into this area are controlled by the discharge rates of, and the biological processes in its major nutrient supplier - the river Elbe. The shape and direction of the nutrient plume are modulated by wind stress. Eutrophication conditions were inferred from oxygen deficiency associated with remineralization in cold bottom layers in areas deeper than 20 m. Thermocline breakup appeared only following 4 days periods of wind forces >7 Bft. By repeated investigations of the same station grid within 4 days, high variability of nutrient gradients is documented.

INTRODUCTION

Increasing concern on the possible effects of eutrophication has given a new impulse to research on various aspects of the ecological situation in the German Bight. Gillbricht (1981 and 1983) discussed the possible influence of increasing phosphate and decreasing salinities on phytoplankton biomass increases in the area since 1962. During the last decades a two to threefold increase in phosphate concentrations was detected in the eastern German Bight (Weichart, 1984). Oxygen deficiency was observed between 1980 and 1983 (Rosenthal et al. 1981, Rachor & Albrecht 1983, v. Westernhagen & Dethlefsen 1983, this paper) in the near bottom waters of the German Bight and adjacent North Sea areas. Observations of dead fish and benthic organisms (v. Westernhagen & Dethlefsen 1983) and the general impoverishment of the original benthic community in this area (Rachor 1980) shows that eutrophication effects in the German Bight have become severe.

We present data on the spatial distribution of nutrients in the German Bight during summer months when thermal stratification restricts exchange of dissolved oxygen between the upper and lower layers. However, stratification only partly affects the sedimentation of particulate material. Special emphasis will be given to the variability of nutrient distribution in the German Bight during different weather situations.

MATERIAL AND METHODS

From 1979 to 1984 several cruises with the research vessels GAUSS and VALDIVIA were undertaken in the German Bight and the North Sea. Samples were collected at station grids with 10 n.m. distance. Water samples were filtered through glass fibre filters (Whatman GF/C) and fixed with mercury chloride (0.01 % w/v). Nutrients were analyzed on an AutoAnalyzer (Eberlein et al. 1983). Oxygen and temperature were measured by a Multisonde (ME, Trappenkamp, FRG): Oxygen measurements were calibrated by the Winkler method. Chlorophyll was analyzed with a Turner fluorometer and calibrated according to Lorenzen (1967).

RESULTS AND DISCUSSION

With the depletion of nutrients in the mixed layer caused by the spring-phytoplankton bloom in the North Sea, the nutrient distribution in the estuaries exhibits clearly detectable horizontal gradients.

Concentrations of more than 0.5 µM phosphate, found during

February, 1984 (Brockmann and Wegner 1985) in the total water column, were detected only in the estuaries of Humber, Thamse, Rhine and Elbe-Weser rivers (Fig. 1). The same situation for nitrate was found in the central North Sea during February with surface-concentrations of more than 5 µM. Such values were found only in the estuaries and along the Dutch and German coasts during May (Fig. 2). It must be mentioned that in February nutrient concentrations were still low in areas above the Doggerbank.

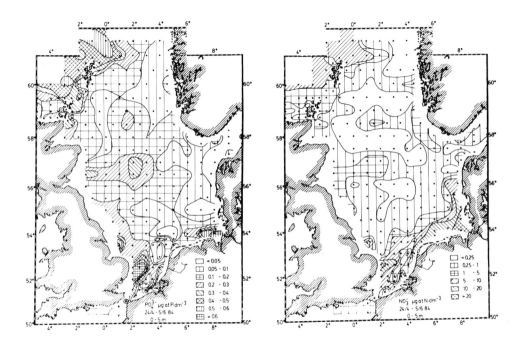

Fig. 1: Concentrations of phosphate (µM dm^{-3}) at the North Sea surface (0-5 m) during a cruise from 24/4 to 5/6 1984 by R.V. GAUSS

Fig. 2: Concentrations of nitrate (µM dm^{-3}) at the surface (see Fig. 1)

Nutrient input into the German Bight is influenced mainly by the freshwater run-off which is dominated by the Elbe river. Significant linear correlations between nitrate and salinity were observed in the Elbe estuary during cruises in June 1979, 1980 and 1981 as well as in August 1981 and September 1983.

Slopes were especially high when the freshwater run-off was
>1000 m^3s^{-1} (Fig. 3). Additionally, nutrient transport to the
estuary is influenced by the biological processes in the river.
For example, nitrification, a temperature-dependent process,
causes a decrease of ammonia and an increase of nitrate in the
outer estuary during spring to autumn (ARGE Elbe, 1983)

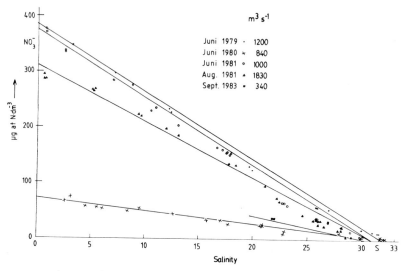

Fig. 3: S/NO_3-diagrams from the Elbe estuary at different
cruises. Freshwater runoff from Neu-Darchau (m^3s^{-1}) is given
for comparison (ARGE Elbe, 1983)

Primary production in the marginal seas during summer time is
not only influenced by the amount of nutrient input, but also
by mixing and exchange processes in the estuary. The expansions and shapes of river plumes are hence important. Our investigations performed at a station grid with 10 n.m. distance
in the German Bight in June 1979, 1980 and 1981, and August/
September 1981, 1982 and 1983 demonstrated that there is a
high variability of the horizontal distributions of the Elbe/
Weser plume in the German Bight. As an example, the nitrate
concentrations at the surface (3 - 5 m) are presented in Fig. 4.
The direction and shape of the plume was formed by wind-induced residual currents. In 1979, two days easterly wind (13m
s^{-1}) followed by a more moderate south-west wind (11 m s^{-1})
forced a surface current to the north-west in the inner German
Bight and to north-west and west in the northern German Bight.

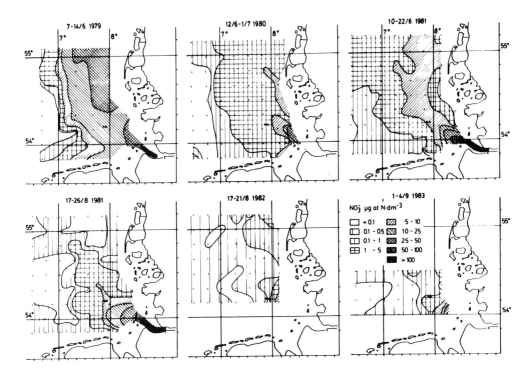

Fig. 4: Surface-nitrate-concentrations in the German Bight during 6 cruises.

As a consequence, a plume with high nitrate concentrations, caused by a river discharge of 1200 m^3 s^{-1}, covered a great part of the German Bight. During June 1980 and 1981 south-westerly changed with north-westerly winds (9 m s^{-1}), retarding the spreading of the plume which had a greater extension in 1981 due to a larger run-off. In August 1981 also the river plume was clearly separated from the North Frisian coast, by continuous north-westerly winds forming a current from north to south along the coast-line. During August 1982 and September 1983 strong continuous south-westerly winds (15 m s^{-1}) formed a surface current which was directed into the German Bight along the East Frisian coast. Due to these currents the river plume was pressed to the coast and could only be detected in the inner German Bight.

The river plume, with high nutrient concentrations, sometimes spread, as was observed in 1979, 1980 and 1981, in the area of

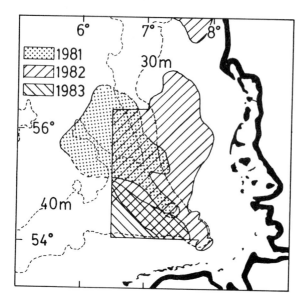

Fig. 5: Areas in which oxygen depletion was observed in 1981 - 1983 (partly after v. Westernhagen & Dethlefsen 1983 and Rachor & Albrecht 1983).

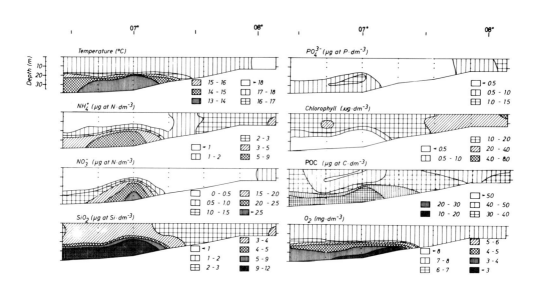

Fig. 6: Vertical distribution of temperature, chemical and biological parameters at an east-west profile (54°30'N) in the German Bight on 19/8 1982. Values are given by isolines. Points indicate sampling positions.

the Helgoland Channel, where the oxygen depletion was observed in 1981, 1982 and 1983 (Fig. 5). As an example, a vertical profile of an east-west section at 54°40'N during August 1982 is given in Fig. 6. A thermocline at a depth of about 20 m separated a colder nutrient-rich bottom layer: 5 - 9 µM ammonia, 2 - 3 µM nitrate, 9 - 12 µM silicate, and 1 - 1.5 µM phosphate. In the mixed layer nutrient concentrations were reduced to below 1 µM ammonia, 0.5 µM nitrate, 0.5 µM phosphate and 4 µM silicate. From the nutrient ratios it can be concluded that nitrogen was the limiting factor of phytoplankton growth. A standing crop of phytoplankton in the mixed layer existed with only about 2 µg chlorophyll dm^{-3} and 50 µg at particulate carbon dm^{-3} (W. Hickel, unpublished). Remineralization of the sedimented or sedimenting organic material, trapped in the bottom layer, caused a drop in oxygen contents below 4 mg dm^{-3}. This is about 40 % of the saturation concentration.

This oxygen depletion was observed in the bottom layer of the deeper parts of the German Bight in 1981, 1982 and 1983 (Rachor & Albrecht 1983, v. Westernhagen & Dethlefsen 1983) (Fig. 5). Since the Elbe-Weser plume spreads also during summer over an area of about 5000 km^2, the nutrient discharge of these rivers will be a direct factor increasing eutrophication conditions in the inner German Bight. Oxygen consumption of about 4 mg dm^{-3} within a 20 m high bottom layer corresponds to an oxidation of about 154000 t biomass in turns of carbon. This corresponds (considering a C/N ratio of 6:1) to 30.000 tons nitrogen, which is 36 % of the annual Elbe-Weser discharge. Thus river discharge (85000 tons/year) can intensify eutrophication effects in the bottom layer, especially during summer when a thermocline has been formed.

The eutrophication situation can only be changed by storms destroying the thermocline. Such situations were observed during August 1982 and September 1983 in the northern German Bight. A stable thermocline was still found even after 4 days wind forces of Bft 7, whereas in the southern part the stratification had already disappeared (Fig. 7).

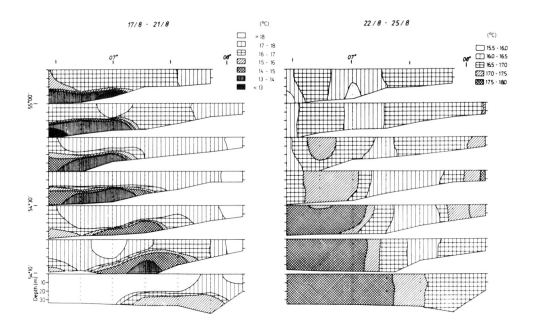

Fig. 7: Temperature profiles in the German Bight from 17/8 - 21/8 and 22/8 - 25/8, 1982. During the first investigation wind forces were between 6 and 9 Bft.

Comparison of vertically integrated silicate concentrations shows clearly the position of the nutrient-rich bottom layer (Fig. 8) which has moved more than 30 miles north after 4 days due to the south-north current (Fig. 9).

At the same time, the phosphate concentrations increased in the deeper part of the German Bight by about 0.3 $\mu M\ dm^{-3}$ (Fig. 10). This rapid increase was probably caused by release of phosphate from the sediment following the exchange of interstitial water. Sudden exchange of water masses with different densities above the sediment could cause this. Thus within a few days the nutrient situation had completely changed in the whole German Bight. The nutrients coming partly from the river discharge and being trapped for a time in the bottom layer, were now redistributed throughout the water column and were again available for biomass production.

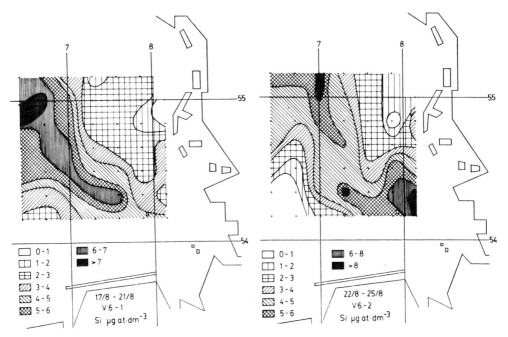

Fig. 8: Horizontal distribution of vertically integrated silicate (μM dm^{-3}) in the German Bight from 17/8 to 21/8, 1982.

Fig. 9: Horizontal distribution of periodically integrated silicate (μM dm^{-3}) in the German Bight from 22/8 to 25/8, 1982.

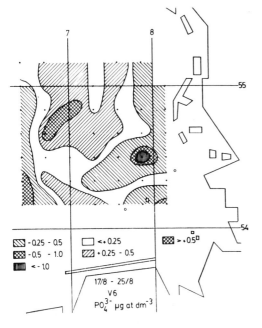

Fig. 10: Differences of vertically integrated phosphate concentrations (μM dm^{-3}) between two investigations in the German Bight (17/8 - 21/8, 1982 and 22/8 - 25/8, 1982).

ACKNOWLEDGEMENTS

For sampling, analyzing and data processing we are grateful to the P 1 crew from the Sonderforschungsbereich 94 "Marine Research Hamburg", for logistical support to the crews of the research vessels and the crew-leaders of GAUSS cruises from the Deutsches Hydrographisches Institut, and to V. Ittekkot for discussions. Financial support given by the Deutsche Forschungsgemeinschaft made these investigations possible.

REFERENCES

Arbeitsgemeinschaft für die Reinhaltung der Elbe (ARGE Elbe) (1983): Wassergütedaten der Elbe von Schnackenburg bis zur See. Hamburg

Brockmann, U.H., G. Wegner (1985): Hydrography, nutrient and chlorophyll distribution in the North Sea in February 1984. Arch.Fisch.Wiss., 36: 27 - 45

Eberlein, K., U.H. Brockmann, K.D. Hammer, G. Kattner, M. Laake (1983): Total dissolved carbohydrates in an enclosure experiment with unialgal Skeletonema costatum culture. Mar.Ecol. Prog.Ser. 14: 45 - 58

Gillbricht, M. (1981): Die langjährige Entwicklung des Phytoplanktonbestandes bei Helgoland. Jber.Biol.Anst.Helgoland 1980: 23 - 24

Gillbricht, M. (1983): Eine "red tide" in der südlichen Nordsee und ihre Beziehungen zur Umwelt. Helgol.Meeresunters. 36: 393 - 426

Lorenzen, C.J. (1967): Determination of chlorophyll and pheopigments. Spectrophotometric equations. Limnol. Oceanogr. 12: 343 - 346

Rachor, E. (1977): The inner German Bight - an ecologically sensitive area as indicated by the bottom fauna. Helgol. Meeresunters. 33: 522 - 530

Rachor, E., H. Albrecht (1983): Sauerstoff-Mangel im Bodenwasser der Deutschen Bucht. Veröff.Inst.Meeresforsch. Bremerh., 19: 209 - 227

Rosenthal, H., P. Hansen, K. Klöckner and H. Krüner (1981): Sauerstoffverhältnisse in der Deutschen Bucht im Frühjahr 1980. Jber. Biol. Anst. Helgoland 1980: 57 - 58

Weichart, G. (1984): Nährstoffe und gelöster Sauerstoff. In: Gütezustand der Nordsee. G.A. Becker (Hrsg.) Deutsches Hydrographisches Institut Nr. 2149/27, Hamburg, 89 - 99

Westernhagen, H. v., V. Dethlefsen (1983): North sea oxygen deficiency 1982 and its effects on the bottom fauna. Ambio, 12, No. 5, 264 - 266

EFFECTS OF FRESH WATER INFLOW ON THE DISTRIBUTION, COMPOSITION AND PRODUCTION OF PLANKTON IN THE DUTCH COASTAL WATERS OF THE NORTH SEA

H.G. Fransz
Netherlands Institute for Sea Research
P.O. Box 59, 1790 AB Den Burg
Texel, The Netherlands

ABSTRACT

The rivers Rhine, Meuse and Scheldt discharge about 75 km^3.y^{-1} of fresh water into the North Sea. The distributions of fresh water, nutrients, chlorophyll and zooplankton during the growing season of plankton indicate that the outflow from the rivers maintains high concentrations of N, P, Si and Fe in a 40 km wide coastal zone, where biomass and production of algae and copepods are enhanced. The highest plankton biomass coincides with the highest temperature and Fe concentration south of the Rhine/Meuse outlet, but not with the lowest salinity and highest N, P and Si concentrations found in the northward flowing river plume. A model of phytoplankton growth predicts a much higher algal production and biomass for the river plume than is actually measured. This indicates the existence of still unknown causes of growth limitation or increased mortality. It also suggests that increased nutrient discharge (eutrophication) multiplied the summer flagellate biomass by a factor 2 to 4 since 1930, and caused a shift in dominance from diatoms to other algae.

INTRODUCTION

According to measurements by the Dutch Institute of Sewage and Industrial Waste Treatment the rivers Rhine, Meuse and Scheldt discharge about 75 km^3 of fresh water per year (van Bennekom et al., 1975) into the Dutch coastal waters of the North Sea (Fig. 1).

Table 1 shows the average amount of important nutrients discharged into the delimited area from 1976 to 1982, as estimated by analysis of data collected by the Dutch government (Rijkswaterstaat) and given by Fransz & Verhagen (1985).

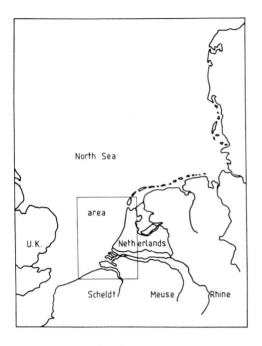

Fig. 1. The Dutch coastal area of the North Sea with main rivers.

Table 1. 1976-1982 river discharges of nutrients.

	10^3 metric tons / year	quarterly percentage 1	2	3	4
Si	198.6	47	17	8	28
$NO_2 + NO_3 - N$	296.9	35	24	16	25
$PO_4 - P$	27.4	29	22	21	28

Since 1930 riverine nutrient input of P has increased by 7 times and N by 5 times (van Bennekom et al., 1975) giving reason for concern about eutrophication (Postma, 1973; van Bennekom et al., 1975). This paper brings together available information related to the effects of the rivers on biomass and production of marine phyto- and zooplankton.

METHODS

The distribution of river water in the area is reflected by the salinity pattern. A first step in the assessment of the river influence is a comparison of the nutrient and plankton distribution with the salinity pattern during the season of plankton growth (April-September). A water quality monitoring programme (Rijkswaterstaat, 1979-1982) provided ample information on nutrient and chlorophyll concentrations during recent years. These data were used to derive isopleths of mean values during the 2nd and 3rd quarters of the years 1979, 1980, 1981 and 1982 for salinity, temperature, and concentration of chlorophyll-a, dissolved N, P, Si and total Fe, all measured by standard methods.

Copepods and, to a much lesser extent, the tunicate Oikapleura dioica are the most abundant pelagic metazoan herbivores in the area. Biomass estimates based on abundance and length-dry weight relationships were obtained by Fransz (1976, 1977) and Fransz & Gieskes (1984). The distribution of these animals was studied in 1973 throughout the year.

Increases in nutrient and biomass levels may indicate enhanced plankton growth, but because the turnover rates can also be affected, direct estimates of plankton production are better measures of growth conditions. Fransz & Gieskes (1984) presented the seasonal variation in estimates of carbon biomass and production of algae and copepods for the coastal zone and more offshore waters.

A comparison of the present situation of nutrient discharge and plankton production with conditions before the eutrophication would be illustrative for the influence of the rivers on the coastal ecosystem. Unfortunately, plankton biomass estimates of periods before 1960 are unreliable and production data are lacking. Fransz & Verhagen (1985) applied a simulation model in a theoretical approach. In this model, phytoplankton biomass, distribution, composition and production are related to nutrient dynamics, advective and dispersive transport, light and temperature conditions and grazing by copepods. Nutrient concentrations depend on river discharge, influx from the south, transport, uptake by algae, excretion and exudation by plankton, and mineralization of suspended and benthic detritus.

RESULTS AND DISCUSSION

The river water is transported mainly in a northerly direction by residual currents along a 40 km wide coastal zone (Fig. 2). Temperature depends on water depth and tends to be relatively high in the shallow and less exposed area near the southern islands. Most nutrients have a pattern identical to the fresh water distribution, indicating the importance of river discharge.

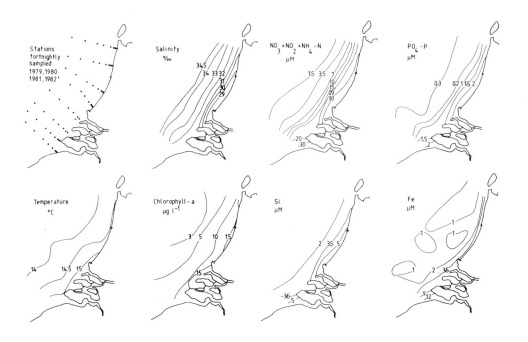

Fig. 2. Mean distributions during the growing season (April-September) of 1979-1982.

Fe has a different distribution and seems to be discharged in large quantities by the Scheldt. The pattern of chlorophyll indicates that algal biomass is produced and maintained mainly in the enriched coastal zone. But the chlorophyll concentration in the river plume area of lowest salinity and highest nutrient concentrations north of the mouths of the Rhine and Meuse is not higher than in the relatively warm and Fe-rich area around the southern islands. De Kroon (1971) measured potential primary production (in an incubator at a light intensity of 4200 lux) and found a summer distribution resembling the chlorophyll distribution in Fig. 2 with the main center in the southern region (see Postma, 1973). There is no doubt that river-induced high food levels stimulate the late spring and summer development of herbivore populations (Fig. 3), but the highest biomasses in 1973 were found near the southern islands.

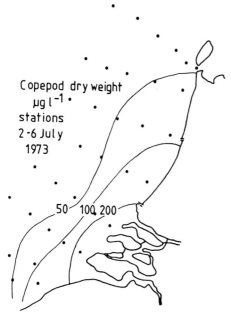

Fig. 3. Distribution of pelagic herbivore biomass during the summer peak in 1973 (Fransz, 1977).

The ratio of plankton production nearshore to offshore during late spring and summer is at least of the same order as the biomass ratio (Fig. 4). Annual primary production, however, is estimated by Fransz & Gieskes to be slightly higher offshore (250 against 200 gC m^{-2} nearshore) because the growing season is reduced in the coastal zone by the higher turbidity. Annual copepod production ranges from 12-23 gC m^{-2} nearshore to 5-15 offshore. Because copepod growth is more restricted to the summer period than primary production (Fig. 4), the annual copepod production very much depends on the summer biomass of algae, which is highest nearshore.

For a number of areal compartments used in the simulation model (Fig. 5) theoretical seasonal cycles of chlorophyll (Fig. 6) and diatom and flagellate biomass (Fig. 7) were computed for different river nutrient discharge levels and southern influx concentrations. The continuous lines correspond with the 1980 conditions, the interrupted lines with the river discharge in 1930 and a southern influx as in 1980. The dotted lines correspond with the lowest possible nutrient loads estimated for 1930, with a southern influx directly from the Channel. The 1930 nutrient load was presumably somewhere between the

conditions represented by the interrupted and dotted lines.
The model confirms on a theoretical basis the observed river effects, but it suggests for the nearshore zone a much higher increase of algal biomass by eutrophication than is actually observed. Apparently algal growth is limited or the mortality is increased by still unknown causes. The influence of nutrient inflow from the south may be equal to the river effects, as indicated by the difference between the interrupted and dotted lines. Eutrophication did not increase the biomass of diatoms, which are limited by Si depletion. Si discharge slightly decreased since 1930 (van Bennekom et al., 1975). The shift from a diatom to a flagellate dominated algae population

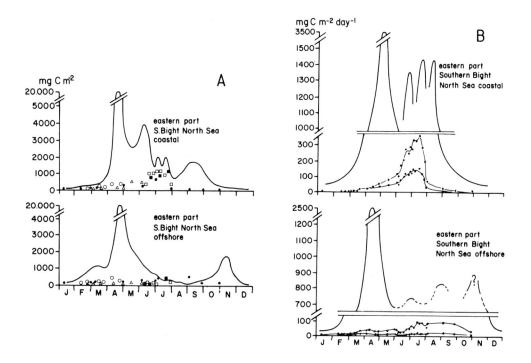

Fig. 4. Seasonal variation of biomass (A) and daily production (B) of algae (continuous lines) and copepods (dots) in 1973-1980. Upper and lower lines of copepod production are based on maximum growth potential and growth rates measured at current food levels.

suggested by the model is not contradicted by observations on species composition (Gieskes & Kraay, 1977). Such a shift is often related to human influences from man-made lakes and eutrophication of rivers, which tend to decrease the input of dissolved silica into the sea (van Bennekom & Salomons, 1980) and may be considered as an undesirable eutrophication effect (Officer & Ryther, 1980).

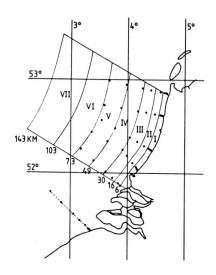

Fig. 5. Compartments used for simulation of transport.

CONCLUDING REMARKS

During the growing season of plankton the river discharge maintains high concentrations of nutrients N, P and Si in a 40 km wide coastal zone. The mean biomass of algae and the summer production are enhanced in this area. However, the highest plankton biomass coincides with relatively high temperature and Fe concentration, not with the lowest salinity and highest nutrient concentrations. The algal biomass in the nearcoastal zone is lower than theoretically expected, indicating possible effects of some unknown growth limiting, growth reducing, or mortality inducing factors, which may be of riverine origin. Eutrophication may have increased the flagellate biomass by a factor 2-4 since 1930, causing a shift from a diatom to a flagellate dominated population.

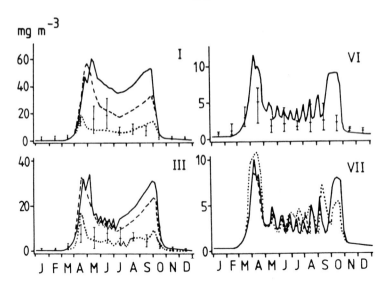

Fig. 6. Computed seasonal variation of chlorophyll-a in some compartments. Bars indicate observed means and extremes.

REFERENCES

van Bennekom, A.J., W.W.C. Gieskes & S.B. Tijssen, 1975. Eutrophication of Dutch coastal waters. Proc. R. Soc. R. B 189:359-374.

van Bennekom, A.J. & W. Salomons, 1980. Pathways of nutrients and organic matter from land to ocean through rivers. In River Inputs to Ocean Systems, ed. J.-M. Martin, J.D. Burton and D. Eisma. Proc. SCOR/UNESCO/IAPSO Workshop on RIOS, Rome, March 26-30, 1979:33-52.

Fransz, H.G., 1976. The spring development of calanoid copepod populations in the Dutch coastal waters as related to primary production. Proc. 10th E.M.B.S., Ostend, Belgium, Sept. 17-23, 1975. Vol. 2:247-269.

Fransz, H.G., 1977. Productiviteit van zoöplankton in de Zuidelijke Bocht van de Noordzee. Vakbl. Biol. 57(17):288-291.

Fransz, H.G. & W.W.C. Gieskes, 1984. The unbalance of phytoplankton and copepods in the North Sea. Rap.-v. Réun. Cons. int. Explor. Mer 183: 218-226.

Fransz, H.G. & J.H.G. Verhagen, 1985. Modelling research on the production cycle of phytoplankton in the Southern Bight of the North Sea in relation to riverborne nutrient loads. Neth. J. Sea Res. 19 (in press).

Gieskes, W.W.C. & G.W. Kraay, 1977. Continuous plankton records: changes in the plankton of the North Sea and its eutrophic Southern Bight from 1948 to 1975. Neth. J. Sea Res. 11(3/4):334-364.

de Kroon, J.C., 1971. Potentiële primaire produktie in het oostelijk deel van de zuidelijke Noordzee. NIOZ Publ. no 1.

Officer, C.B. & J.H. Ryther, 1980. The possible importance of silicon in marine eutrophication. Mar. Ecol. Progr. Ser. 3(1):83-91.

Postma, H., 1973. Transport and budget of organic matter in the North Sea.

In North Sea Science, ed. E.D. Goldberg, MIT Press, Cambridge, Massachusetts: 326-334.
Rijkswaterstaat, Project group WAKWON. Monitoring Water Quality North Sea. Time series 2nd and 3rd quarter 1979, 1980, 1981 and 1982. Staatsuitgeverij, 's-Gravenhage.

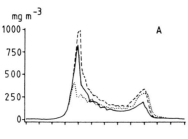
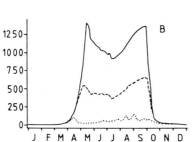

Fig. 7. Computed seasonal variation of diatom biomass (A) and biomass of other phytoplankton (B) in compartment I.

FRESHWATER RUNOFF CONTROL OF THE SPATIO-TEMPORAL DISTRIBUTION OF PHYTOPLANKTON IN THE LOWER ST.LAWRENCE ESTUARY (CANADA)

J.-C. THERRIAULT and M. LEVASSEUR
Centre Champlain des Sciences de la Mer,
Ministère des Pêches et des Océans,
C.P. 15500, 901 Cap Diamant, Québec
(Québec), Canada, G1K 7Y7

ABSTRACT

Sampling carried out in the Lower St.Lawrence Estuary reveals a strong influence of freshwater runoff on the spatio-temporal distribution of phytoplankton. At high discharge rates (spring and fall) the whole Lower Estuary forms a single freshwater plume and phytoplankton distribution is mainly controlled by dilution and advective processes. At low discharge rates (summer) the two main freshwater sources form two geographically localized plumes with different physical and biological characteristics.

INTRODUCTION

Natural or man-made alterations of the pattern of freshwater discharge into the marine system are of great concern for those who study or manage the marine environment, because of the potential impact of such alterations on the marine biota and particularly on the commercially exploited species (e.g. Sutcliffe, 1973; Neu 1976; Skreslet 1976).

Taking the Gulf of St.Lawrence system as an example, Bugden et al. (1982) have reviewed the various interactions between freshwater runoff and the physical, chemical and biological processes, and they have proposed a number of mechanisms by which freshwater runoff can affect the marine environment. Their review suggests that the main physical processes that are directly affected by runoff regulation are the circulation and mixing patterns and the ice formation and distribution. In turn, these changes in the physical environment have a strong influence on primary production by altering nutrient transport and enrichment processes, and by changing light and stratification conditions in the system. A further inference is that variations in phytoplankton production in the Gulf system have important consequences, a number of years later, for the commercial fish and invertebrate yields, through larval survival.

In a recent study, we identified freshwater runoff as one of the most important forces controlling phytoplankton production in the St.Lawrence Estuary (Therriault and Levasseur 1985). For the purpose of the present paper, we use the same extensive data set and focus exclusively on the freshwater runoff aspect, in order to show how freshwater discharge controls the spatio-temporal distribution of phytoplankton in the Lower Estuary.

MATERIAL AND METHODS

Collection of the data and technical details of the field and laboratory analyses have already been fully described (Therriault and Levasseur 1985). To summarize, sampling was carried out at approximately monthly interval on 16 occasions during 1979-1980, on a grid of 29 stations in the Lower St.Lawrence Estuary (Fig. 1). Sampling at each station consisted of measures of water transparency, vertical profiles of temperature and salinity, and a cast of 5 l Niskin bottles at 5 depths in the photic zone and 5 depths under the photic zone. Phytoplankton biomass and nutrient concentrations were measured at each depth, while production measurements and cell counts were carried out in the photic zone only.

RESULTS AND DISCUSSION

Freshwater runoff influence on the annual cycle of phytoplankton production and biomass.

Figure 2a shows the annual cycle of freshwater discharge into the Lower Estuary in 1979-1980. Freshwater runoff originates mainly from two sources, the combined Upper Estuary

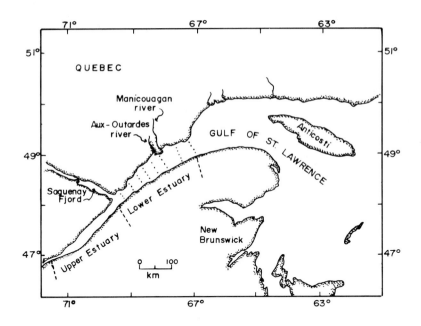

Figure 1. Map of the Gulf of St.Lawrence system showing the location of the sampling stations (dots) and of the main freshwater sources (Upper Estuary, Saguenay Fjord, Manicouagan and Aux-Outardes rivers) in the Lower Estuary.

and Saguenay Fjord discharge (UE-SF) upstream, and the combined Manicouagan and Aux-Outardes (M-O) rivers on the North shore of the lower Estuary (Fig. 1). The runoff from the UE-SF is characterized by strong seasonal peaks in spring and fall, while that from the M-O rivers is highly controlled throughout the year (Fig. 2a).

A drastic decrease in mean surface salinity (average of 29 sampling stations: Fig. 2b) coincides with the spring runoff peak. However, no significant decrease in mean surface salinity parallels the fall peak in runoff. Strong stratification of the water column is established starting with the spring peak in freshwater discharge in April-May. The end of the stratification period corresponds to the fall increase in freshwater discharge and wind velocity (Fig. 2a, e) in October-November. At the onset of stratification, the mean (of 29 stations) light availability in the mixed layer (<I> estimated as in Riley 1957) rapidly increases above the critical threshold, Icr, (which has been set empirically at 7.6 Ein $m^{-2}.d^{-1}$: Riley 1957), and stays well

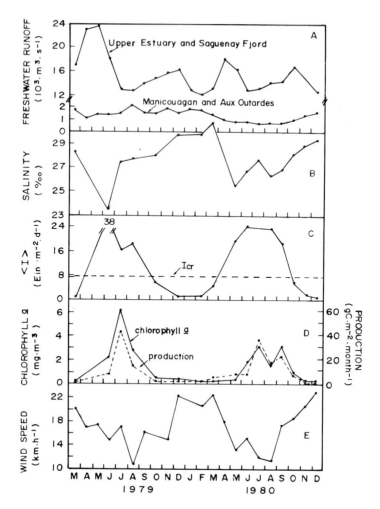

Figure 2. Temporal variations for the whole Lower Estuary in 1979-80 of a) freshwater discharge, b) mean (of 29 stations) surface salinity, c) mean light intensity <I> in the mixed layer (Icr is the light limitation level: Riley 1957), d) integrated photic zone phytoplankton production and mean photic zone chlorophyll a, and e) monthly mean wind velocity. (Figure adapted from Therriault and Levasseur 1985).

above limitation levels until late fall, corresponding to the end of the stratification period (Fig. 2c).

Phytoplankton growth and biomass accumulation (Chlorophyll a) reflecting mainly diatom growth in the Lower Estuary only occurs when mean light levels above Icr are found in the surface mixed layer (Fig. 2c, d). However, diatom growth is not initiated before June, even if mean light conditions are adequate in the mixed layer, starting in April. Levasseur et al. (1984) and Therriault and Levasseur (1985) have attributed this retardation of the diatom bloom to the increased freshwater discharge during the spring which, by decreasing the eddy exchange between the surface and the deeper layers due to increased stratification (Bugden 1981), delays seeding of the surface mixed layer by diatom cells. This interpretation is consistent with the conceptual model of a decreased vertical mixing at higher runoff levels, rather than that of a continuous increase, albeit at a decreasing rate, with runoff increases (see p. 45 in Bugden et al. 1982).

The termination of phytoplankton growth in the fall, on the other hand, also corresponds to a period of increased freshwater runoff, but it is evident from Fig. 2 that the end of the growth period is not solely due to freshwater runoff. The large increase in wind mixing during the fall (Fig. 2e) is probably the cause of the observed deepening and/or destruction of the mixed layer, which consequently imposes a halt on biological production processes, due to light limitation. The absence of a relationship between freshwater runoff and surface salinity during the fall supports this interpretation.

Freshwater runoff influence on the spatial distribution of phytoplankton

In the first section we discussed the effects of freshwater runoff at the scale of the whole Lower Estuary. However, considering the fact that major freshwater discharges occur at discrete geographical locations (Fig. 1), spatial differences in the response of phytoplankton to river discharge were certainly to be expected in this system. Different runoff effects on phytoplankton, at periods of higher river discharge than at period of lower discharge, were also expected.

High freshwater runoff periods. - May and October 1980 are periods of high freshwater discharge into the system (Fig. 2a) and freshwater transport processes dominate the hydrodynamics at the scale of the whole Lower Estuary. This is well illustrated in Figure 3 where clear downstream increases of the mean photic

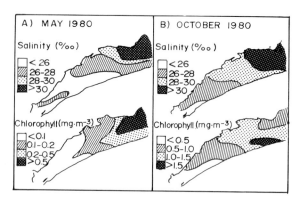

Figure 3. Spatial distribution of mean photic zone salinity and chlorophyll a in the Lower St. Lawrence Estuary for a) May and b) October 1980.

zone salinity are observed for May and October. Chlorophyll a shows a spatial distribution pattern which closely parallels the salinity distribution and which is characterized by a downstream accumulation. Chlorophyll concentrations are relatively low since μ-flagellates are essentially the only phytoplankton cells found in significant numbers in the water column in May and October (Therriault and Levasseur 1985). By comparing the flushing rate or the residence time of phytoplankton cells in the surface waters with their doubling time, Therriault and Levasseur (1985) came to the conclusion that phytoplankton biomass could only accumulate in the downstream region of the Estuary during these periods of high freshwater discharge rates.

An appreciation of the relative importance of freshwater dilution and advective processes versus other processes during May and October 1980, can be obtained by examining the shape of the relationship between mean photic zone salinity and the integrated photic zone chlorophyll a (Fig. 4 a,b). A clear difference is observed between those two months which probably reflects the dominance of different processes during spring and fall. In May, a highly significant (at $P < 0.01$) linear regression is observed between salinity and chlorophyll (Fig. 4a), which sugests that advection and dilution processes alone can explain the relationship. Data from a few stations in an upwelling zone (see Fig. 3; o in Fig. 4a; and Therriault and Levasseur 1985) were excluded from this regression analysis. In October, the regression between salinity and chlorophyll (Fig. 4b) is still significant (at $P < 0.01$), but the relationship is non-linear, suggesting that phytoplankton growth is acting in conjunction with dilution and other physical mixing processes during the fall. The deepening and destruction of the mixed layer by a large increase in wind mixing, following a period of intense phytoplankton growth (Fig. 2) is probably the reason underlying the particular relationship between salinity and chlorophyll in October (Fig. 4b).

Low freshwater runoff period. - During the summer (June to September), freshwater discharge in the Lower Estuary is drastically reduced (Fig. 2d) and a relationship can no longer be found between salinity and chlorophyll (Fig. 4c). At this reduced freshwater discharge rate, the extent and influence of the UE-SF plume declines and other circulation and mixing processes become relatively more important (Therriault and Levasseur 1985). This is illustrated by Figure 5a which shows a strong spatial heterogeneity of mean summer photic zone salinity distribution (June to September) in the Lower Estuary. The two main freshwater sources into the system are clearly evidenced in this figure by the two well defined freshwater plumes. The UE-SF plume is essentially restricted to the upstream region of the Estuary during the summer months and high turbidity is associated with this freshwater source (Fig. 5b). On the other hand, the M-O plume gains considerably in importance, even though the freshwater discharge from this source ($\sim 1,500$ $m^3.s^{-1}$) during the summer remains much lower than that of the UE-SF plume ($\sim 10,000$ $m^3.s^{-1}$). This can be explained by the fact that the water from the North shore source enters the system at salinity 0 $^o/_{oo}$ compared to waters at salinity ~ 26 $^o/_{oo}$ from the upstream UE-SF source.

Figure 5 also shows that the chlorophyll distribution during the summer is closely related to the different physical processes in the Lower Estuary. For example, high chlorophyll values are associated with the upwelling zone (Therriault and Levasseur 1985) in the upstream region, but also with the plume of the Manicouagan and Aux-Outardes rivers. On the other hand, very low phytoplankton biomass is associated with the UE-SF plume. Therriault and Levasseur (1985) attributed the low

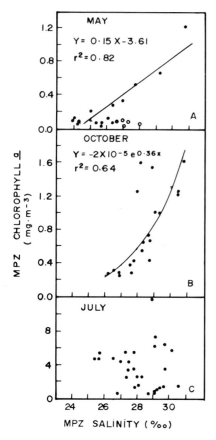

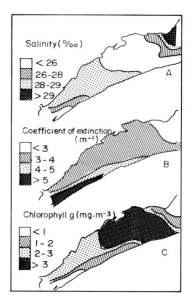

Figure 4. Mean photic zone salinity versus the mean photic zone chlorophyll a for a) May, b) October and c) July 1980 in the Lower St. Lawrence Estuary.

Figure 5. Mean summer (June to September 1980) spatial distribution of a) mean photic zone salinity, b) coefficient of extinction (corrected for chlorophyll a: Lorenzen 1972) and c) mean photic zone chlorophyll a in the Lower St. Lawrence Estuary.

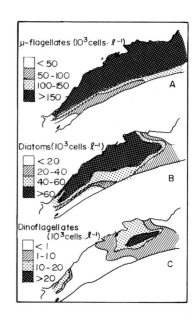

Figure 6. Mean summer (June to September 1980) spatial distribution of a) microflagellates, b) diatoms and c) dinoflagellates in the Lower St. Lawrence Estuary.

productivity of this plume, to high turbidity, high flushing rates and strong tidal mixing, whereas nutrient enrichment in the upwelling zone and increased stability in the M-O plume were invoked to explain the higher productivity of these latter two zones.

In terms of community structure, Figure 6 shows that the upwelling zone favors the growth of µ-flagellates and diatoms in large abundance, that dinoflagellates as well as diatoms and µ-flagellates can be found in large abundance in the M-O plume and that the UE-SF plume does not favor the growth of any phytoplankton group. Finally, it is of interest to note that the toxic species Protogonyaulax tamarensis largely dominates the dinoflagellate community in the M-O plume during the summer, such that a relationship between freshwater discharge and paralytic shellfish poisoning distribution can be inferred.

CONCLUSION

To conclude, our data clearly show that at the scale of the whole Lower St.Lawrence Estuary, there exists a strong influence of freshwater runoff on the temporal and spatial distribution of phytoplankton. Our data also indicate different influences of freshwater runoff during periods of high discharge than during periods of low discharge. At high discharge rates the whole Lower Estuary forms a single freshwater plume (a typical estuary) whose effects are felt much lower in the Gulf of St.Lawrence and even along the Atlantic coast (Bugden et al. 1982). During periods of flow reduction (summer), the impact of freshwater runoff becomes more geographically localized to the Lower Estuary. It is of interest to point out that at this time of the year, the two freshwater plumes of the Lower Estuary have opposite effects on phytoplankton growth: the UE-SF and M-O plumes showing detrimental and beneficial effects on phytoplankton growth, respectively. Similarly, during high runoff periods, detrimental effects on phytoplankton growth might be observed in the Lower Estuary at the same time as beneficial effects can be observed simultaneously much farther in the Gulf (the phytoplankton bloom occurs in April-May in the Gulf: Steven 1975). Thus, temporal as well as spatial scale effects should be considered when studying freshwater influence on the marine environment.

REFERENCES

Bugden, G.L., 1981. Salt and heat budgets for the Gulf of St.Lawrence. Can. J. Fish. Aquat. Sci., 38: 1153-1167.

Bugden, G.L., B.T. Hargrave, M.M. Sinclair, C.L. Tang, J.C. Therriault, and P.Y. Yeats. 1982. Freshwater runoff effects in the Marine Environment: The Gulf of St.Lawrence example.- Can. Tech. Rept. Fish. Aquat. Sci. No 1078, 89 pp.

Levasseur, M., J.-C. Therriault and L. Legendre, 1984. Hierarchical control of phytoplankton succession by physical factors. Mar. Ecol. Prog. Ser. 19: 211-222.

Lorenzen, C.J., 1972. Extinction of light in the ocean by phytoplankton. J. Cons. int. Explor. Mer. 36: 262-267.

Neu, H.J.A. 1976. Runoff regulation for hydro-power and its effects on the ocean environment. Hydrol. Sci. 21: 433-444.

Riley, G.A., 1957. Phytoplankton of the north central Sargasso sea. Limnol. Oceanogr., 2: 252-270.

Skreslet, S., 1976. Influence of freshwater outflow from Norway on recruitment to the stock of Arctic-Norwegian cod (Gadus morhua), In Fresh Water on the Sea, S. Skreslet, R. Leineb, J.B.L. Mattews and E. Sakshaug (Eds), The association of Norwegian Oceanographers, Oslo, p. 233-237.

Steven, D.M., 1975. Biological production in the Gulf of St.Lawrence. p. 229-248. In T.W.M. Cameron (ed)., Energy flow its biological dimensions. A summary of the IBP in Canada 1964-1974. Ottawa, Royal Society of Canada. 319 pp.

Sutcliffe, W. H. Jr., 1973. Correlations between seasonal river discharge and local landings of american lobster (Homarus americanus) and Atlantic halibut (Hippoglossus hippoglossus) in the Gulf of St.Lawrence. J. Fish. Res. Bd. Can. 30: 856-859.

Therriault, J.C. and M. Levasseur, 1985. Control of phytoplankton production in the Lower St.Lawrence Estuary: Light and Freshwater runoff. Naturaliste Can. 112: 77-96.

BIOLOGICAL AND PHYSICAL CHARACTERISTICS OF A FRONTAL REGION ASSOCIATED WITH THE ARRIVAL OF SPRING FRESHWATER DISCHARGE IN THE SOUTHWESTERN GULF OF ST. LAWRENCE

Côté, B.[1], M. El-Sabh[2] et R. de la Durantaye[3]

Laboratoire Océanologique de Rimouski, INRS[1], UQAR[2]
310 Avenue des Ursulines
Rimouski (Québec) G5L 3A1

Pêches et Océans Canada[3]
Centre Champlain des Sciences de la Mer
C.P. 15500, 901 Cap Diamant
Québec (Québec) G1K 7Y7

ABSTRACT

The preliminary results of a study carried out in the southwestern gulf of St. Lawrence to examine the relationship between the physical and biological characteristics of the area and larval fish distribution are presented. It will be shown that the arrival of the spring discharge from the St. Lawrence estuary significantly affects the hydrographic characteristics of the area and that through the formation of fronts may result in increased aggregation of food organisms and fish larvae.

INTRODUCTION

The Magdalen shallows within the Gulf of St. Lawrence is one of the richest fishing areas in the northwest Atlantic (Dickie & Trites 1983). Recruitment to the various fisheries (cod, herring, plaice, mackerel, etc) has varied significantly over the years. However we still know relatively little about the factors regulating their year-class success.

Mesoscale features associated with early life stages are now recognized as important in the life history of marine fish. In this respect the Magdalen shallows is characterized by marked fluctuations in surface current patterns, by the presence of eddies and fronts (El-Sabh 1976; Pingree & Griffiths 1980) and by the discharge of the St. Lawrence Estuary (Sutcliffe *et al.* 1976; Bugden *et al.* 1982). Each of these features can affect the early life history stages of fish.

We are presently studying the southwestern region of the Magdalen shallows, Miscou-Shediac Valley (Fig. 1). This region seems to be a preferred site for spawning and larval development of a number of species including mackerel, herring, cod and plaice. It is influenced to the north by the outflow of Chaleurs Bay and the Gaspé current (a strong coastal jet flowing seaward along the Gaspé peninsula of Québec) and to the south, by the outflow of Miramichi Bay. Data will be presented on the physical and biological characteristics of the region. It will be shown, that the arrival of the St. Lawrence estuary spring discharge contributes to the formation of a front and that the latter strongly influences the biological characteristics of the area.

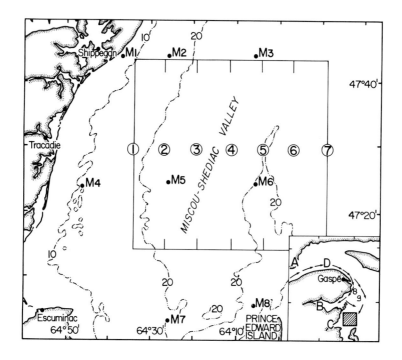

Figure 1. Map of survey area: (M) current mooring positions, ① transect number and location of stations referred to in Fig. 6 and 7. Inset: location of sampling area relative to gulf and position of sampling station in (#8) and out (#9) of Gaspé current, from El-Sabh (1976); a) Pointe-des-Monts, b) Chaleur Bay, c) Miramichi Bay, d) Gaspé current.

METHODS

The preliminary results of our July 1984 cruise are presented here. Seven transects were covered within a 4-day period between 27 and 31 July aboard C.S.S. L.M. Lauzier. The cruise was designed as a joint biological and physical investigation of the area. The underway sampling system consisted of a fluorometer (Turner design model III), a thermistor and an acoustic profiling system (Datasonic DSF 2100 High frequency echosounder, Transducer 200 KHz, 3° beam). Seawater was supplied by the ship's in-line seawater system (hull inlet at 3 m depth). Water from the outflow was used to calibrate the fluorometer. Samples were also taken every 20 minutes for nutrient and particle size determinations, particulate carbon, particulate nitrogen, chlorophyll, phaeopigments, DNA and RNA analyses. Simultaneously the fire intake system was used to collect zooplankton samples (153 μm mesh net). A number of oblique and horizontal Tucker hauls were also taken. Temperature and salinity were profiled using a Guildline Mark IV CTD system. An array of current meters were deployed in the area between 16 June and 20 September (Fig. 1).

RESULTS

Surface salinity and density distributions (Fig. 2b & c) show the presence of a front separating offshore water from coastal waters. West-east temperature and salinity isopleths, constructed from a series of stations positionned midway along the transects show that for instance the 27.5‰ isohaline and 13.5°C isotherm rise from about 20 to 10 m over a distance of 10 km (Fig. 3). The corresponding changes in surface chlorophyll a, DNA and silicate concentrations show higher values of all three variables in coastal water (Fig. 4b & d). This finding holds as well for changes in zooplankton biomass (Fig. 4c). In the latter case, sharp changes in the species composition of the zooplancton community occurred across the front (Fig. 4c). Fish larvae were also found to be unevenly distributed across the front, for instance capelin larvae were found almost exclusively in offshore water while mackerel, cod, plaice were found in coastal water (data not shown).

It is apparent from the temperature-salinity data obtained from our current meters (Fig. 5) that the sampling period coincided with a period of change in the hydrographic characteristics of the area. An abrupt rise in temperature and fall in salinity start to occur on July 17 at all current meter moorings positionned between longitudes 64°25.00' and 64°45.00' but not at moorings positionned more offshore. As will be discussed below, the major source of this low salinity water, is believed to be the outflow from the St. Lawrence estuary (Fig. 6). In addition, an east-westward oscillating movement of the outer part of the Gaspé current may explain the high frequency variability observed at 24 m depth at M_5 between 1 and 23 August. A T-S diagram (not shown) comparing data collected within and outside the Gaspé current (see inset Fig. 1) with data collected within the study area shows a close affinity between sector A (southwestern) profiles and Gaspé current data on the one hand and sector B (northeastern) profiles and offshore water on the other.

DISCUSSION

The study area receives freshwater discharge mainly from rivers flowing into the St. Lawrence estuary, the Chaleur Bay and the Miramichi Bay. The seasonal character of freshwater

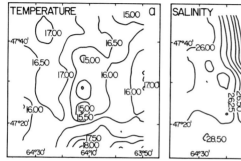
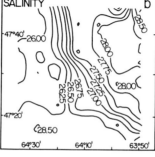
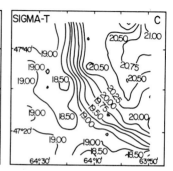

Figure 2. Surface distribution a) temperature, b) salinity and sigma-t derived from CSTD data.

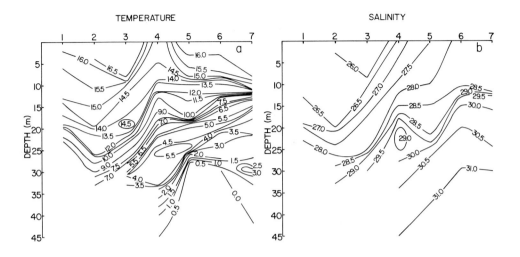

Figure 3. West-east a) temperature and b) salinity isopleths (see Fig. 1 for station locations).

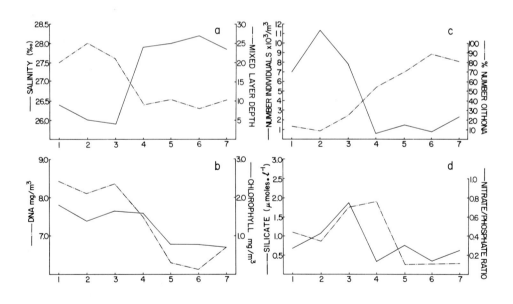

Figure 4. a) Surface salinity and mixed layer depth at stations shown in Fig. 1; b) DNA and chlorophyll concentration; c) zooplankton concentration and % of zooplankton sample comprised of *Oithona sp.*; d) silicate concentration and nitrate-phosphate ratio.

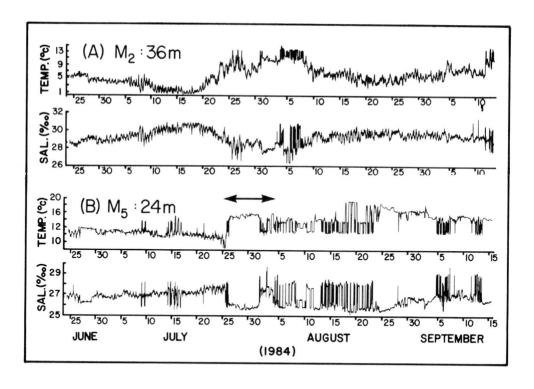

Figure 5. 20-minute a) temperature and b) salinity time-series observed at current meter morrings M_2 and M_5 (see Fig. 1). The double arrow indicates sampling period.

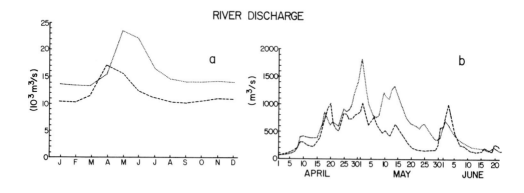

Figure 6. a) Monthly freshwater inflow from the St. Lawrence River drainage basin at Québec City (-----), at Pointe-des-Monts (.....), average (1949-1981) from El-Sabh (1985); b) Daily values of freshwater discharge from Restigouche River (.....) and southwest Miramichi River (-----).

discharge at the mouth of the St. Lawrence Estuary, near Pointe-des-Monts, reveals marked monthly fluctuations (Fig. 6a) (El-Sabh, 1985). The data show that 26% of the yearly drainage at Pointe-des-Monts is usually produced during May and June, the two flood months of the year. From studies by El-Sabh (1976), and Sutcliffe et al. (1976), we expect that the effect of peak discharge should be felt in the study area around mid-July. On the other hand the outflow of Restigouche and south west Miramichi River after 1 June is insignificant (Fig. 6b). The changes observed in the temperature-salinity data obtained from our current meters (Fig. 5) in late July are thus believed to reflect the arrival of the St. Lawrence River spring discharge.

The water of low salinity produced in the estuary gradually becomes associated with the Gaspé current (Lauzier et al. 1957). This current is buoyancy driven and its properties are strongly influenced by the seasonal discharge from the St. Lawrence estuary (Benoit et al. 1985). During the period of peak discharge, the Gaspé current once passed Cap Gaspé maintains a south, south westerly direction, i.e. it is oriented in the general direction of the study area whereas during the fall it is directed southeasterly and it is spread more uniformly over the Magdalen Shallows (Lauzier et al. 1957). The close affinity between T-S profiles in sector A, and the Gaspé current data indicate that the same water mass is being sampled. The importance of the Gaspé current as a mecanism controlling the biological activity of the Magdalen Shallows has been discussed by Lauzier and Marcotte (1965) and Bugden et al. (1982). It is apparent from these studies that we know very little about how freshwater affects plankton dynamics.

While it is somewhat speculative to postulate mechanisms of possible freshwater effects on fish populations (Bugden et al. 1982), nevertheless it is interesting to note for the Atlantic mackerel, that peak spawning precedes the arrival of the spring freshet by a couple of weeks. At the temperatures prevailing in the gulf at this time of the year (11-14°C), eggs would hatch in approximately a week. Transition from endogenous to exogenous feeding would therefore coincide in the study area with the arrival of the spring freshet and the formation of a front. Studies of larval fish distributions have recently indicated that larvae as well as adult fish may be influenced by the hydrographic and/or biological activity occurring in frontal areas (e.g. Sherman *et al.* 1984). The higher temperatures, increased habitat volume (mixed layer depth), greater food concentrations observed in the southwestern sector of the study area would all favor the rapid growth requirements characteristic of mackerel (Fig. 4). Furthermore the sharp change in ichthyoplankton species composition across the front suggests that the front acts as a barrier and hence reduces cross-shelf mixing. This may help maintain the aggregation of fish larvae. In the case of mackerel, larvae, this could be particularly important for the development of their schooling behavior.

It is apparent from our study that the spring discharge from the St. Lawrence estuary significantly affects the hydrographic characteristics of the Magdalen Shallows and that through the formation of fronts may result in increased aggregation of food organisms as well as larvae. Dickie and Trites (1983) noted that in the gulf the success of higher trophic levels did not appear to depend solely on a rich broth of primary and secondary producers. They concluded that geographic and temporal aggregations on relatively small scales may be involved in the success of year-class production of both fish and invertebrates and may need to be understood in explaining year-to-year differences. The results of our study suggest that this may indeed be the case. Much more needs to be known, however, of the processes involved in the dispersion of the St. Lawrence estuary spring freshet through the shallows before mechanisms of possible freshwater effects on vertebrate and invertebrate populations can truly be understood (Bugden *et al.* 1982).

ACKNOWLEDGEMENTS

We thank E. Bonneau, M. Gagnon, I. Guay, C. Girouard, E. Laberge, M. Morissette, G. Ouellet, J. Plourde, M. Roberge for their excellent help at sea and in the laboratory, A. Roy for drafting the figures and M. Cogné for typing the manuscript. This study was supported by grants to B. Côté and/or M. El-Sabh from the following organisations: Fonds FCAC, Québec (EQ2685); Department of Fisheries and Oceans; Institut National de la recherche scientifique; Fondation de l'Université du Québec à Rimouski; and the NSERC, Ottawa (grant A-0073).

REFERENCES

Benoît, J., M.I. El-Sabh and C.L. Tang, 1985: Structure and seasonal characteristics of the Gaspé Current. J. Geophys. Res., 90, 3225-3236.

Bugden, G.L., B.T. Hargrave, M.M. Sinclair, C.L. Tang, J.-C. Therriault and P.A. Yeats, 1982: Freshwater runoff effects in the marine environment: The Gulf of St. Lawrence Example. Can. Tech. Rept. Fish. Aquat. Sci., 1078, 89 p.

Dickie, L.M. et R.W. Trites, 1983: The Gulf of St. Lawrence. p. 403-425. In: Ketchum, B.H. (ed.), Ecosystems of the world. 26. Estuaries and Enclosed Seas. Elsevier Sci. Publ.

El-Sabh, M.I., 1976: Surface circulation pattern in the gulf of St. Lawrence. J. Fish. Res. Bd. Can., 33, 124-138.

El-Sabh, M.I., 1985: Oceanography of the St. Lawrence estuary. In: B. Kjerfve (ed.), Hydrodynamics of estuaries, CRC Press (in press).

Lauzier, L., R.W. trites and H.B. Hachey, 1957: Some features of the surface layer of the Gulf of St. Lawrence. Fish. Res. Bd. Can., MS Rept. Biol. Sta. 417, 1-19.

Lauzier, L. et A. Marcotte, 1965: Comparaison du climat marin de Grande-Rivière (Baie des Chaleurs) avec celui d'autres stations de la côte Atlantique. J. Fish. Res. Bd. Can., 6, 1321-1334.

Pingree, R.D. and D.K. Griffiths, 1980: A numerical model of the M_2 tide in the Gulf of St. Lawrence. Oceanol. Acta, 3, 221-225.

Sherman, K., W. Smith, W. Morse, M. Berman, J. Green and L. Ejsymont, 1984: Spawning strategies of fishes in relation to circulation, phytoplankton, production, and pulses in zooplankton off the northeastern United States. Mar. Ecol., prog. ser., 18, 1-19.

Sutcliffe, W.H. Jr., R.H. Loucks and K.F. Drinkwater, 1976: Coastal circulation and physical oceanography of the Scotian shelf and the Gulf of Maine. J. Fish. Res. Bd. Can., 33, 98-115.

RELATIONSHIPS OF ST. LAWRENCE RIVER OUTFLOW WITH SEA SURFACE TEMPERATURE AND SALINITY IN THE NORTHWEST ATLANTIC

J.A. Koslow[1], R.H. Loucks[2], K.R. Thompson[1], R.W. Trites[3]
1 - Oceanography Department, Dalhousie University, Halifax, N.S. B3H 4J1, Canada
2 - Loucks Oceanology, 24 Clayton Park Drive, Halifax, N.S., Canada
3 - Marine Ecology Laboratory, Bedford Institute of Oceanography, Dartmouth, N.S. B2Y 4A2, Canada

ABSTRACT

We examine the influence of St. Lawrence River runoff on sea-surface temperature (SST) and salinity along the eastern seaboard of Canada and the United States. To avoid spurious statistical relationships, we take into account the direct influence of meteorological forcing, which is closely linked to St. Lawrence discharge.

The low-salinity pulse of the St. Lawrence spring freshet has been traced to the tip of southwest Nova Scotia, but not into the Gulf of Maine (e.g. at Boston). Multiple regression analysis indicates Boston salinity is significantly influenced by meteorological forcing and possibly local runoff but not St. Lawrence discharge.

Freshwater outflow is correlated with SST on the northwest Atlantic shelf, at 0-1 season lag, from the Gulf of St. Lawrence to Georges Bank due to large-scale meteorological forcing and local runoff effects. The influence of St. Lawrence outflow on SST is not traced beyond the upper half of the Scotian shelf by our seasonal, lagged correlation analysis.

INTRODUCTION

The problem of understanding fluctuations in recruitment to North Atlantic fish stocks has been the subject of renewed interest (Sutcliffe et al. 1977, 1983; Loucks and Sutcliffe 1978; Koslow 1984a,b). The large spatial extent of the recruitment patterns (Koslow 1984a) has led to a search for causal relationships with large-scale environmental variables (e.g. air pressure, sea surface temperature). Several statistically significant relationships between recruitment and environmental signals have been obtained (Koslow 1984b), but the underlying mechanism(s) remain unclear. Part of the problem of interpretation stems from the complex couplings among the environmental signals. We therefore undertook a preliminary study of the coupled environmental signals. This paper is thus part of an ongoing study of recruitment/environmental interactions and takes as its focus the influence of freshwater runoff on the climatic and hydrographic conditions of the northwest Atlantic.

It is recognised that the input of freshwater (from land runoff and ice melt) plays an important role in determining the circulation and hydrographic conditions off eastern Canada. To illustrate, Figure 1 clearly shows the influence of freshwater input through reduced coastal salinity. The Nova Scotian current (Figure 1) has a strong seasonal component associated with St. Lawrence discharge (Drinkwater et al. 1979). Further north, the inshore branch of the Labrador Current is a mixture of the Baffin Land Current and Hudson Strait outflow. Recent

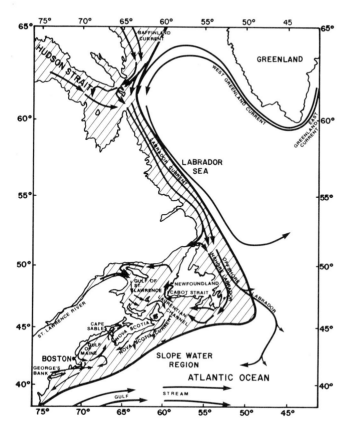

Figure 1. The northwest Atlantic showing circulation patterns and the mean spring 32.5% surface isohaline (U.S. Naval Office 1967) to indicate the approximate extent of low-salinity runoff.

studies have defined as well an intense, southward current trapped to within about 15 km of the Labrador coast that is thought to be driven by local runoff (Fissel and Lemon, unpubl. mss.) and possibly Hudson Strait outflow.

In this paper we will examine the role of St. Lawrence outflow on the sea surface temperature and salinity southwest of the Gulf of St. Lawrence. In contrast to previous authors (e.g. Sutcliffe et al. 1983), we decided not to use St. Lawrence discharge as a surrogate for Hudson Strait outflow and thus extend our analysis north of the Laurentian Channel into the Labrador Sea, because recent analyses of North American precipitation patterns and variabilty suggest that conditions over the St. Lawrence drainage area may differ considerably from those further north (Fortin et al. 1983; Vines 1984). Also, ice melt is presumably a major contributor to the fresh water flux through Hudson Strait and should be considered in any attempt to determine its variability (Jordan 1976).

Table I. Data and sources used in this paper.

	Period	Region	Reference
Air pressure	1950-75	North Atlantic 15-65°N, 0-90°W	Thompson et al. (1983)
Geostrophic wind	1950-80	Scotland Shelf (44°N, 63°E)	Thompson and Hazen (1983)
Salinity	1956-72	Boston Lightship	Sutcliffe et al. (1976)
River discharge	1950-80	St. Lawrence	Sutcliffe et al. (1976), Drinkwater pers. commn. Bedford Institute of Oceanography, Dartmouth, N.S.
Sea surface temperature	1950-80	Eastern Seaboard U.S. & Canada	Loucks et al. (in press)

We first explore the relationship between river discharge and meteorological forcing in order to illustrate the degree of coupling between two main drivers of the system. The river influence on salinity is then discussed and compared to that of meteorological forcing. Finally, we examine the river influence on sea surface temperature along the eastern seaboard of Canada and the United States. All data used in this study, and their sources, are listed in Table 1. The basic averaging period was three months giving seasonal means (Winter=Dec-Feb, etc). The number of independent data points (N*) in the correlation analyses are corrected for autocorrelation in the time series (Garrett and Toulany 1981), and significance levels are based upon two-tailed Student t-tests with N* - 2 degrees of freedom.

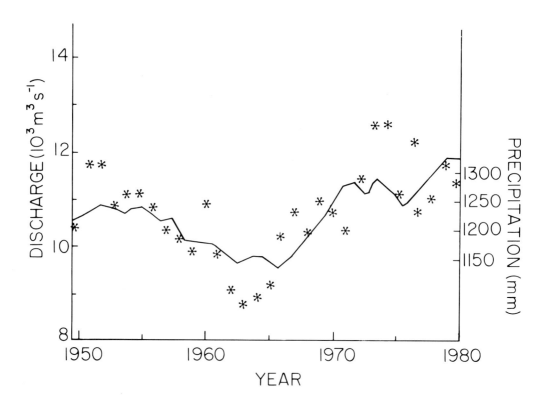

Figure 2. Five-year running mean (solid line) of mean precipitation for selected stations including Montreal, Quebec, and St. John's (Saulesleja and Phillips 1982) and annual mean discharge of the St. Lawrence River (1950-1980) (points).

RESULTS

River Discharge and Meteorology

The annual discharge record for the St. Lawrence River (1950-80) is dominated by low-frequency changes, which are closely related to precipitation variations over the drainage area (Figure 2) and thus underlying atmospheric conditions. An immediate problem is now obvious in trying to identify the role of river discharge on hydrographic conditions along the eastern Canadian seaboard - atmospheric conditions can also influence temperature and salinity directly (through air sea exchanges and wind-forced advection) and could be misinterpreted as a river influence.

We have examined the degree of coupling between discharge and atmospheric conditions using two types of index for the latter. The first type of index was simply the seasonal geostrophic wind over the Scotian Shelf which, as noted above, could be expected to have a direct influence on oceanographic

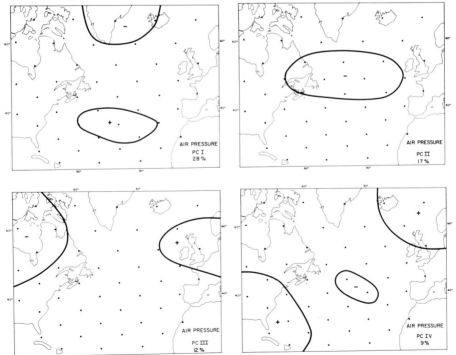

Figure 3. Spatial patterns based upon significant loadings onto the first four principal components (PC) for atmospheric pressure on a grid over the North Atlantic and eastern North America, 1950-75. The variance explained by each PC is shown.

conditions. The second type of index was derived from large-scale atmospheric conditions over the whole North Atlantic and eastern North America and could be related to advection by Ekman drift and wind-forced ocean circulation. (These indices were principal components from an analysis of the monthly air pressure data, based on the correlation matrix. The first 4 modes explained 66% of the total standardised variance and are shown in Figure 3).

The correlation between spring river discharge and the two types of meteorological index are shown in Table 2. We have included the meteorological indices for both spring (corresponding to the time of discharge) and the previous winter, which we also considered could be potentially important. High runoff is generally associated with northward, onshore airflow. But perhaps the most important point to note from Table 2 is that statistically significant relationships exist between both types of index and spring discharge. (A separate multiple regression analysis showed that over 50% of the spring discharge variance could be accounted for by a linear combination of just 3 of the principal components). In other words river discharge is significantly correlated with atmospheric conditions prevailing locally over the eastern seaboard of Cananda, as well as the whole North Atlantic.

Table II. Product-moment correlation coefficients (r) of spring St. Lawrence River discharge with winter and spring air pressure principal components (PC) 1-4 and the north and east components of geostrophic wind on the Scotian Shelf.

* – $p \leq 0.05$; ** – $p \leq 0.01$; *** – $p \leq 0.001$.

	PC1	2	3	4	Wind E	N
Winter	0.20	-0.20	-0.31	-0.48*	-0.07	0.38*
Spring	-0.33	-0.15	0.40*	-0.23	-0.23	0.36*

Another complication in any attempt to identify the river influence on the hydrography without first removing the direct meterological effect is that _winter_ atmospheric conditions are also significantly correlated with _spring_ discharge.

River Discharge and Salinity

Previous studies have clearly shown the effect of the St. Lawrence on salinity can be traced through the Gulf of St. Lawrence, Cabot Strait and along the Scotian Shelf to Cape Sable (e.g. El Sabh 1977, Drinkwater et al. 1979, Smith 1983). The seasonal pulse of fresh water peaking in May, reaches the southwest tip of Nova Scotia in November to February, implying a probable advection time of 7-8 months with a range of 6-9 months.

In a study of monthly St. Lawrence discharge and coastal salinity, Sutcliffe et al. (1976) were unable to trace the river influence into the Gulf of Maine. We have also tried to detect the St. Lawrence river influence into the Gulf of Maine using Boston salinity. The major difference in our approach however was that we also _simultaneously_ modelled the effect of local meteorology, which we have shown to be significantly correlated with the river signal itself and which could mask its effect. A stepwise regression analysis (Draper and Smith, 1966) was performed using seasonal mean Boston salinity as the dependent variable and wind in the Gulf of Maine and river discharge as the "independent" variables (lagged back 0-4 seasons). We hoped that this approach would allow the relative importance of both local meteorology and St. Lawrence discharge to be correctly determined. The results (Table 3) show that spring is the only season in which St. Lawrence runoff significantly enters the regression. However the lag is 0 seasons, which indicates that the correlation presumably results from local runoff into the Gulf of Maine area, which is correlated with St. Lawrence discharge. Thus our analysis agrees with Sutcliffe et al (1976) - the effect of St. Lawrence discharge on inshore coastal salinity cannot be traced beyond the southwest tip of Nova Scotia.

In every season except winter, the eastward wind for that season enters the regression. The general association of offshore winds with higher salinity may be related to offshore

Table III. Stepwise multiple regression analysis of seasonal salinity (S) at Boston with seasonal geostrophic winds (northerly (N) and easterly (E) component(s) in the Gulf of Maine. Wind variables were available to enter the regression for the season in question and preceding seasons of that year; seasonal river discharge for the preceding year (-1) was available to enter as well. Variables entered if significant at $p \leq 0.05$ and subsequently removed if $p \leq 0.10$. The variables in the regression in order of entry are shown below, the standardized regression coefficients (Beta) and their significance, and the variance explained by the regression and the significance of the regression.
* - $p \leq 0.05$; ** - $p \leq 0.01$; *** - $p \leq 0.001$.

Dependent variable		Variables in regression	Beta	R^2 and significance
Boston S	Wi	None		
	Sp	River (Sp)	-0.51*	0.58**
		E Wind (Sp)	0.44*	
	Su	E Wind (Sp)	0.69***	0.80***
		N Wind (Sp)	-0.55**	
		E Wind (Su)	0.35*	
	Fa	E Wind (Fa)	0.63**	0.40**

transport of surface water and upwelling of higher salinity, subsurface water. However, the influence of wind on Boston salinity is particularly pronounced in summer (80% of the variance can be related to local winds; Table 3), when there is an additional contribution from spring winds. The regression analysis selects energetic, spring on-offshore winds as the most effective in changing summer salinity. This could result from the increased precipitation associated with onshore winds and hence, presumably, local runoff into the Gulf of Maine which is trapped above the summer thermocline. The influence of high-salinity slope water intrusions into the Gulf of Maine are a further important factor (Sutcliffe et al. 1976, Loucks and Trites in press).

River Discharge and Sea Surface Temperature

Recently Loucks et al. (in press) analysed monthly sea surface temperatures from 24 areas in the northwest Atlantic.

They found that sea surface temperatures were coherent over large spatial scales (~1000km) and long time scales (>1 year). It is generally accepted that river discharge can influence sea surface temperature through changes in the stability of the upper water column and hence the rate of vertical mixing with the deeper layers. In this section we explore the role of St. Lawrence discharge in causing large sea surface temperature patterns. Our approach is similar to that employed by Sutcliffe et al (1976), who reported a relationship between lagged sea surface temperature and the advected seasonal pulse of fresh water from the St. Lawrence onto the Scotian Shelf and into the Gulf of Maine. The major differences are two-fold: our study is based on the sea surface temperatures of Loucks and Trites (in press) rather than nearshore surface and subsurface temperatures; and whereas Sutcliffe et al. (1976) used monthly data, we use three-month seasonal averages to

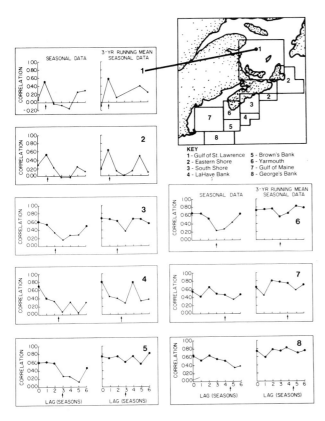

Figure 4. Correlations of spring discharge of the St. Lawrence River with SST at 8 areas of the northwest Atlantic from the Gulf of St. Lawrence to Georges Bank with 0-6 seasons lag. Correlations at left are based upon unmodified data; those at right are based upon 3-point running means over three years. Large dots indicate the correlation is significant ($p \leq 0.05$).

reduce the meteorological effect, which smooths peaks in lagged correlations over a five-month period with a symmetric triangular weighting (Fig. 4).

The locations of 8 sea-surface temperature subareas used in this study are shown in Figure 4. We calculated 32 separate time series of winter, spring, summer and fall temperatures for each area (1950-80). Another 32 time series were generated by applying a 3-point, between-years running mean to each seasonal series to suppress the higher frequency variability. This smoothing approach, which was used by Sutcliffe et al. (1976), presumably reduces the meterological influence. All temperature series were then correlated with spring discharge, and the resulting lagged correlation functions, both with and without smoothing, are plotted in Figure 4.

There is evidence for an advected river signal in areas 1 and 2 but it is difficult to trace along the rest of the Scotian Shelf and into the Gulf of Maine (Figure 4). It is clear that the effect of the 3-point running mean is to generally increase the level of correlation at all lags, but it does not help isolate an advective signal. The increased correlation presumably results from accentuation of the trends in the discharge and sea surface temperature series.

DISCUSSION

St. Lawrence discharge is a major influence on the inshore circulation and hydrography of the Gulf of St. Lawrence and Scotian Shelf. However, in agreement with Sutcliffe et al. (1976) we were unable to trace its influence into the Gulf of Maine through examination of coastal salinity (Boston). Presumably the fresh water pulse enters the Gulf of Maine during winter when wind and tidal mixing combine to homogenise the water column and thereby attenuate the salinity signature of the fresh water pulse. Regression analysis suggested that local meteorology was the dominant influence on salinity during seasons with a stratified water column, perhaps the result of local upwelling and runoff trapped in the near surface layer. Warm, high-salinity, intrusions into the Gulf of Maine would also confuse the relationship between low-salinity runoff and surface warming.

In contrast with Sutcliffe et al. (1976), we could not trace the St. Lawrence influence on sea surface temperature beyond the northeast half of the Scotian shelf. It is possible that the difference is due to their use of monthly nearshore data and our use of shelfwide seasonal means. The areal data are subject to averaging errors associated with uneven spatial and temporal coverage over the designated subareas. Sutcliffe et al. (1976) also made use of subsurface temperature data at several stations, where the river's influence was clearer than in the SST data. However, the major sea surface temperature mode of variation in our data is a simultaneous rise and fall of temperature over extensive space (~1000km) and time (>1 year) scales. This encourages us to think that the dominant driving mechanism is probably related to large-scale atmospheric conditions.

ACKNOWLEDGEMENTS

This research was supported by a Canadian Natural Sciences and Engineering Research Council grant and Department of Fisheries and Oceans Subvention to J.A. Koslow.

REFERENCES

Drinkwater, K.F., B. Petrie, and W.H. Sutcliffe, Jr. 1979. Seasonal geostrophic volume transports along the Scotian Shelf. Estuarine Coastal Mar. Sci. 2: 17-27.

El-Sabh, M.I. 1977. Oceanographic features, currents, and transport in Cabot Strait. J. Fish. Res. Board Can. 34: 516-528.

Fissel, D.E. and D.D. Lemon. Unpubl. mss. Analysis of physical oceanographic data from the Labrador shelf, summer 1980. Arctic Sciences, Ltd.

Fortin, J.-P., G. Morin, and L. Dupont. 1983. La rationalisation du reseau meteorologique du Quebec: strategie d'intervention et methodes d'analyse des donnees. Atmosphere-Ocean 21: 365-386.

Garrett, C.J.R. and B. Toulany. 1981. Variability of the flow through the Strait of Belle Isle. J. Mar. Res. 39: 163-189.

Jordan, F. 1976. Canadian run-off contribution to the Labrador Current. Unpubl. mss. Coastal Oceanogr. Div., Bedford Institute of Oceanography.

Koslow, J.A. 1984a. Recruitment patterns in northwest Atlantic fish stocks. Can. J. Fish. Aquat. Sci. 41: 1722-1729.

Koslow, J.A. 1984b. Recruitment variability - biologically or environmentally driven? Presented to MEES Workshop. Bedford Institute of Oceanography, Dartmouth, N.S.

Loucks, R.H. and W.H. Sutcliffe Jr. 1978. A simple fish-population model including environmental influence, for two western Atlantic shelf stocks. J. Fish. Res. Board Can. 35: 279-285.

Loucks, R.H., K.R. Thompson, and R.W. Trites. In press. Sea-surface temperature in the northwest Atlantic - time and space scales, spectra and spatially-smoothed fields. Can. Tech. Report Fish. Aquat. Sci.

Loucks, R.H. and R.W. Trites. In press. Variability of sea surface temperatures in the Northwest Atlantic. Can. Tech. Report Fish. Aquat. Sci.

Saulesleja and Phillips. 1982. Meteorological conditions in the decade 1970-79 and their impacts over the northwest Atlantic. NAFO Sci. Coun. Studies 5: 17-32.

Smith, P.C. 1983. The mean and seasonal circulation off southwest Nova Scotia. J. Phys. Ocean. 13: 1034-1054.

Sutcliffe, W.H. Jr., K. Drinkwater, and B.S. Muir. 1977. Correlations of fish catch and environmental factors in the Gulf of Maine. J. Fish Res. Board Can. 34: 19-30.

Sutcliffe, W.H. Jr., R.H. Loucks, and K.F. Drinkwater. 1976. Coastal circulation and physical oceanography of the Scotian Shelf and the Gulf of Maine. J. Fish. Res. Board Can. 33: 98-115.

Sutcliffe, W.H. Jr., R.H. Loucks, K.F. Drinkwater, and A.R. Coote. 1983. Nutrient flux onto the Labrador shelf from Hudson Strait and its biological consequences. Can. J. Fish. Aquat. Sci. 40: 1692-1701.

Thompson, K.R. and M.G. Hazen. 1983. Interseasonal changes of wind stress and Ekman upwelling: North Atlantic, 1950-1980. Can. Tech. Report Fish. Aquat. Sci. 1214.

Thompson, K.R., R.F. Marsden, and D.G. Wright. 1983. Estimation of low-frequency wind-stress fluctuations over the open ocean. J. Phys. Ocean. 13: 1003-1011.

U.S. Naval Office Publications. 1967. Oceanographic Atlas of the North Atlantic Ocean, Section II, Physical Properties.

Vines, R.G. 1984. Rainfall patterns in the eastern United States. Climate Change 6: 79-98.

WATER RETENTION OVER FLEMISH CAP

S. A. Akenhead
Department of Fisheries and Oceans
Fisheries Research Branch
P. O. Box 5667
St. John's, Newfoundland A1C 5X1

ABSTRACT

Current meters, T-S analysis, and surface drifters indicate fresh water is retained in a slow gyre over Flemish Cap. Fitted annual cycles of salinity in the top 100 m are used in a simple compartment model to provide a salt flux estimate. About 50% of Cap surface waters are replaced each month. North Atlantic Current water does not influence Cap water composition. Computing heat fluxes from the salt flux predicts the correct phase and appropriate amplitude and mean for the solar heating cycle.

INTRODUCTION

Flemish Cap (47°N 45°W) is a steep-sided plateau of about 150 km diameter, separated from the Grand Bank of Newfoundland by the 1 km deep Flemish Pass (Fig. 1). It is bounded on the west and north by branches of the Labrador Current. The south and east sides are Slope Water regions, strongly influenced by the eddy energy of the Gulf Stream.

The water over the Cap has been referred to as "mixed water", meaning that it was formed from both Labrador Current and North Atlantic Current water (e.g. Hayes et al. 1977; Anderson 1984; Grimm et al. 1980). This paper contends that Flemish Cap water with salinity over 35 does not derive from North Atlantic Current water, rather it is derived exclusively from Labrador Current water. The T-S plot of the 45°W transect north of Flemish Cap (Fig. 2) shows salinities entirely under 35, and no indication of shallow source waters with salinity and temperature typical of Gulf Stream influence. The waters of Flemish Cap could be created by warming the Labrador Current.

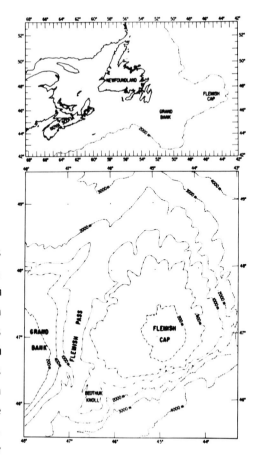

Fig. 1. Study area.

USSR reports consistently identify an anti-cyclonic gyre over Flemish Cap, using dynamic height maps (Kudlo and Burmakin 1972; Kudlo et al. 1983). Using current meters, Ross (1981) confirmed a slow gyre from January to April 1979, but no net motion from April to July 1979. The moorings were on the 200 m isobaths crossing 45°W, and showed mean currents of 3.9 to 5.0 cm $\cdot$ s^{-1} aligned with the isobath. The roughly 260 km perimeter of the 200 m isobath suggests a gyre rotation time of about 2 months. This is slow compared to the water turnover rates (see below), and to biological time-scales. Ross (1981) speculated that the slow gyre was due to a wind induced four-day period internal wave, apparent in the current records.

Six drogued satellite-tracked buoys left on the Cap (Ross 1980) moved slowly with lengthy stalls. Evidence for an anticyclonic gyre within the 400 m shelf-break was weak during summer. The buoys exited mainly from the steep south-east region, where strong eddy activity removed them quickly. Average residence time on the Cap for the 6 buoys was 55 days.

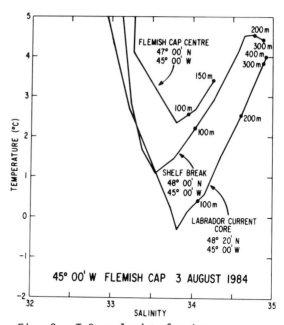

Fig. 2. T-S analysis of water masses on the North of Flemish Cap.

If Flemish Cap water is derived solely from the Labrador Current, then annual cycles of temperature and salinity on Flemish Cap are the result of similar cycles in the Labrador Current. Cap salinity must be entirely predictable whereas Cap temperatures are due to exchange with the Labrador Current but also to heat exchange through the surface. If an exchange coefficient can be derived from salinity, it can be applied to temperature. Residuals from predicted temperature will lead to estimate of surface heat flux. This flux, through comparison to similar studies, allows a test of the model.

MATERIALS AND METHODS

Depth-averaging of historical data gave 1,873 values of mean T and S for the top 100 m ($\overline{T}_{100}$ and $\overline{S}_{100}$) within the 1 km isobath of Flemish Cap. No annual cycle could be seen in the 100-200 m layer. This is verified by the six month T and S records from the 180 m deep current meters.

Stations were averaged by months within years by 5 zones (defined by the 400 m isobath, 47°N and 45°W) to avoid bias from different sampling effort between years. The 576 means resulting were averaged into months across years, weighted by the inverse of the standard error of the mean. For months with a single observation, the mean of the standard deviations of $\overline{T}_{100}$ and $\overline{S}_{100}$ within months was used as the standard error. ANOVA was used to examine the variance partitioned into error, months, and years. Because of bias between instrument types only reversing thermometer data were used for the analysis in this paper (Wyrtki and Uhrich 1982).

We wish to know the salinity cycle over Flemish Cap, driven by the salinity cycle of the adjacent Labrador Current through some constant exchange coefficient, k. The model is

$d/dt \ (S_{cap}) = k \ (S_{lab} - S_{cap})$ or

1) $S_{cap} = e^{-kt} \ (C + \int k \ S_{lab}(t) \ e^{kt} dt)$.

If $S_{lab} = \bar{S}_{lab} + A_{lab} \sin b \ (t + \omega)$, then

2) $S_{cap} = C \ e^{-kt} + \bar{S}_{lab} + A_{lab} k^2/(k^2+b^2) \ [\sin b(t+\omega) - (b/k) \cos b \ (t+\omega)]$

$= \bar{S}_{lab} + A_{lab} \ b(k^2+b^2)^{\frac{1}{2}} \sin b(t+w+\phi)$

The initial salinity, C dies away with time. $\bar{S}_{lab}$ is the mean salinity of both the compartment and the external environment, at steady state. $b/(k^2+b^2)^{\frac{1}{2}}$ indicates the amplitude of the salinity cycle in the shallows of the Cap depends on the extent of exchange and upon the frequency of the exterior cycle. If the Cap is weakly linked to the cycles of the exterior environment, then if the frequency of the outside cycles is too high, if will not be 'felt' by the Cap. A phase shift, ϕ indicates Flemish Cap salinity lags Labrador Current.

Three different analyses were considered for the model. First, the question could be "Are the Flemish Cap-Labrador Current salinities sufficiently described a single sine wave and an exchange coefficient"? This leads to fitting a four parameter model to combined Flemish Cap and Labrador Current salinities. Because lack of fit for Flemish Cap salinities would influence the Labrador Current fit, the results would be

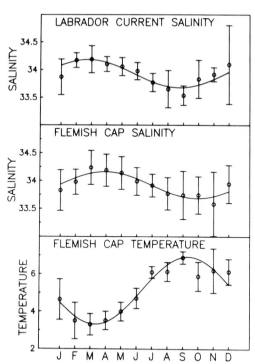

Fig. 3. Monthly means and standard errors with fitted annual cycles.

unclear. Second, the question could be "Is Flemish Cap salinity sufficiently described by linkage only to the Labrador Current"? This involves parameterizing the Labrador Current salinity cycle, and using those parameters to fit one free parameter, k, to Flemish Cap salinity. Only measures of fit are available to reject the model, and any annual cycle that leads the Cap by 0 to $\pi/2$ cycles can explain the Cap cycle. Third, the question could be "Can independently fitted cycles lead to falsifying a presumed linkage"? Since there is, by hypothesis, only 1 free parameter for the Flemish Cap salinity, k, then there are two redundant parameters that can be falsified. The linkage coefficient can be determined from either gain or phase. This third analysis is addressed in what follows.

The linkage of the Flemish Cap salinity cycle to the Labrador Current can be predicted from the lag,

3) $S_{cap} = \bar{S}_{lab} + A_{lab} \cos b \circ \sin b (t + \omega - \phi)$.

RESULTS

For Flemish Cap centre, $\bar{S}_{100}$ was partitioned by ANOVA into year, month and error components. Error accounted for 16% of the total sum of squares. Months (the annual cycle) accounted for 56% of the variance due to model, leaving 44% of model variance due to interannual and year x month interactions. The data were concluded to be sufficient to proceed.

In the flat, shallow centre region of Flemish Cap, less than 400 m deep, the salinity ($\bar{S}_{100}$) varies from 33.74 in September to 34.24 in April, an amplitude of 250 parts per million. The area of this region is 26,130 km^2, and the volume to 100 m deep is then 2,613 km^3.

Monthly means (± one standard error) of $\bar{S}_{100}$ for the NW slopes and $\bar{T}_{100}$ and $\bar{S}_{100}$ for the bank of Flemish Cap are plotted in Fig. 4. Month means and standard errors were used to fit the equation $S = \bar{S}_{100} + A \sin(b(t+\omega))$, with t in days of the year. The coefficients and their standard errors, as determined using the Marquardt algorithm, are in Table 1. The fitted sine waves are compared to the monthly data in Fig. 3.

The utility of an expression for k as a function of ϕ is that merely identifying the phase difference in the two fitted salinity cycles indicates the linkage coefficient. This linkage is essentially the loss rate of Flemish Cap water, resupplied solely from the Labrador Current.

From Table 1, ϕ = 25 -(-13)=38 days, so k=-b tan^{-1} (b38)=-0.0225 d^{-1}.

This exchange is 50% per month, or a flow of k · $V_{CAP}/86400$ = .68 Sv. The

Table 1. Sine wave coefficients for the annual temperature and salinity cycles of five compartments on Flemish Cap, averaged over the surface 100 m.

	Mean (± 1SE)	Amplitude (± 1SE)	Phase (± 1SE)
Temperature			
CR	5.11±.123	1.80±.163	188.±5.73
NW	3.94±.172	1.80±.226	183.31±8.33
SW	4.45±.175	1.86±.23	187.07±7.40
NE	4.72±.313	1.80±.32	175.96±10.84
Salinity			
CR	33.923±.024	0.2375±.035	-13.00±7.86
NW	33.930±.028	0.253±.039	+25.03±9.11
SW	33.875±.019	0.132±.024	+2.60±8.58
NE	34.068±.032	0.172±.0295	+15.15±19.23
SE	34.142±0.19	0.197±.033	+ 7.26±12.95

gain of the Flemish Cap salinity cycle predicted from ϕ is $\cos b\phi = 0.79$. The observed gain (A_{cap}/A_{lab}) from Table 1 is 0.94. Using the square of the standard error of the amplitudes of the salinity cycles for computing the variance of a ratio from the variance of independent parts,

$$\text{Var}(\frac{a}{b}) = (\frac{d(a/b)}{da})^2 \text{Var}(a) + (\frac{d(a/b)}{db})^2 \text{Var}(b),$$

gives an estimate of .145 for the standard error of the observed gain. This shows that the predicted gain is within one standard error of the observed, so there is no basis for rejecting the model based on this discrepancy.

The other test of the model is that the mean salinity of the Labrador Current be the same for Flemish Cap. Since $\bar{S}_{100}$ for the NW slope = 33.93 ± 0.28 and $\bar{S}_{100}$ for the centre region is 33.92 ± 0.24, the means are indistinguishable and the model is apparently sufficient.

Horizontal Heat Flux

The salinity-based model for residence over Flemish Cap predicts the concentrations of conservative tracers. For validation, only temperature observations are available. The horizontal flux of heat will correspond to salt, but there is an important flux of heat through the ocean surface.

The horizontal heat flux was calculated by the formula used to predict salt flux, equation 3, but with the fitted cycle of Labrador Current 100 m mean temperatures. The phase lag, and hence transfer coefficient, of the salinity model was used. The predicted temperature of Flemish Cap is

therefore independent of the observed temperatures, allowing a comparison. From Table 1 and equation 3 the predicted temperature is
T=3.94+.795(1.80) sin (b(t+183.3-38)) = 3.94+1.43 sin (bt+2.50). The predicted temperature on the Cap is compared with the fitted temperature cycles of the Labrador Current and Flemish Cap in Fig. 4, demonstrating that the horizontal flux is not the only component of the observed temperature cycle on Flemish Cap. The Cap is on average warmer than the mean Labrador Current temperature (4°C) by about 1.1°C. The maximal rate of increase in the temperature anomaly occurs near the spring equinox whereas solar heating is maximum at the summer solstice. This is because the retention of solar heat input is no better than that of the salt input.

Calling the surface heat flux L, the heat anomaly from that predicted from the Labrador Current, T', changes as $dT'/dt = L/f - kT'$, so the surface flux is $L/f = dT'/dt + kT'$, where f relates heat content to temperature and is effectively a constant.

The coefficients of the predicted and observed temperature cycles were used to calculate the temperature anomaly cycle (the difference of two sine waves), T' = 1.17 + 1.211 sin (b(t + 241.1)).
The heat flux L, is computed from the derivative of the heat anomaly, and a correction for the horizontal flux of heat corresponding to that of salt, L/f =0.02085 cos (b(t+241.1))+(0.0225) (1.17+1.211 sin (b(t+241.1))=.0263-.0342 sin (b(t+85.9)).

Since f has a value of 4758 Watts $m^{-2} \cdot °C^{-1} \cdot day^{-1}$, that is, a change of 1 degree per day throughout a 100 m water column represents an average input of 4758 W · m^{-2}, the heat flux through the surface is L=125.1+162.8 sin (b(t-96.6)). The temperature anomaly and surface heat flux cycles are shown in Fig. 4. Because the surface heat flux has the same phase as the solar cycle (the 5 day difference is surely within estimation errors) and because the surface fluxes are reasonable values (-38 W · m^{-2} at winter solstice, + 288 at summer solstice), the original model is apparently validated. The temperature and salinity cycles on Flemish Cap show connections only to the Labrador Current.

DISCUSSION

The analysis of salinity and temperature annual cycles at North Atlantic ocean weather stations by Taylor and Stephens (1980) showed a sine wave described of the sea surface salinity at OWS B. The amplitude of the salinity cycle fitted to OWS B surface data was 0.20, less than Flemish Cap

(0.52 for 0-20 m). This lower amplitude is expected since the mean salinity at OWS B is 34.5, higher by 1.04 then the mean surface salinity on Flemish Cap and indicative of the much reduced freshwater influences in the centre of the Labrador Sea compared to the western boundary coastal current.

Smith and Dobson (1984) detailed the heat budget at OWS B(56°N 51°W), and found a net flux from the ocean of 28 W·m^{-2}, and an annual cycle amplitude of roughly 170 W·m^{-2} (their Fig. 3). OWS B is in the centre of the Labrador Sea gyre, and if there is any advective heat flux, it is warming due to the Irminger Current component of West Greenland Current. Flemish Cap is 1000 km south, and very greatly of the icy Labrador Current. Smith and Dobson (1984) produce revisions of Bunker (1976) that seem to preclude further comparisons to Bunker's result at this time.

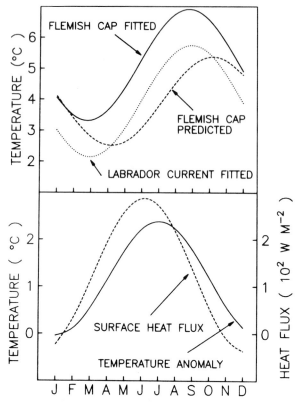

Fig. 4. Components of surface heat flux derivation.

Meehl (1985) presented the heat storage rates for OWS C (52°N 35°W) and for the zonal means of the North Atlantic. Both support the Flemish Cap observation that storage is maximal at the summer solstice, and that the surface heat flux amplitude is in the range 150-200 W·m^{-2}. Zonal values of heat storage in Land and Bunker (1982) have an amplitude similar to this Flemish Cap calculation (about 130 W·m^{-2}).

The North Sea (56°N) heat budget was examined by Becker (1981), who determined the that amplitude of annual heat content cycles generally ranged about 75·10 J m^{-2}. The mean heat budget indicated a release of 5 W m^{-2} from the North Sea to the atmosphere. Because the heat budget is nearly balanced by considering only surface fluxes, horizontal advection of heat can be ignored and the surface heat flux, L, will be the derivative of the heat content $dH/dt = bA\cos(b(t+w))$. A is given as 0.5 to $1.0 \cdot 10^9 Jm^2$, so the amplitude of the surface flux is 100-200 W·m^{-2}. This corresponds to the amplitude of the surface flux the surface heat flux amplitude observed on Flemish Cap. The phase of North Sea heat content is given by Becker (1981) as 250 days, and the phase of the temperature anomaly on Flemish Cap is 244 days; a satisfactory agreement.

Loder et al. (1982) examined water retention over Georges Bank (41°N). The analysis was complicated by having to deal with a mixed layer that extended to a sea bottom which varied from 10 to 43 m. Their radial diffusion coefficient (about 250 $m^2 \cdot s^{-1}$) is not directly comparable to a compartment linkage coefficient. The time for 50% loss a patch at the centre of Georges Bank through a perimeter 45 km away is given as 20 days. Their model was fitted by the inverse of this paper's approach, from empirical solar heating to a predicted salt pattern. The solar heating sinusoid used, had the phase (90 days) and amplitude (200 W·m^{-2}) expected from Flemish Cap results.

A salinity depression of 0.1% was predicted from precipitation over Georges Bank by Loder et al. (1982). Precipitation effects are expected to be of the same order or smaller on Flemish Cap because of the deeper mixed layer. Rainfall at St. John's has an annual cycle with a phase of about 270 days, and cannot be argued to account for either the fitted annual cycle or to the slight discrepancy in observed amplitude compared to expected.

Having confirmed that the model is correct, and that Flemish Cap water is derived solely from Labrador Current water, a linkage coefficient was fitted to Flemish Cap salinity, given the description of the Labrador Current salinity cycle (the second analysis as discussed in Methods). The estimate for k was .025, which compared well with the .0225 estimated from phase alone. Fitting 3 parameters of a sine wave to Flemish cap salinity left a sum of squared residuals of 0.197. Using just a single parameter, k, left only 0.214. This is further confirmation of the model, since part of the 8% discrepancy is due to error in the fitted parameter of the Labrador Current.

A tentative extrapolation from water turnover rates to source and sink terms in budgets of pelagic biology can provide valuable contraints for biologists. From Table 2 of Anderson (1984), the mortality rate during spring of redfish larvae over Flemish Cap is .047 ± .0047 d^{-1}. Under the assumption that ichthyoplankton are passive drifters and will be lost from the Cap in the same way water is, half this mortality can be ascribed to transport losses. Since the water exchange of .025 d^{-1} is computed for the top 100 m, if ichthyoplankton are mainly associated with the mixed layer and thermocline, the transport loss rate may be higher and the biological sources of mortality correspondingly reduced.

REFERENCES

Anderson, J. T. 1984. Early life history of Redfish (Sebastes spp.) on Flemish Cap. Can. J. Fish. Aquat. Sci. 41(7): 1106-1116.

Becker, G. A. 1981. Beiträge zur Hydrographic and Wärmebilanz der Nordsee. Deutsche Hydrographische Zeitchrift, 34, H. 5, pp. 167-262.

Bunker, A. F. 1976. Computations of surface energy flux and annual air-sea interaction cycles of the North Atlantic Ocean. Mon. Weather Rev. 104: 1122-1140.

Hayes, R. M., D. G. Mountain, and T. C. Wolford. 1977. Physical oceanography and the abiotic influences on cod recruitment in the Flemish Cap region. ICNAF Res. Doc. 77/54, Ser. No. 5107.

Grimm, S., A. Furtak, J. Wysaki, M. Baranowski. MS 1980. Distribution and abundance of redfish larvae against thermal conditions on Flemish Cap in April 1978. NAFO SCR Doc. 80/VI/62, Ser. No. N101.

Kudlo, B. P., and V. V. Burmakin. 1972. Water circulation in the South Labrador and Newfoundland areas in 1970-71. ICNAF Redbook 1973, Part III, p. 27-33.

Kudlo, B. P., V.A. Borokov, and N. G. Sapronetskaya. 1983. Results of the Soviet oceanographic investigations in accordance with the Flemish Cap Project in 1977-1982. NAFO SCR Doc. 83/VI/41, Ser. No. N697.

Meehl, G. A. 1985. A calculation of ocean heat storage and effective ocean surface layer depths for the Northern Hemisphere. J. Phys. Oceanogr. 14(11): 1747-1761.

Land, P. J., and A. F. Bunker. 1982. The annual March of the heat budget of the north and tropical Atlantic Oceans. J. Phys. Oceanogr. 12: 1388-1410.

Loder, J. W., D. G. Wright, C. Garrett, and B. Juzko. 1982. Horizontal exchange in central Georges Bank. Can. J. Fish. Aquat. Sci. 39: 1130-1137.

Ross, C. K. 1981. Drift of Satellite-tracked buoys on Flemish Cap, 1979-80. NAFO Scientific Council Studies 1: 47-50.

Ross, C. K. 1980. Moored current meter data from Flemish Cap January-July 1979. NAFC SCR Doc. 80/IX/128, Ser. No. N200.

Smith, S. S., and F. W. Dobson. 1984. The heat budget at Ocean Weather Station Bravo. Atmosphere-Ocean 22(1): 1-22.

Taylor, A. H., and J. A. Stephens. 1980. Seasonal and year to year variations of surface salinity at the nine North Atlantic ocean weather stations. Oceanologica Acta 3(4): 420-430.

Wyrtki, K., and L. Uhrich. 1982. On the accuracy of heat storage computations. J. Phys. Oceanogr. 12: 1411-1416.

THE SCOTTISH COASTAL CURRENT

J. H. Simpson and A. E. Hill

Department of Physical Oceanography
Marine Science Laboratories
Menai Bridge
Gwynedd   LL57 5EY
U.K.

ABSTRACT

Low salinity water flows northward along the Scottish coast forming a persistent coastal current whose mean speed (~ 5 kmd^{-1}) contrasts with the generally low net flows in the U.K. shelf seas. The short period and seasonal variability of the current is assessed on the basis of recent current meter measurements and the monthly mean flow is compared with variations in freshwater runoff from the principal coastal sources.

A striking feature of the current is its interaction with the island chain of the Hebrides. Tracer distributions suggest than on entering the Minch (the channel between Scotland and the islands), part of the flow turns westward, crosses the Minch and then flows southward along the western side of the channel before proceeding northwards again up the west coast of the Hebrides. This bifurcation and meandering of the current, which is confirmed by current meter observations, is discussed in relation to a quasi-geostrophic model of topographic influence on the flow.

INTRODUCTION

Long period mean flows in the shelf seas around the U.K. are generally weak with typical velocities. 1kmd^{-1} or less when averages are taken over periods of 30 days. Booth (1979) has shown this directly by averaging high quality current meter data from the northern Irish Sea but the general weakness of flow may be deduced from the application of continuity arguments to distributions of tracers such as salinity and Caesium. For example Wilson (1974) has used the distributions of both these tracers to determine that the mean flow through the Irish Sea on the St. Georges Channel section is only 300-450 md^{-1} in reasonable agreement with earlier estimates by Bowden (1950) based on salinity along a section from

It has also been argued that the consistency of the positions of the shelf sea fronts testifies to the weakness of residual flows. An argument (Simpson 1981) based on depth uniform advection indicates that fronts would be significantly displaced from their positions predicted by the tidal mixing model if the mean flow normal to the front is 1 kmd^{-1}.

A striking exception to this pattern of generally sluggish mean currents, is the flow of low salinity water northwards along the west coast of Scotland. The outflow of brackish water from the Irish Sea supplemented by freshwater discharge from sources on the Scottish coast, notably the Clyde and the Firth of Lorne maintain this substantial salinity anomaly. The associated buoyancy forces must play a part in driving the current which under the influence of the Coriolis forces turns northward along the Scottish coast to form a classical coastal current (Griffiths and Linden, 1982). This flow is readily identified in published tracer distributions (fig 1) and see McKinley et al (1980) and there are references to northward flow in this area in early fisheries studies (Craig 1959), but in spite of its potential importance the current has not been the subject of intensive study by modern methods until a joint Menai Bridge - SMBA programme commenced in 1982. In this paper we shall make an initial report of results from this programme and suggest a dynamical explanation from the pronounced meandering and bifurcation of the current where it interacts with the Hebridean island chain.

STRATEGY AND LOCATION

The observational programme consisted of a series of five

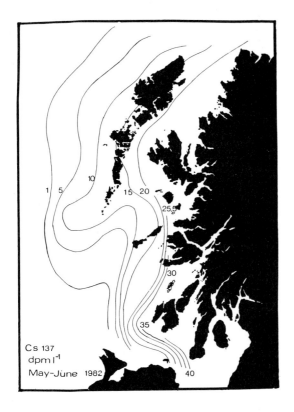

fig 1 Distribution of the radioiosotope Cs-137 (d min^{-1}l^{-1}) West of Scotland in 1982.

extensive surveys of the Scottish shelf to determine the extent and intensity of the current at different seasons of the year. Measurements were made of temperature, salinity and nutrient distributions and in the May '82 survey samples were also taken for the determination of Cs concentrations. During the surveys which were typically of two weeks duration, short term current meter deployments were made to investigate particular aspects of the flow. The station grid mooring positions are shown in fig 2.

To investigate the variability in time of the current and provide data spanning the intervals between cruises, a second important aspect of the programme was to deploy long term moorings at the positions M2 and HS1 for a continuous 12 month period. This was achieved with only one small lacuna at HS1 but a severe storm in January '83 caused the loss of mooring M2 with a consequent break of five months in the data record. It is hoped to complete the annual cycle at M2 with further measurements this year, but in any case valuable supplementation of our own data has been provided by data from an SMBA mooring in the Tiree Passage.

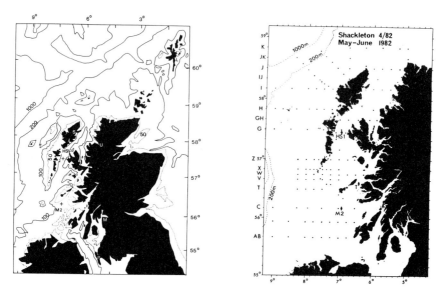

fig 2 Bottom topography of the continental shelf west of Scotland. Barra Head is denoted by 'B' and 'S' indicates the island of Skye both of which are referred to in the text. Positions of the long term current meter moorings M2, HS1 and the Tiree Passage mooring (T) which were maintained in 1983-84. Also shown are the positions of CTD stations from the May 1982 survey.

The Scottish shelf exhibits a complex topography largely on account of the numerous islands of the Inner and Outer Hebrides (fig 2). The latter form a more or less continuous island chain which together with the large island of Skye may be expected to affect the course of the current.

The tidal streams on this part of the shelf are generally rather weak so that strong seasonal stratification develops in most areas except for a mixed region inshore of the Islay front in the south and a number of localised mixing zones around islands and headlands such as that associated with Barra Head (Simpson and Tett 1985).

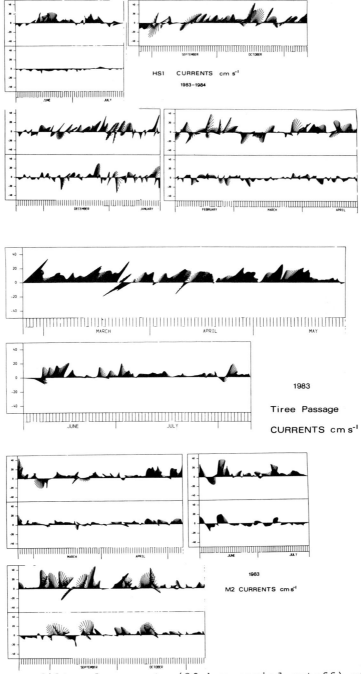

fig 3 Low pass filtered currents (30 hrs period cutoff) at (a) M2 (b) Tiree Passage (c) HS1. The upper panels in (a) and (b) are near surface currents and the lower panels show the near bottom surface.

CURRENT METER RESULTS

Data from the long period current meter deployments has been filtered (cut-off 30hrs) to remove tidal and higher frequency motions. Samples of the resulting mean current components for the M2 position (depth = 60m) are shown in figure 3(a). For the meter in the upper part of the water column (20m below the surface) the flow is predominantly northward with speeds of typically 5-10 cms^{-1}. There is, however, considerable variability with peaks of up to 40 cms^{-1} lasting several days and a few episodes where the flow reverses.

For the lower current meter (10m above the seabed) the currents again exhibit a mean northward flow with a similar pattern of variability but speeds are generally lower. Reverse flow is again apparent, particularly in June and July, when the mean current was southward for much of the time.

The net transport, however, is generally northward and this is emphasised by the data from the Tiree Passage mooring (fig 3b). At this station which is in relatively shallow mixed water (h 50m) the mid-depth flow is strongly northward throughout the period March-July except for a few short periods of reversal. The flow is noticeably weaker in the summer months being at a level of only a few cms during much of July.

fig 4 Summary of mean current vectors on continental shelf west of Scotland. Northward flow is confined to the coastal zone. Open arrow heads indicate near bottom currents and full arrow heads denote near surface currents.

At the HS1 mooring in the central Minch (h = 150 m) near surface flow (20 m below surface) is predominantly northward and more

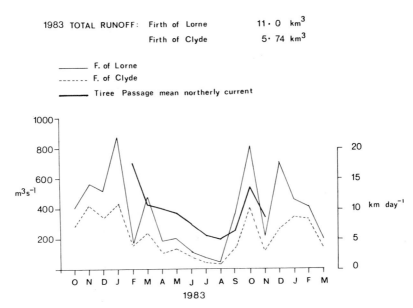

fig 5 A comparison between the monthly mean current speed in the Tiree Passage and the monthly mean fresh water runoff from the two principal fresh water sources, the Firth of Lorne and the Firth of Clyde.

strongly so in the autumn. The currents 20m above the seabed, however, show frequent reversals with a weak but sustained southward flow in June. Currents in both surface and bottom layers show evidence of quasi-periodic variations. It is of interest to note that at certain times these fluctuations at the M2 and HS1 stations are partially coherent.

We have averaged the data from these recent deployments together with all other available current meter data for the Scottish shelf. The result (fig 4) confirms the existence of a mean northward flow of 5 cms in the coastal region. Currents at the few positions outside the salinity deficit region (high Cs in fig 1), are generally to the east and are presumably supplying the demand for offshore water which is entrained into the current.

The seasonal cycle in the mean flow is emphasised by fig 5 which illustrates the monthly mean flow in the Tiree Passage plotted alongside the runoff data from the Firth of Lorne and the Clyde. There is reduction in the residual flow in late

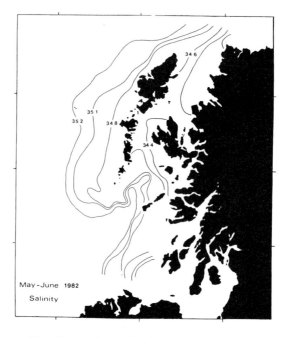

fig 6 Surface (5m depth) salinity in May 1982. Low salinity water is confined to a broad coastal band. Low salinity water is present west of the outer Hebridean island chain and there is a pronounced meander in the isohalines in the entrance to the Minch.

summer to about one third of the highest winter values with a rapid recovery in September and October following the increase in run-off. This picture is strongly suggestive of the idea that the current is largely buoyancy driven but it should not be forgotten that the wind field shows a similar seasonal pattern.

MEANDERING OF THE CURRENT

A most striking feature of the current is its behaviour as it enters the channel (the 'Minch') between the Outer Hebrides and the Scottish mainland. Salinity distributions (fig 6) as well as the Caesium data (fig 1) suggest that the current meanders and bifurcates in the Minch, part of the flow turning westwards and then flowing southwards along the western margin of the channel before passing around Barra Head and proceeding northwards again up the west coast of the island chain.

The results of all our surveys and many of the published salinity distributions indicate behaviour of this kind but the meander system is clearly developed in the April and May surveys when a section across the entrance to the Minch (fig 7) shows welldefined wedges of low density water on either side of the channel separated by almost depth uniform fully saline water occupying the central channel. Such a density field implies a southward flow on the western side of the channel and we have undertaken a series of current meter deployments to detect this flow. Meters deployed at stations SU1 and BA3 showed persistent southward flow in two deployments in July and November both of about 10 days duration.

A BAROTROPIC MODEL

We were tempted to seek an understanding of this phenomenon in terms of the response of a density current to the presence of a gap (the Minch) in its supporting land barrier. Tank experiments undertaken in collaboration with Dr P F Linden suggested that gap crossing and meander development would occur for the case where the gap width is of the same order as the internal Rossby radius. The experiments were, however, undertaken for currents flowing along vertical walls in deep water which does not correspond well to the real current in which there is little vertical structure and gradients are predominantly horizontal on account of the shallow water and strong mixing near the coast (see fig 8).

It seems that in these circumstances the influence of the bottom topography will be predominant and we have therefore looked at a simple model of the flow based on the quasi-geostrophic dynamics of a homogeneous fluid following the approach of Laegerloef (1983) who uses the formulation of Pedlosky (1979, p. 91).

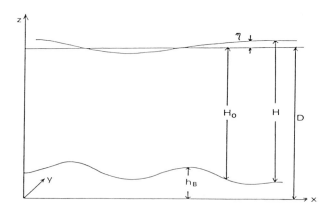

fig 7   Definition figure for the quasi-geostrophic model.

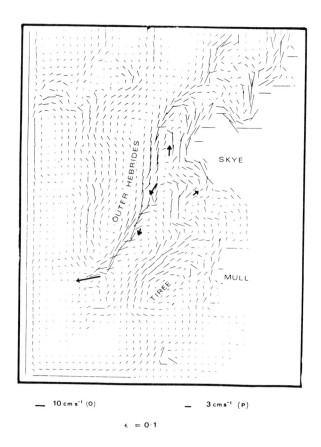

fig 8   Current vectors predicted from the model. Superimposed are mean currents determined from a 10 day current meter-deployment in July 1983.

Using a systematic expansion of the equations of motion in powers of the Rossby number it can be shown that for inviscid, steady state flow at small , the conservation of potential vorticity may be written in terms of non-dimensional variables as

$$\left(\frac{\partial \psi}{\partial x} \frac{\partial}{\partial y} - \frac{\partial \psi}{\partial y} \frac{\partial}{\partial x} \right) \left(\nabla^2 \psi - F\psi + \eta_B \right) = 0$$

where ψ is the stream function which is equivalent, in this case, to the (non-dimensional) sea level. The parameter η_B is defined in terms of the height h_B of the bottom above a reference depth D (see fig 8) as

$$\eta_B = \frac{h_B}{\epsilon D}$$

The quantity $F = (L/R)^2$ represents the ratio of the geometrical length scale L to the Rossby radius $R = \sqrt{gD}/f$

For pronounced topography the η_B term dominates so that equation 1 has the limiting form

$$J(\psi, \eta_B) \equiv \left(\frac{\partial \psi}{\partial x} \frac{\partial \eta_B}{\partial y} - \frac{\partial \psi}{\partial y} \frac{\partial \eta_B}{\partial x} \right) = 0$$

in which geostrophic flow is exactly parallel to the bottom contours.

Generally equation 1 implies that

$$\nabla^2 \psi - F\psi + \eta_B = G(\psi)$$

which can be written

$$\nabla^2 \psi + \eta_B = G(\psi) + F(\psi) \equiv K(\psi)$$

Once the quantity $K(\psi)$ is specified for each streamline the solution for may be obtained numerically.

We have applied this approach to the flow of the coastal current as it enters the Minch using a 5km grid and the simplified coastline. Values of $K(\psi)$ on streamlines entering the southern boundary are fixed by specifying a uniform current of width 30km. The remaining boundaries are streamlines (ψ = constant) except for the northern limit where we have set $\frac{\partial \psi}{\partial y} = 0$ which implies a northward flow.

The numerical solution was obtained using a Gauss-Seidel iteration with successive over-relaxation.

Results for $\epsilon = 0.1$ are shown in fig 9. The flow is seen to meander and bifurcate in the Minch with a strong southerly flow on the western side of the channel in accord with the observations. There is rather close agreement between the direction of the current vectors (fig. 9b) predicted by this model and the observed residual currents. In particular the model correctly predicts the location of the split in the current between north and south flowing branches at about 57° 20'N. The current speeds predicted are of the right order but are generally in less favourable agreement with the observations.

DISCUSSION

The principal conclusion to be drawn at this stage is that the Scottish coastal current while probably largely driven by buoyancy input is strongly steered by the bottom topography. In this respect it contrasts with the Norwegian coastal current (Mork 1981) which conforms more closely to the classical picture of a density current flowing parallel to the coast in deep water uninfluenced by bottom configuration. The differences arise from the much greater depth of water along the Norwegian coast. The relatively shallow water on Scottish coast is subject to rather strong tidal stirring which prevents the maintenance of vertical structure and ensures that the gradients are mainly horizontal. Control by the topography would also account for the relatively stable and consistent path of the current in contrast to the Norwegian coastal current which is subject to severe time dependant meandering and instabilities.

An interesting question arising from the kinematics of the current inferred here concerns the proportion of the current passing to the west of the hebrides. McKay et al (1985) have recently estimated on the basis of the Cs_{137} distribution that only 20% of the flow passes around Barra Head.

ACKNOWLEDGEMENTS

We would like to thank Dr. M. S. Baxter of the Department of Chemistry, Glasgow University, for the analysis of the Caesium data. The Tiree Passage current meter data was kingly provided by Mr David Ellett of S.M.B.A. We are also grateful to Mr. P. Taylor of Research Vessel Services for his assistance with current meter deployments. A.E.H. acknowledges the support of an N.E.R.C. research studentship. This work was funded by a grant from N.E.R.C.

REFERENCES

BOOTH, D.A. (1970) The evaluation of a Lagrangian current system and its application to water movement in the Irish Sea. Ph.D thesis. University of Wales.

BOWDEN, K.F. (1950) Processes affecting the salinity of the Irish Sea. Mon. Not. Roy. Astron. Soc. Geophys. Suppl. $\underline{6}$ 63-89.

CRAIG, R.E. (1959) Hydrography of Scottish coastal waters. Marine Research, Scottish Home Department 1958 (2) 30pp.

GRIFFITHS, R.W. and LINDEN, P.F. (1982)

Laboratory experiments on fronts. Part 1 : density driven boundary currents. Geophys, and Astrophys. Fluid Dynamics $\underline{19}$, 159-187.

LAGERLOEF, G. (1983) Topographically controlled flow around a deep trough transecting the shelf of Kodiak Island, Alaska. J Physical Oceanography $\underline{13}$ 139-146.

McKINLEY, I.G., BAXTER, M.S., ELLETT, D.J. and JACK, W (1978) Tracer Applications of Radiocaesium in the Sea of the Hebrides.
Estuarine, Coastal and Shelf Science $\underline{13}$ 69-82.

McKAY, W.A., BAXTER, M.S., ELLETT,D.J., and MELDRUM, D.T. (1985)

Radiocaesium and circulation patterns west of Scotland (in press).

MORK, M. (1981) Circulation phenomena and frontal dynamics in the Norwegian coastal current. Phil. Trans. Roy. Soc. Lond. $\underline{A}$ 302, 635-647.

PEDLOSKY, J. (1979) Geophysical Fluid Dynamics. Springer-Verlag.

SIMPSON, J.H. (1981) Shelf sea fronts : implications for their existence and behaviour. Phil. Trans. Roy. Soc. Lond. $\underline{A}$ $\underline{302}$ 531-546.

SIMPSON, J.H. and P.B. TETT (1985) To appear as a chapter of "Tidal Mixing and Plankton Dynamics" eds. Bowman, Yentsch, Petersen.

WILSON, T.R.S. (1974) Caesium-137 as a water movement tracer in St. George's Channel. Nature 248, 125-127.

RUNOFF DRIVEN COASTAL FLOW OFF BRITISH COLUMBIA

P.H. LeBlond[1], B.M. Hickey[2], R.E. Thomson[3].

1. Dept. of Oceanography, Univ. of British Columbia, Vancouver B.C. Canada
2. School of Oceanography, Univ. of Washington, Seattle, WA, USA
3. Institute of Ocean Sciences, Sidney, B.C., Canada

ABSTRACT

First results from a recent measurement program on the west coast of Vancouver Island, Canada, have demonstrated the existence of a narrow northward flowing current within 20 km from the shore. This current is associated with the strong freshwater runoff flowing into the sea along that coastline.

INTRODUCTION

The Pacific coast of Canada is an area of strong precipitation and hence large fresh water runoff (Fig. 1). As on the continguous Alaskan coast (Royer, 1981), runoff is expected to have a significant influence on coastal currents and water properties. We review here, briefly, local runoff characteristics and historical evidence for a northward runoff-induced near-shore flow off Vancouver Island and present preliminary results from a recent observational program in that area.

Two sources of runoff affect the area of interest: 1- Direct discharge from coastal streams along the Vancouver Island and Washington state coasts, with freshet in the fall-winter rainy season; and 2- stored discharge, mainly from the Fraser River which peaks in late spring-early summer (Fig. 2). The Columbia River is also an important contributor to runoff further south.

The two types of runoff sources differ not only in their timing, but in their spatial distribution: direct runoff along the coast is a distributed line source, whereas the summer-time stored runoff arrives at the coast as a point source, mainly via Juan de Fuca Strait. Finally, whereas direct runoff reaches the shelf relatively undiluted, the stored runoff is mixed with about ten times its volume of salt water in the Strait of Georgia, Puget

Sound and Juan de Fuca Strait. In view of these differences, one would expect considerable seasonal variations in any near shore flow associated with runoff. A recent review of historical data by Freeland et al. (1984) has provided evidence of that current and of its variability. The Vancouver Island Coastal Current, as it is now called, is found within 20 km of the shore and appears strongest in late fall and early summer. The evidence for the current has nevertheless been gathered rather coincidentally and remains rather fragmentary. A recent intensive measurement program was designed to identify that current and relate its characteristics to coastal runoff. Preliminary results from that field program are discussed here.

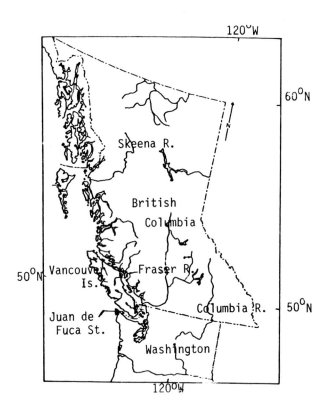

Figure 1. The Pacific Coast of Canada and of the northwestern USA, showing the principal rivers.

DATA COLLECTION

A combination of moored current meters, tide gauges and meteorological stations were deployed from June to October 1984 in a pattern focussing on the mouth of Juan de Fuca Strait and the coast of Vancouver Island (Fig. 3). Three one-week hydrographic cruises (in June, July and October 1984) provided dense sampling of temperature, salinity and dissolved oxygen. Winds were monitored and runoff measured at a number of locations. In addition, a number of satellite-tracked drifters were released along the west coast of Vancouver Island in a simultaneous program undertaken jointly by the Canadian Coast Guard and the Institute of Ocean Sciences in a effort to improve the efficiency of their air-sea rescue operations. The entire field program shall be referred to as the Coastal Current Experiment (CCE).

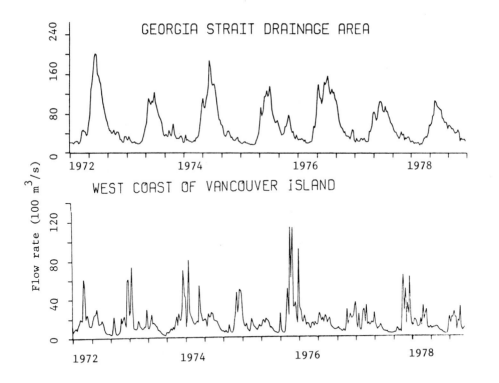

Figure 2. Time series of weekly averaged runoff into the Strait of Georgia (mainly the Fraser River) and along the west coast of Vancouver Island.

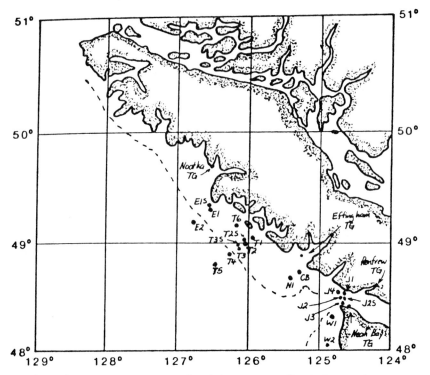

Figure 3. Current meter and tide gauge (TG) locations for the Coastal Current Experiment. The dotted line shows the positions of the 100-meter contour.

Preliminary results from CCE have clearly shown the presence of a narrow band of current along the coast of Vancouver Island during the summer of 1984. This northwestward flow is confined to the shallow shelf region and is unambiguously identifiable in direct current measurements, drifter tracks and dynamic topography. Runoff data for the observation period not being available at the time of writing, our comments will be restricted to an initial description of the data sets mentioned above.

RESULTS

Velocity components at two locations on the Vancouver Island shelf and at a mooring at the mouth of the Strait of Juan de Fuca are shown in Fig. 4. Tidal fluctuations clearly dominate the record at all three locations. The diurnally varying currents which dominate the variance on the shelf are probably diurnal

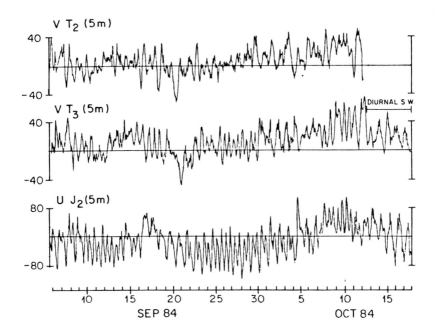

Figure 4. Near surface (5m) current time series during Sept-Oct. 1984 over the 50m (T2) and 80m (T3) isobaths along the T-line and mid-channel at the exit of the Strait of Juan de Fuca (J2). The north-south component of velocity V (positive northward) is shown over the shelf; the east-west component U (positive up-strait, i.e. eastward) is shown at mooring J2. Velocity scales are in cm/sec.

shelf waves, as described by Crawford and Thomson (1984). Currents at the mouth of the Strait of Juan de Fuca are more nearly semi-diurnal. Mean flow on the shelf is mainly northward, with varying intensity, but with a typical strength of 20 cm/sec. At mooring J2, the flow is mainly out of the strait, but with occasional intrusive reversals; there appears to be a connection between such reversals and a reduction of the northward current on the shelf, but this relation could also be attributed to southward wind events, as was already noted by Cannon (1972) and Frisch et al. (1981).

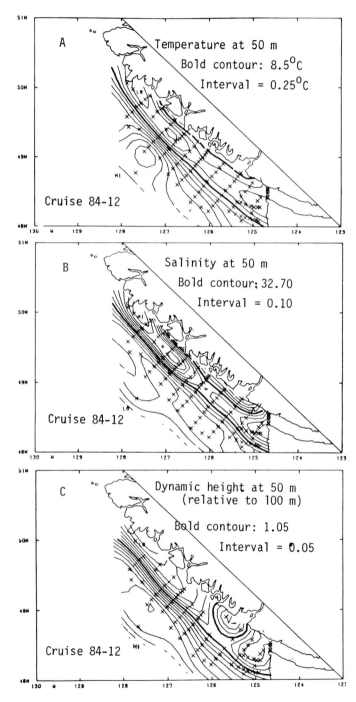

Figure 5. Distribution of temperature (A), salinity (B) and dynamic height (C) at a depth of 50m during cruise 84-12 (June 18-24, 1984). Crosses indicate sampling locations.

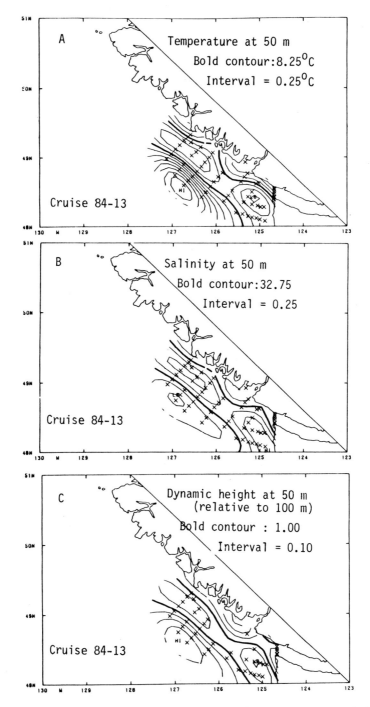

Figure 6. Distribution of temperature (A), salinity (B) and dynamic height (C) at a depth of 50m during cruise 84-13 (July 23-29, 1984). Crosses indicate sampling locations.

Distributions of temperature, salinity and dynamic height off southern Vancouver Island are shown in Fig. 5 and 6 for data obtained during the hydrographic cruises, on June 18-24 and July 23-29 respectively. Property maps at a depth of 50m illustrate best the distinction between near-shore and shelf-edge regimes. As seen in Fig. 5a, b, a ridge of cold, high salinity water parallels the shore.

A maximum in dynamic height roughly follows the crest of that ridge, separating a northwestward nearshore geostrophic tendency from the southeastward regime which prevails over the shelf break. This situation is found again later in the season (Fig. 6). Severe weather conditions during the mooring recovery cruise in late October made it impossible to sample widely and densely enough to establish the presence (or absence) of similar conditions in the fall.

Satellite-tracked drifters released along a line normal to the coast at various times during the summer also confirmed the presence of the inshore northwestward current. Drifters released within 20 km of the shore follow that current, moving northward even against an adverse wind, while those launched offshore (>20 km) travelled southwards.

CONCLUSIONS

Preliminary examination of CCE data has confirmed the presence of the Vancouver Island Coastal Current, substantiating local folklore and earlier measurements (Thomson, 1981, p. 233). Further analysis will concentrate on the dynamics of the current and its relation to winds and freshwater runoff variations. In particular, the relative importances of the line source of runoff along Vancouver Island, and of the point source at the mouth of Juan de Fuca Strait, will be investigated. Another relevant question is that of the importance of currents south of Juan de Fuca Strait on the flow along the coast of Vancouver Island.

ACKNOWLEDGEMENTS

In addition to the logistic support provided by the Institute of Ocean Sciences, funding for this project was provided by grants from the US Office of Naval Research to B.M. Hickey and from the Canadian Natural Sciences and Engineering Research Council to P.H. LeBlond.

REFERENCES

Cannon, G.A., 1972. Wind effects on currents observed in Juan de Fuca submarine canyon. J. Phys. Oceanogr. 2, 281-285.

Crawford, R.W. and R.E. Thomson, 1984 Diurnal-period shelf waves along Vancouver Island: a comparison of observations with theoretical models. J. Phys. Oceanogr., 14, 1629-1646.

Freeland, H.J., W.R. Crawford and R.E. Thomson, 1984. Currents along the Pacific coast of Canada. Atmosphere-Ocean 22. 151-172.

Frisch, A.S., J. Holbrook and A.B. Ages, 1981. Observations of a summertime reversal in the circulation in the Strait of Juan de Fuca. J. Geophys. Res. 86, 2044-2048.

Royer, T.C., 1981. Baroclinic transport in the Gulf of Alaska: fresh water driven coastal current. J. Marine Research, 39, 251-266.

Thomson, R.E., 1981. Oceanography of the British Columbia Coast. Canadian Special Publication of Fisheries and Aquatic Sciences No. 56, Ottawa, Ont., 291 pp.

THE SEASONAL INFLUENCE OF CHESAPEAKE BAY PHYTOPLANKTON TO THE CONTINENTAL SHELF

Harold G. Marshall
Department of Biological Sciences
Old Dominion University
Norfolk, Virginia   23508   U.S.A.

ABSTRACT

High concentrations of phytoplankton characterize the Chesapeake Bay entrance area and the Bay plume during spring and fall. A tendency for vertical homogeneity and low species diversity for the phytoplankton were also associated with the plume. As the plume moves along the shelf it is continually seeding these waters with an estuarine and coastal mixture of species, many of which appear more common over the shelf, and possibly indicative of increased shelf eutrophication.

INTRODUCTION

Phytoplankton composition in the lower Chesapeake Bay has been characterized by Wolfe et al. (1926), Patten et al. (1963), McCarthy et al. (1974), Marshall (1980), and Marshall and Lacouture (In Press), among others. Phytoplankton from shelf waters in this vicinity have been described by Cowles (1930), Mulford and Norcross (1971), Marshall (1982, 1984a, 1984b), and Marshall and Cohn (1985), with the productivity of the area described by McCarthy et al. (1974) and O'Reilly and Busch (1984). These studies indicate a productive and diverse composition of both nanoplankton (<20 μm) and net plankton (>20 μm) components for this region. Seasonal maxima are common during late winter-early spring and in fall. Considerable fluctuations in density occur in the lower Bay, often producing periods containing multiple pulses during the year. Marshall (1984a) has noted similar dominant species common to the lower Bay and coastal waters to those in the high growth belt along the outer shelf margin. The Chesapeake Bay plume is easily distinguished and tends to flow southward as a narrow band along the coast. It will also show seasonal differences in its extent and movement over the shelf associated with changes in wind stress and the amount of water discharged from the Bay (Boicourt, 1973; 1981). Although the mid and outer shelf waters of this region also have a southward drift, the nearshore surface waters are subject to summer changes in wind forcing events. These may produce a slight northward flow that has been noted to influence zooplankton concentrations in this area (Johnson et al., 1984). This same wind stress condition will contribute to the high summer phytoplankton concentrations associated with this region (Marshall and Lacouture, In Press). Once cells enter the Bay they are influenced by a two layered net flow system and local tidal patterns. Tyler and Seliger (1978) have described subsurface phytoplankton transport of a bloom species to the upper Bay (240 km) during late winter and early spring. Surface flow to the Bay entrance would later return these populations in a seasonally cyclic pattern. In response to all of these activities, the Bay entrance is a dynamic region that produces phytoplankton abundance and high productivity. It is also the exchange region where estuarine and shelf phytoplankton assemblages are in continual interaction. The purpose of this paper is to discuss the seasonal influence of the Chesapeake Bay phytoplankton to the distribution and composition of phytoplankton of the near shelf.

METHODS

Water samples utilized in this study came from two major data sets from the vicinity of the Chesapeake Bay entrance. One represents seasonal surface collections from nine cruises between Cape Cod and Cape Hatteras aboard vessels in the NOAA NARMAP and Ocean Pulse Programs (between 1979-1981). The other collections came from monthly samples taken between 1980-1984 in the lower Chesapeake Bay and in near shore and outer shelf waters beyond the Bay entrance. Cell concentrations and biomass levels between November 1983 and September 1984 for near shore stations south of the Bay entrance are given in Figures 1 and 2. Water samples (500 ml) were preserved with buffered formalin, with occasional replicate samples taken for comparisons and preserved with Lugol's solution. A modified Utermohl method was followed in the microscopic analysis (Marshall, 1982) with cell volume determination used as a basis for estimating cell biomass. Clustering analysis was utilized in distinguishing plume and shelf populations, with species diversity determined using the Shannon-Weaver diversity index.

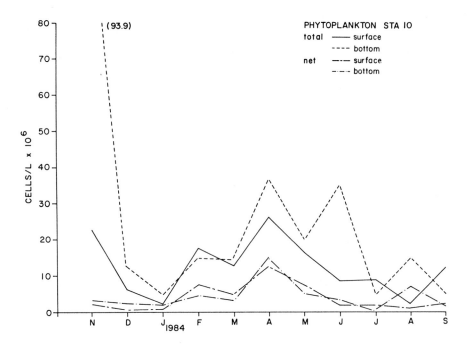

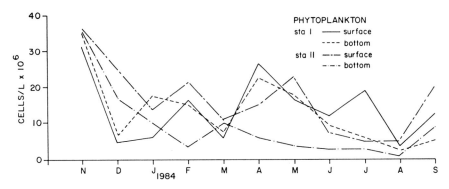

Figure 1. Net and nanoplankton abundance from Station 10 at the Chesapeake Bay entrance with a bottom depth of 22 M. Total phytoplankton concentrations are given for coastal Stations 1 and 10 downstream in the Chesapeake Bay plume, with bottom depths of 17 and 18 M.

RESULTS

Composition

The phytoplankton of the lower Bay and coastal region is dominated by a diatomaceous flora and a pico-nanoplankton complex. The major diatoms consist of small centric, chain-forming species having characteristic periods of high abundance extending from late winter through early spring and during fall. Cyanobacteria, Chlorophyceae, and various phytoflagellates composed the dominant pico-nanoplankton, with the most common sizes <5 µm. These cells were ubiquitous throughout the year, but had higher concentrations from late summer through fall, with occasionally modest pulses during spring. Other chlorophytes were common in spring and summer. More constant levels of concentration were associated with the dinoflagellates, with greater summer abundance for the cryptophyceans and the larger cyanobacteria. Also common, with large concentrations from spring through summer, were the chrysophyceans and haptophyceans. In waters outside the Bay entrance the winter-spring outburst was composed mainly of the diatoms *Skeletonema costatum, Leptocylindrus danicus, Ceratulina pelagica, Cyclotella* spp., *Rhizosolenia delicatula, Asterionella glacialis, Nitzschia pungens,* and *Thalassiosira* sp. The fall development is dominated by *Leptocylindrus minimus, Skeletonema costatum, Cyclotella* sp., *Thalassionema nitzschioides, Rhizosolenia fragilissima,* and *Nitzschia pungens*. A bimodal pattern of seasonal dominance for most of these species is also found in the mid-Atlantic bight (Marshall, 1984a). Lower abundance and other dominants are common to the mid-shelf region and periods between seasonal maxima, with increasing concentrations found along the shelf break (Marshall, 1984b).
Phytoplankton concentrations at surface and bottom depths for near shore stations, inside (Sta. 10, 22 m depth) and outside (Sta. 1 and 11, 17 and 18 m depths) the Bay plume are given for an eleven month period in Figures 1 and 2. Within the plume there is little difference between surface and bottom concentrations for the net phytoplankton. Composed mainly of diatoms, they had a spring maximum, summer and winter lows, and a modest rise in September 1984. The influence of the pico-nanoplankton is indicated by the difference between the net and total phytoplankton abundance at Station 10 in Figure 1. The pico-nanoplankton maxima occur mainly during fall, spring, and summer, with higher concentrations during fall in the bottom depths. The

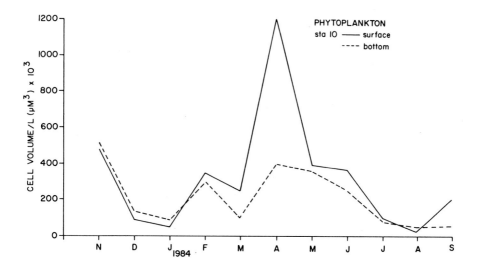

Figure 2. Phytoplankton biomass based on cell volume values at the Chesapeake Bay entrance.

Table 1. Dominant species, number cells/1 x 10^4, and species diversity at shelf stations seaward of the Chesapeake Bay entrance in June 1980.

Location	Surface	3-5 Meters	7-12 Meters	13-15 Meters
Plume	S. costatum 363.4 1.128	S. costatum 377.8 0.844	S. costatum 424.3 0.592	---
Plume	S. costatum 687.9 1.111	S. costatum 740.0 1.227	S. costatum 217.6 1.269	---
Off Shore	S. costatum 77.1 2.810	Mixture 153.2 2.096	Mixture 73.4 2.754	Mixture 121.5 2.450
Off Shore	Mixture 56.7 2.517	E. huxleyi 26.8 2.090	Mixture 61.6 2.430	Mixture 39.0 3.019
Mid Shelf	E. huxleyi 18.9 2.311	E. huxleyi 39.5 2.869	---	---

biomass peak coincides with the vernal outburst of diatoms from late winter through early summer (Figure 1). Another fall peak occurs, with early winter and summer having lower biomass levels. Seaward of the plume, total phytoplankton concentrations have less variability during fall and summer. The dinoflagellate spatial concentrations within the plume were similar throughout the year, but not abundant until the spring pulse.

Spatial composition and concentrations patterns from the Bay entrance across the shelf are given in Table 1. These were June collections that coincided with the decline of the spring growth and were prior to thermocline formation. *Skeletonema costatum* dominated the plume and was well distributed throughout the entire water column. Away from the Bay entrance and seaward, the assemblages became more diverse and cell concentrations decreased. The mixed populations characteristic of these stations contained more coccolithophores and a different assemblage of neritic species. The mid-shelf stations were dominated by *Emiliania huxleyi* and other coccolithophores, large centrales (e.g. *Coscinodiscus* spp., *Rhizosolenia* spp., *Chaetoceros* spp.), and a variety of phytoflagellates that formed an assemblage different than what was found inshore. Continuing seaward to the shelf break, the dominant species again change, with *S. costatum, L. danicus, A. glacialis* and other near shore species dominant (Marshall, 1984a).

DISCUSSION

Estuarine-coastal habitats represent active and dynamic regions that enhance phytoplankton development and the productivity of the inner shelf waters. Along the northeastern coast of the United States mid-Atlantic bight the major estuary systems are connected by a narrow coastal belt of high productivity. Annual levels of concentration for this inner shelf range from 10^5-10^7 cells/l before decreasing seaward to 10^3-10^4 cells/l for the mid-shelf region. Although numerous physical factors and nutrient entry from other sources contribute to this high productivity, the estuarine systems represent a major influence to this growth.

From these coastal spigots an out flow of water will continually enhance the nutrient levels of the inner shelf and at the same time seed these waters with an estuarine-coastal phytoplankton assemblage. Once these waters leave the estuary, the general productivity pattern is to mimic the broad seasonal and more predictable cycles of the shelf. Major outbursts of growth will occur during late winter-early spring and in fall. This is generally in contrast to the frequent pulses and sporadic growth patterns common to the lower Chesapeake Bay. Within the Bay plume vertical homogeneity is greatest near shore and decreases seaward. This pattern breaks down as the plume loses its identity and is mixed with shelf waters. As this occurs shelf species begin to dominate the water column and species diversity increases. Over the past 60 years there has also been an apparent change in the composition and concentrations of phytoplankton in the lower Chesapeake Bay and the water flowing to the shelf. The pattern consists of the continual dominance of *Skeletonema costatum*, which is accompanied by a changing assemblage of sub-dominant diatoms and other phytoplankters. The earlier Bay studies reported a dominant diatomaceous flora with peak growth periods during spring and fall (Wolfe et al., 1926; Cowles, 1930). Concnetrations at the Bay entrance averaged 4.1×10^4 cells/l, with a maximum count noted in April 1916 of 43.8×10^4 cells/l. In contrast, average diatom counts during 1982-1983 were 1.97×10^6 cells/l during spring and fall blooms. Lowest concentrations occurred in winter with average counts of 0.6×10^6 cells/l. Unfortunately, the initial Bay studies either did not include or cover adequately representatives from many of the taxonomic categories common in recent collections. These include the euglenophyceans, cryptophyceans, chlorophyceans, chrysophyceans, and coccolithophores, among others. The omission of these groups may have been due to either differences in methodology followed by the early investigators, or the absence (or low concentrations) of these forms in the samples. It is feasible that these investigators did not record pico-nanoplankton present, but is it not likely that they would omit any of the larger forms from these groups (e.g. euglenoids and cryptomonads). Associated with the existing assemblages in the Chesapeake Bay is evidence for increased eutrophication. Brush (1984) studied chlorophyll degradation products from core samples in a Bay tributary with sediment layers up to 200 years old. She associated increased levels of algal productivity over this period with the introduction of sewage effluents and fertilizer application from agricultural practices. A comparable increase in nutrient levels would be expected in other drainage systems along the Bay. In addition, there is apparently a reduction of erosion and sediment load in the Atlantic drainage systems (Meade, 1972; Smith and Shoemaker, 1983). This pattern would lower the compensation depth, and enhance photosynthesis in the estuaries. Present data also indicates high cell concentration in the Bay during 1982-1984 occurred during seasonal periods of low outflow and reduced sediment load.

These patterns of changing phytoplankton assemblages are not limited to the Chesapeake Bay. Collections over the shelf, specially from the near shore waters will contain numbers of euglenoids, cryptomonads, chlorophytes, chrysophytes (e.g. Calycomonas spp.) and a variety of cyanobacteria, among others. These groups are common to large estuarine systems like the Chesapeake Bay and the shelf waters. Many of these shelf species may also have been overlooked in earlier studies, or they are becoming established in greater numbers in these inner coastal waters. This similar phytoplankton composition for many of the larger Bay systems and the shelf waters may indicate that the seeding of Bay phytoplankters to the ocean not only has been taking place, but the species are becoming more established in time over the shelf. The presence of such assemblages could then be interpreted as indices of increased eutrophication for both the Chesapeake Bay and the continental shelf.

ACKNOWLEDGEMENT

Portions of this study were supported by the NOAA National Marine Fisheries NEMP/Ocean Pulse and the Norfolk Division of the U. S. Army Corps of Engineers.

REFERENCES CITED

Boicourt, W. C. 1973. The circulation of water on the continental shelf from Chesapeake Bay to Cape Hatteras. Ph.D. Thesis, The Johns Hopkins University, Baltimore, Maryland. 183 pp.
Boicourt, W. C. 1981. Circulation in the Chesapeake Bay entrance region: Estuary-shelf interaction. In: J. W. Campbell and J. P. Thomas, Eds.: Chesapeake Bay Plume Study, Superflux 1980. NOAA/NEMP III 81 ABCDEFG 0042, NASA Conf. Publ. 2188. p. 61-78.
Brush, G. S. 1984. Stratigraphic evidence of eutrophication in an estuary. Water Resources Res. 20(5): 531-541.
Cowles, R. 1930. A biological study of the offshore waters of the Chesapeake Bay. Fishery Bulletin 46: 277-381.
Marshall, H. G. 1980. Seasonal phytoplankton composition in the lower Chesapeake Bay and Old Plantation Creek, Cape Charles, Virginia. Estuaries 3: 207-216.
Johnson, D. R., B. S. Hester and J. R. McConaugha. 1984. Studies of a wind mechanism influencing the recruitment of blue crabs in the middle Atlantic bight. Cont. Shelf Res. 3: 425-437.
Marshall, H. G. 1982. The composition of phytoplankton within the Chesapeake Bay plume and adjacent waters of the Virginia coast, U.S.A. Estuarine, Coastal and Shelf Science 15: 29-43.
Marshall, H. G. 1984a. Phytoplankton distribution along the eastern coast of the USA. Part V. Seasonal density and cell volume patterns for the northeastern continental shelf. J. Plankton Res. 6: 169-193.
Marshall, H. G. 1984b. Meso-scale distribution patterns for diatoms over

the northeastern continental shelf of the United States. In: D. B. Mann (Ed.), Proc. 7th Inter. Diatom Symp., Otto Koeltz, Koenigsten. pp. 393-400.

Marshall, H. G. and M. S. Cohn. 1985. Seasonal phytoplankton composition in the mid-Atlantic Bight off the Delmarva peninsula. In press.

Marshall, H. G. and R. Lacouture. Seasonal variations in phytoplankton assemblages from the lower Chesapeake Bay and vicinity. In press.

McCarthy, J. J., W. R. Taylor and J. Loftus. 1974. Significance of nanoplankton in the Chesapeake Bay estuary and problems associated with the measurement of nanoplankton productivity. Marine Biology 24: 7-16.

Meade, R. H. 1972. Transport and deposition of sediments in estuaries. Geol. Soc. Amer. Mem. 133: 81-120.

Mulford, R. A. and J. J. Norcross. 1971. Species composition and abundance of net phytoplankton in Virginia coastal waters, 1963-1964. Chesapeake Science 12: 142-155.

O'Reilly, J. and D. Busch. 1984. Phytoplankton primary production on the northwestern Atlantic shelf. Rapp. P. -v. Reun. Cons. Int. Explor. Mer. 183: 255-268.

Patten, B., R. Mulford and J. Warinner. 1963. An annual phytoplankton cycle in the lower Chesapeake Bay. Chesapeake Sci. 4: 1-20.

Smith, J. A. and L. L. Shoemaker. 1983. The role of sediment in nonpoint pollution in the Potomac River Basin. Sp. Rep. Interstate Commission on the Potomac River Basin. Rockville, MD, 8 p.

Wolfe, J. J., B. Cunningham, N. Wilkerson and J. Barnes. 1926. An investigation of the microplankton of Chesapeake Bay. Journal of the Elisha Mitchell Scientific Society 42: 25-54.

RIVER RUNOFF AND SHRIMP ABUNDANCE IN A TROPICAL COASTAL ECOSYSTEM - THE EXAMPLE OF THE SOFALA BANK (CENTRAL MOZAMBIQUE)

A.Jorge da Silva
Instituto Hidrográfico
Rua das Trinas, 49
P-1296 Lisboa Codex
Portugal

ABSTRACT

Abundance indices of the shallow-water shrimp Penaeus indicus in Central Mozambique were analyzed by length groups, and related to Zambezi runoff on a year-to-year basis for the period 1974-1983. The Zambezi outflow seems to influence the recruitment strength of P.indicus either by directly affecting the number of recruits or by inducing changes in the size at migration (or a combination of both). Too high effort levels during the period of recruitment since 1979 appear, however, to be exerting a detrimental effect on annual abundance.

INTRODUCTION

The shallow-water penaeid shrimp stocks are strategically important in the economy of Mozambique. The main concentrations, exploited by industrial fishing, occur in the Sofala Bank area, particulary to the north of the Zambezi delta, at depths of 5-25 m (Figure 1). 90% of the shrimp caught in that area consist of two species - Penaeus indicus and Metapenaeus monoceros (BRINCA and SOUSA, 1984a).

Catches have decreased from 10,000-12,000 tons in 1974-76 to 8,000 tons in 1983, apparently as a result of a decrease in abundance, as reflected by catch rates (Figure 2). According to GARCIA and LE RESTE (1981), annual catch depends almost entirely on recruitment during the year. There is no indication that fishing effort has caused the spawning stock to drop below a critical level at which the recruitment is affected, and BRINCA and SOUSA (1984a) suggest that fluctuations in recruitment may be due to climatological factors, namely to the Zambezi runoff.

This contribution is an attempt to elucidade some of the possible links between freshwater outflow and abundance of shrimp, by reviewing the existing data on river runoff and catch and effort of the shrimp fishery.

THE DATA

Zambezi runoff data were obtained from the National Directorate for Waters, Maputo, Mozambique, and refer to the Tete gauge station (Figure 1). These data have been evaluted by the author in previous papers and represent an average underestimation of about 16% relative to the runoff at the river mouth (SAETRE and

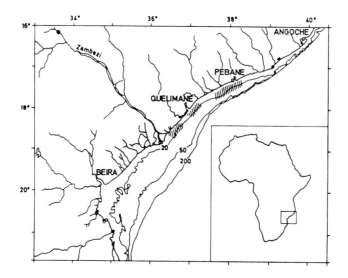

Figure 1. Sofala Bank. Bathymetry (m), rivers and main shrimp fishing areas (shaded). Upper dot denotes Tete gauge station

JORGE DA SILVA, 1982; JORGE DA SILVA, 1984).

Catch and effort data were obtained from the Institute of Fisheries Research, Maputo, Mozambique, and originate from two differente sources:

- Total annual catches for all fleets were extracted from production statistics of the Sate Secretariat for Fisheries.

- Monthly values of effective fishing effort and catch by species were supplied by one single fishing company and relate to a single fleet. Length composition of the P.indicus catch for the same fleet was obtained from commercial size groups after conversion to length groups.

Estimates of catch in numbers per unit of effort (numerical abundance index) were worked out on a monthly and length group basis for the only fleet supplying relevant information. Catch rates in mass units (mass abundance index) were also calculated on a monthly basis, regardless of length composition.

Catch rates calculated as above were assumed to be adequate abundance indices for the whole fishing area. Annual abundance indices were calculated in a similar way. An estimate of total annual effort was then obtained from total catch and average annual catch rate.

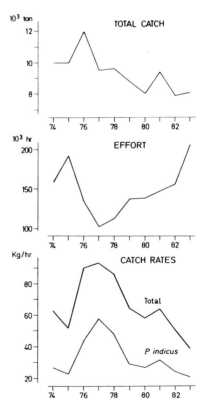

Figure 2. Evolution of the shrimp fishery at the Sofala Bank, 1974-1983 (adapted from BRINCA and SOUSA, 1984a)

ROLE OF THE ZAMBEZI RIVER IN THE SOFALA BANK

JORGE DA SILVA (1984) identified four hydrologic regimes in the Sofala Bank area and showed evidence of different fish species preferring different regimes. The Zambezi was found to be the main influencing factor (and the main source of freshwater) at the inner shelf area between Pebane and Beira. This river influence is well illustrated by the distribution of organic matter in the surface sediments (Figure 3). Comparison of Figures 1 and 3 reveals that shrimp abundance (as indicated by the distribution pattern in the fishery) is highest in the areas influenced by the Zambezi runoff.

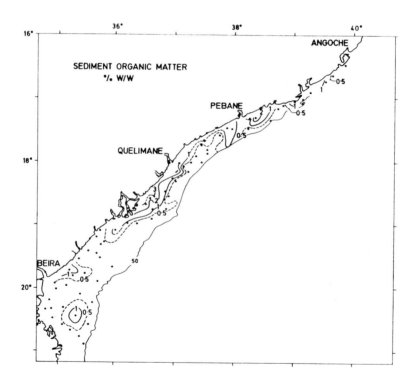

Figure 3. Organic matter in surface sediment (from JORGE DA SILVA, 1984)

The distributional area of shallow-water shrimp is inferred to be related to a persistant coastal current which is partly fed by Zambezi water. On the synoptic scale, the longshore extent of such current appears to be determined by lateral oscillations of the Mozambique current core (JORGE DA SILVA, 1984). These oscillations seem to influence, on the same time scale, the establishment of important shrimp concentrations within the distributional area (BRINCA, SILVA and SILVA, 1981; BRINCA, REY, SILVA and SAETRE, 1981; BRINCA, BUDNICHENKO, JORGE DA SILVA and SILVA, 1983).

The interannual differences in the dynamics of the coastal current have not been specifically followed. They are, however, beleived to depend on the regulation constraints that have been imposed since late 1974 at the Cahora-Bassa dam. The major features of the Zambezi pattern before and after regulation are well illustrated in Table 1 and Figures 4 and 5, and can be summarized as follows:

a) interannual variability did not decrease with the regulation;

b) mean runoff in January-June (flood) decreased, although mean annual runoff increased;

c) mean runoff in July-December (drought) almost doubled;

d) runoff variability for January-June is now twice as high as for July-December.

SHRIMP LIFE CYCLE

Schematically, GARCIA and LE RESTE (1981) described the life cycle of penaeid shrimps as follow: spawning takes place at sea; the larvae and first postlarvae are planktonic. The postlarvae later migrate inshore and very often into estuaries, lagoons and mangroves. The shrimps return to sea when they have reached a

Table 1. Average runoff ($\bar{Q}$) and variability coefficient (C_V) for the Zambezi river at Tete before and after the regulation at Cahora-Bassa.

($C_V = (s/\bar{Q}) \times 100\%$, with s - standard deviation)

YEARS	JAN - JUN		JUL - DEC		JAN - DEC	
	$\bar{Q}(Km^3)$	$C_V(\%)$	$\bar{Q}(Km^3)$	$C_V(\%)$	$\bar{Q}(Km^3)$	$C_V(\%)$
1961-74	56.8	42	20.9	46	77.7	40
1975-83	50.3	58	40.2	28	90.5	42

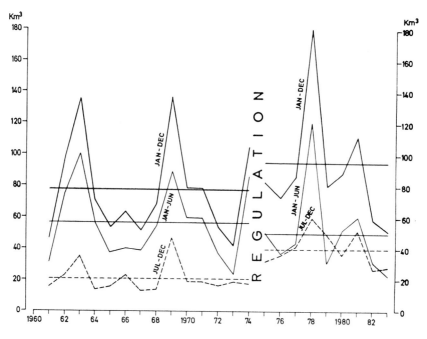

Figure 4. Evolution of the Zambezi runoff at the Tete gauge station, 1960-1983. Horizontal lines represent mean values. (Notice the changes imposed by the regulation, since 1975)

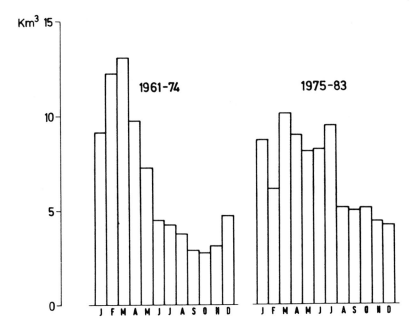

Figure 5. Monthly means of the Zambezi runoff at Tete gauge station, before and after the regulation at the Cahora-Bassa dam

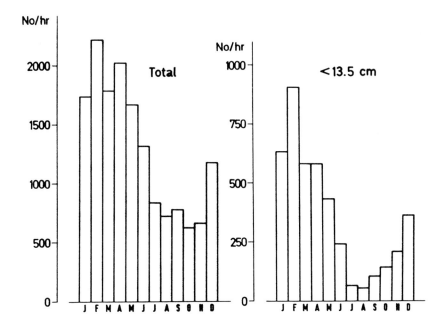

Figure 6. Monthly mean catch of Penaeus indicus

length of about 10 cm. The end of the cycle takes place at sea. The life span is approximately one year.

Different factors can be expected to trigger the shrimp migration from the nurseries to be offshore areas. This process is often connected with the establishment of unfavorable conditions in the estuaries (GARCIA and LE RESTE, 1981). In central Mozambique these conditions are likely to be strongly influenced by the seasonal cycle in freshwater runoff (BRINCA and SOUSA, 1984a).

The average seasonal pattern of P.indicus abundance in the Sofala Bank (Figure 6) suggests that the recruitment cycle begins in August-September, reaches a maximum in February and remains high until April. The strongest recruitment occurs in January-April, i.e. four months after peak spawning, as pointed by BRINCA and SOUSA (1984a).

Growth curves of M.monoceros in Maputo Bay, Mozambique, are shown in Figure 7. A striking aspect is the absence of males in the larger size groups, which is characteristic of penaeids. The general trend of Figure 7 also holds for P.indicus at the Sofala Bank, as suggested by preliminary results of on-going investigations (LÍLIA BRINCA, personnal communication).

SHRIMP ABUNDANCE

In this and the following sections catch rates of P.indicus will be used as abundance indices of the whole shrimp stock in the fishing area, a procedure that is supported by Figure 2. Catch rates of M.monoceros are not considered as this species is often mixed with carids in the same commercial designation and thus the data are suspect.

Figure 8 shows annual abundance indices (catch rates) of P.indicus by length groups together with the catch rates in mass units. Careful examination of the figure will reveal interesting

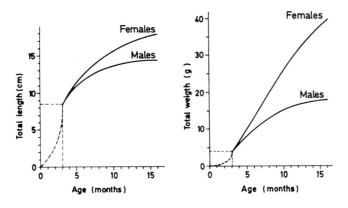

Figure 7. Growth curves of Metapenaeus monoceros at Maputo Bay: a) in total length; b) in total weight (adapted from BRINCA and SOUSA, 1984b)

aspects of the evolution of the P.indicus stock during the decade 1974-83:

i) Abundance in mass units, although related to total numerical abundance, exhibits the best relation with length

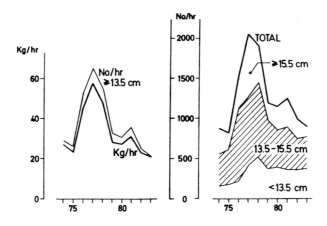

Figure 8. Evolution of Penaeus indicus abundance by length groups. Shaded area represents abundance of the dominant group

groups bigger than 13.5 cm; the catch (in weight) of P.indicus will thus be mainly determined by those groups.

ii) The dominant length group (13.5-15.5 cm, shaded in Figure 8) became progressively less abundant since 1978, and reached the level of the smaller length group (<13.5 cm) in 1982-83. Similar, but even more dramatic, was the decrease in abundance of the bigger length group (⩾15.5 cm) which, as suggested by the growth curves of Figure 7, is likely to be composed only of females.

iii) The abundance of the smaller length group (<13.5 cm) approximately doubled in 1977 and since then has maintained a rather constant level. If size at recruitment did not change much from year to year, this would mean that annual recruitment has kept approximately constant since 1977 (with the exception of 1978) at a level which is the double of that observed in 1974-76.

How can the low catch rates of large size classes of shrimp be interpreted if the recruitment has not declined substantially since 1978? The answer to this question is attempted in relation to the influence of freshwater runoff and fishing effort.

FACTORS AFFECTING SHRIMP ABUNDANCE

The resemblance is striking between the average seasonal distribution of P.indicus abundance (figure 6) and that of the Zambezi runoff prior to the regulation at the Cahora-Bassa dam (Figure 5, left). This resemblance can be plausibly interpreted. The abundance in the fishing areas increases when the shrimps migrate from the estuaries, and this migration occurs mainly during the flood season. On the other hand, the interannual trends shown in Figures 4 and 8, although similar, do show important discrepancies.

Having in mind the considerations made above on the recruit

Table 2. Correlation coefficients between Zambezi runoff at Tete and abundance of Penaeus indicus

Zambezi Aug - Jul vs. P.indicus in	number per hour			Kg/hr
	1<13.5 cm	1≥13.5 cm	total	
Aug - Jul	0.758	0.645	0.786	0.686
Sep - Aug	0.759	0.624	0.761	0.648

Zambezi Aug - Mar vs. P.indicus in	number per hour			Kg/hr
	1<13.5 cm	1≥13.5 cm	total	
Aug - Mar	0.845	0.656	0.826	0.714
Aug - Apr	0.894	0.608	0.803	0.667
Sep - Apr	0.898	0.561	0.775	0.627

ment of P.indicus, Zambezi runoff and shrimp abundance values were analyzed in more detail. Lagged and unlagged correlations were attempted both for one-year and for shorter periods, these corresponding to recruitment. Figure 9 illustrates the best

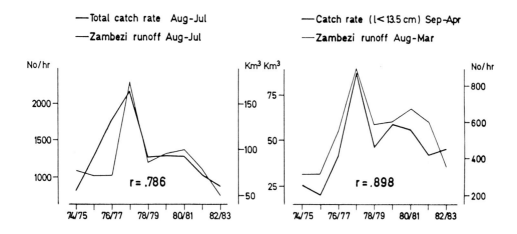

Figure 9. Relation between abundance of Penaeus indicus (thick line) and Zambezi runoff (thin line)

relations. The most significant correlation coefficients are summarized in Table 2. The best ones were obtained for the recruitment period with a one-month lag between runoff and catch rate, and were always better for smaller length groups (< 13.5 cm) than for the total number or for catch weight of P.indicus.

In these circumstances, the Zambezi runoff is inferred to affect, either directly or indirectly, the recruitment strength of P.indicus to the fishing areas. The interaction mechanism requires however, some clarification. Even though the number of recruits may be influenced by freswater runoff, the interannual trend observed in the larger size groups cannot be accounted for by runoff alone. Explanation for it has to be sought in the evolution of fishing effort.

Effective fishing effort doubled from 1877 to 1983 (Figure 2). On the other hand, the abundance of small P.indicus during the recruitment period did not vary much since 1978-79, in good agreement with the Zambezi runoff (Figure 9). Given the absence of seasonality in fishing effort (Table 3), a change in the size structure of the stock towards smaller lengths is to be expected, due to a significant increase in effort during the recruitment peak.

A rather different mechanism, involving salinity induced variations of size at migration, is also an interpretation of the present findings. In this case one could assume, with LE RESTE (1979), that when runoff increases the salinity decreases leading to a shorter residence period in the estuaries, decreasing the age at first capture and the catchable biomass through the interaction of growth and mortality. ULTANG, BRINCA and SILVA (1980) provided a good suport for this mechanism. They claimed that P.indicus smaller than 13.5 cm were not fully recruited to the fishing area in 1974-76, when their abundance was comparati vely low (Figures 8 and 9). The considerations that were made above on the evolution of fishing effort will also apply in this case, providing additional justification to the changes in the length composition of the stock.

Table 3. Percent distribution of fishing effort by quarters for 1974 - 1983

	74	75	76	77	78	79	80	81	82	83
JAN-MAR	22.2	25.0	21.8	21.5	24.5	25.0	16.3	26.1	21.4	25.4
APR-JUN	29.1	23.2	27.9	28.5	22.1	29.0	22.4	24.7	26.5	29.1
JUL-SEP	31.2	33.9	24.1	28.8	33.2	20.2	30.7	25.7	29.4	16.8
OCT-DEC	17.5	17.9	26.2	21.2	20.2	25.8	30.6	23.5	22.7	28.7

DISCUSSION AND CONCLUSION

Making use of yield per recruit curves, ULLTANG, BRINCA and SOUSA (1985) concluded that recruitment has been at a low level since 1979, and can be linearly related to the catch rate of P.indicus in January-March. Such conclusions disagree with the present findings.

ULLTANG et al. (1985) assumed a constant value of natural mortality when calculating recruitment indices for 1977-83. If one takes into account the different estimates of natural mortality that were arrived at by those authors, it is no longer possible to conclude that the recruitment has been low since 1979. Neither is it possible to relate recruitment with catch rates in January-March. As shown in Figure 10, even during the first quarter, when length groups smaller than 13.5 cm are most abundant, the catch rate of P.indicus still reflects the abundance of the larger size groups, as do the annual catch rates.

The apparently increased abundance of the smaller length groups could also be due to a modification of the fishing pattern (e.g. a change in the mesh size of the gear, or a displacement of the fleet towards other areas). There is, however, no indication that such modification has occured.

The author's main conclusion is that the Zambezi influences the recruitment process of Penaeus indicus. It should, however, be borne in mind that, in spite of the high correlation coefficients that were obtained, the relation between the Zambezi runoff and the abundance of P.indicus may not be direct. It is possible that changes in size at first capture will reflect major changes in runoff, while small fluctuations around a given runoff level will induce variations in the number of recruits. Additional complexity may also be introduced by the duration of the floods and the moments they are triggered. The examination of further details is, however, regarded as premature requiring a longer time series and more biological information.

ULLTANG et al. (1985) calculated natural mortality during periods of "no recruitment", when the population was formed by

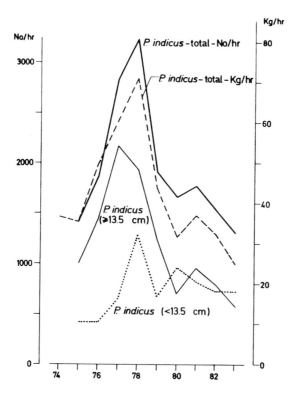

Figure 10. Evolution of catch rates in January-March, 1974-1983

older individuals. Their figures may, therefore, be overestimates influenced by death after spawning (BRINCA and SOUSA, 1984c). Mortalities of the smaller length classes, during the main recruitment period, may be lower than the estimates of ULLTANG et al. (1985). In these circumstances, it seems worthwhile to reduce the effort during the first quarter, allowing for shrimp to reach larger sizes before being caught.

In the future it may also be possible to regulate the Zambezi runoff in an appropriate manner to induce higher yields, if the mechanisms by which runoff controls recruitment is completely understood.

REFERENCES

BRINCA L., V.A.BUDNICHENKO, A. JORGE DA SILVA and C. SILVA (1983) A report on a survey with the R/V "Ernst Haeckel" in July-August 1980 - Shallow-water shrimp and by catch; oceanography. Revista de Investigação Pesqueira, No. 6, Maputo, 105 pp.

BRINCA L., F. REY, C. SILVA and R. SAETRE (1981) A survey on the marine fish resources of Mozambique, Oct-Nov. 1980. Reports on surveys with the R/V Dr. Fridtjof Nansen, Instituto de Desenvolvimento Pesqueiro, Maputo, Institute of Marine Research, Bergen, 58pp.

BRINCA L. and E. SANTOS (1984) Síntese dos resultados e métodos de análise utilizados na avaliação do recurso de camarão de águas pouco profundas do Banco de Sofala. Revista de Investigação Pesqueira, No. 9, Maputo, 189-215 (in Portuguese)

BRINCA L., C. SILVA and A. SILVA (1981) Relatório do cruzeiro realizado no Banco de Sofala pelo arrastão "Muleve" em Julho -Agosto 1979. Informação No. 4, Instituto de Desenvolvimento Pesqueiro, Maputo, 53pp. (in Portuguese)

BRINCA L. and L.P. SOUSA (1984a) O recurso de camarão de águas pouco profundas. Revista de Investigação Pesqueira No. 9, Maputo, 45-60 (in Portuguese)

BRINCA L. and L.P. SOUSA (1984b) A study on growth of Metapenaeus monoceros (Fabricius) of Maputo Bay. Revista de Investigação Pesqueira, No. 11, Maputo, 77-101

BRINCA L. and L.P. SOUSA (1984c) Mortality rates estimates for Metapenaeus monoceros (Fabricius) of Maputo Bay. Revista de Investigação Pesqueira, No. 11, Maputo, 41-76

GARCIA S. and L. LE RESTE (1981) Life cycles, dynamics, exploitation and management of coastal penaeid shrimp stocks. FAO Fisheries Technical Paper, 203, 215pp

JORGE DA SILVA A. (1984) Hydrology and fish distribution at the Sofala Bank (Mozambique). Revista de Investigação Pesqueira No. 12, Maputo, 5-36

LE RESTE L. (1979) The relation of rainfall to the production of the penaeid Penaeus duorarum in the Casamance estuary (Senegal). Paper presented to the fifth International Symposium of Tropical Ecology, Kuala Lumpur, 5 pp. (cited in GARCIA and LE RESTE, 1981)

SAETRE R. and A. JORGE DA SILVA (1982) Water masses and circulation of the Mozambique Channel. Revista de Investigação Pesqueira, No. 3, Maputo, 83 pp.

ULLTANG Ø, L. BRINCA and C. SILVA (1980) A preliminary assessment of the shallow water prawn stocks off Mozambique, north of Beira. Revista de Investigação Pesqueira, No. 1, Maputo, 69 pp

ULLTANG Ø, L. BRINCA and L. SOUSA (1985) State of the stocks of shallow-water prawns at Sofala Bank. Revista de Investigação Pesqueira, No. 13, Maputo, 97-126

TIMING AND DURATION OF SPRING BLOOMING SOUTH AND SOUTHWEST OF ICELAND

Thórunn Thórdardóttir
Marine Research Institute, Reykjavík

ABSTRACT

Primary production measurements carried out during spring for several years in the waters S and SW of Iceland are dealt with. On average the blooming starts in nearshore waters and is delayed with increasing distance from the coast. In the shallowest part of Faxaflói the onset of blooming is in late March, whereas in offshore waters the start is not until after the middle of May. For individual years the results reveal great yearly fluctuations in the onset of blooming as well as lasting of established blooms. Of the factors causing these variations the stability and grazing are the most important ones. Within the shelf region the interactions between runoff and wind regime greatly affects stability conditions and thus the pattern of phytoplankton growth development each year.

INTRODUCTION

The main spawning grounds for Icelandic fish stocks, exploited commercially in Icelandic waters are found in the shelf area south of Iceland. During the first stages of their life the larvae remain in the waters off the south and southwest coasts where their survival will largely depend on the availability of food. There are indications that early stages of copepods provide an important food source for newly hatched cod larvae (Bainbridge and McKay 1968, Friðgeirsson 1984). Timing and duration of the phytoplankton bloom may be of key importance for the spawning success of the zooplankton and consequently, also of fish larvae which feed on the zooplankton.

Due to the biological importance of the area south and southwest of Iceland it has been surveyed in spring during several years. Preliminary results have revealed great yearly variability in conditions for vernal blooming and its continuation (Thórdardóttir 1976, Fridgeirsson et al. 1979).

In this paper main results of primary production studies in this area will be outlined and an attempt made to relate onset of blooming and its duration to environmental variables.

HYDROGRAPHY

The shelf waters south and southwest of Iceland consist essentially of Atlantic water, relatively warm and saline. In spring its salinity ranges from 35.0-35.2, and its temperature from about 6-8 °C. In the coastal region, however, this water may be appreciably diluted by fresh water from land, so that at near-shore stations salinity as low as 33.0 is frequently found.

Studies made in the Faxaflói area have indicated that the extension and accumalation of fresh water inside the coastal region is largely controlled by the wind regime (Stefánsson and Gudmundsson 1978) and presumably, the same applies to the shelf area south of Iceland (Thórdardóttir, 1976).

The stratification of the surface waters in spring will depend on a) the offshore extension of fresh water and b) the development of a thermocline. In the early spring, the near-shore water may attain considerable stability due to fresh water dilution of the surface water, whereas the formation of thermocline will normally not take place until the second half of May. Thus in general, the shallow near-shore area becomes stratified considerably earlier than the deeper more saline region.

Direct and indirect current measurements (Hermann and Thomsen 1946, Malmberg 1968) have indicated a mean residual surface current that flows clockwise around Iceland, with a speed of 3-5 miles a day in the area south and west of Iceland. However, the rate of the coastal current may show large variation depending upon wind conditions and fresh water runoff.

MATERIAL AND METHODS

The investigations in the water south and southwest of Iceland (Fig. 1) have included routine measurements of temperature, salinity, nutrients, zooplankton and in most surveys primary production measurements by the ^{14}C technique (Steeman Nielsen 1952). Samples for the primary production measurements have usually been collected at 0, 10, 20 and 30 m and illuminated for 4 hours in an air temperature controlled incubator. The illumination used, about 250 $\mu Em^{-2}s^{-1}$ has been found to be at the saturation level for different phytoplankton communities in Icelandic waters.

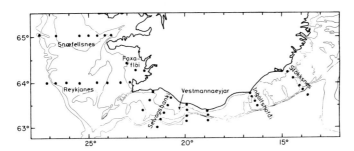

Fig. 1. Main sections and stations occupied during hydrobiological surveys in the area south and southwest of Iceland in the years 1971-1981. Depth contours: 100, 200, 400 and 600 m.

RESULTS

In the years 1958-1980 the blooming starts at near-shore stations in Faxaflói (Faxabay) in late March, whereas nearshore off the south coast the start is at later date (Fig. 2). With increasing distance from the coast the blooming is retarded and it does not start in offshore waters until late May. At that time the primary production values are similar all over the area (average mostly in the range 5-10 mg C m^{-3} h^{-1}) and the zonation in production, clearly appearant during 1-15 May averages (Fig. 2) has vanished. There is however a marked zonation in the nitrate concentrations (Fig. 3) which increases in offshore direction. By late May the near-shore primary production has decreased whereas that of the deeper area has increased.

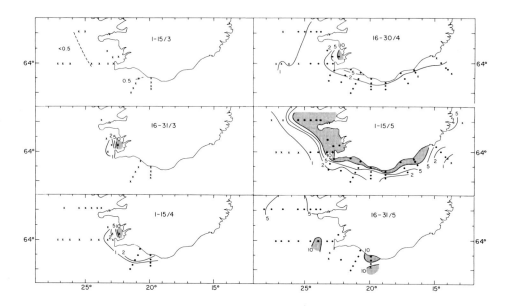

Fig. 2. Mean primary production (mg C m^{-3} h^{-1}) at 10 m in the years 1958-1980 within 14 days periods from March to May. x: available observations ≤ 2.

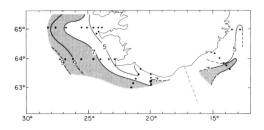

Fig. 3. Average nitrate concentrations (μM) 0-20 m in May-June for the years 1970-1975.

When individual years are considered the general picture of growth development in these waters becomes much more complex. Comparison of the primary production during April-June of 1976-1981 (Fig. 4) reveals that there are great yearly fluctuations both with respect to the location of the main production areas and the timing of the blooming.

During the first half of April 1981 an appreciable blooming had only started at near-shore stations in Faxaflói. In the other 5 years measurements were not available until the second half of April. In 1976 and 1981 the blooming clearly started off the eastern part of the south coast, and the same applied to 1980, but in more limited areas. In the years 1977, 1978 and 1979, on the other hand, the blooming seems to have started off the southwest coast and in Faxaflói.

During the first half of May 1976 very high productivity values were measured on the Snæfellsnes section and over the outer part of Faxaflói, while in 1978 high values occurred both north of Snæfellsnes, in Faxaflói and inside a narrow belt along the south coast. In 1977, 1979 and 1980 the early May values were appreciable lower than in 1976, with the highest production occurring off the southwest coast in 1977 and 1979, while in 1980 the highest values were off the southeast coast. In 1981 the main production during the first half of May was found on both sides of Reykjanes, west of Reykjanes and within a small area farthest to the east in the study area.

Usually, the primary production was lower in late April than during the first half of May. During the second half of May and in early June there were large differences between years. Within the shallower part of the region the locations of the production areas had at this time been markedly displaced in all the years from what they were about 2 weeks earlier. The near-shore values were now frequently lower than during the preceding periods. At the same time the production had moved farther offshore to the deeper regions.

The growth period is of longest duration inside Faxaflói and becomes gradually shorter with increasing distance away from the coast (Fig. 5). In Faxaflói and west of Reykjanes the productivity remains high on the average throughout the growth

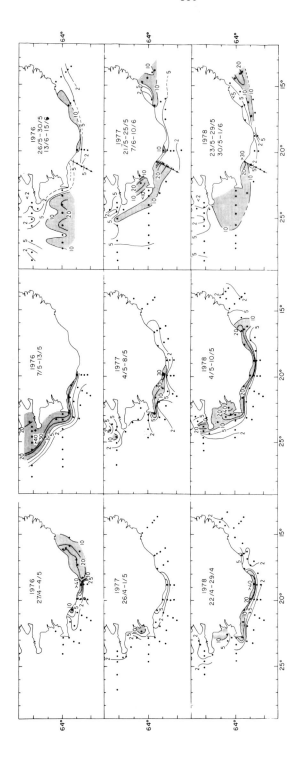

Fig. 4. Primary production (mg C m^{-3} h^{-1}) at 10 m from April to early June in the years 1976-1981.

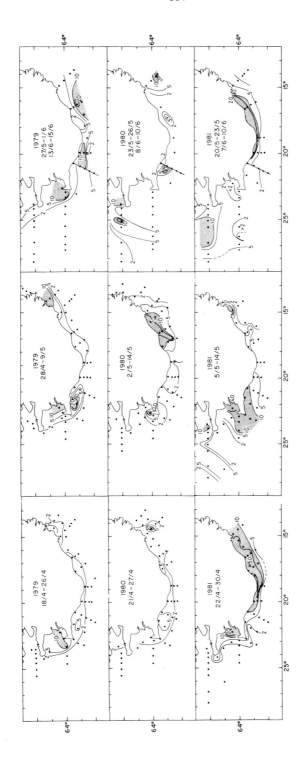

Fig. 4 cont.

season, but year to year fluctuations are conspicuous. In this region fresh water influence is considerable but varying. In the deep ocean area (St. 3 in Fig. 5) there is a well defined spring bloom, but the growing season starts at a later date than in the coastal region. Then there is a marked difference between years in the productivity level at the late May peak. This difference may be explained by yearly fluctuation in the establishment of a thermocline as well as yearly differences in zooplankton densities (Thórdardóttir 1976, Ástthórsson et al 1983). The data from the Selvogsbanki region reveal essentially the same features as those found for the shelf west of Reykjanes. However, the Selvogsbanki values appear to be slightly lower than

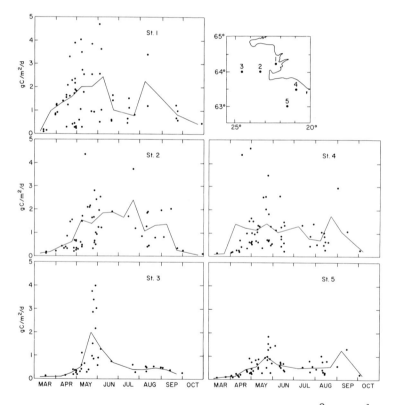

Fig. 5. All primary production values (g C m^{-2} day^{-1}) in the years 1958-1982 at a few selected stations the position of which is shown on the inset map. The curve gives the average annual cycle based on 14 days means. Dots indicate individual observations.

those at comparable positions on the section off Reykjanes. At the same time it seems that in some years relatively intensive production started somewhat earlier at St. 4 than at St. 2 (see Fig. 5). At the offshore station on the Selvogsbanki section (St. 5) the productivity peaks appear to be much less conspicuous than at St. 3 and the productivity seems to remain at a similar level throughout summer.

DISCUSSION

Having described large irregularities in the spatial pattern of phytoplankton growth within individual years and marked year to year variations, it seems appropriate to ask what may cause these observed anomalies and fluctuations. Without doubt we are dealing with a most complicated system governed by an interplay of several variables. The most important of these are probably light conditions, stratification, nutrients and grazing. These variables will now be considered.

There is no reason to believe that shortage of nutrients have caused the observed variations in the primary production in early spring south and southwest of Iceland. Data on nutrient concentrations (see Fig. 6) have revealed that in April the utilization is generally negligible except possibly in the innermost part of Faxaflói, and in early May the concentrations are still relatively high. Even in late May there are normally considerable nutrient stocks available (Fig. 3) not only in the open ocean beyond the shelf, but also over the coastal region south of Iceland.

Studies of the global radiation at different northern latitudes and the available information on phytoplankton cycles in Norwegian coastal waters led Smayda (1959) to present the theory that spring bloom will not start until the global radiation has surpassed 120-150 ly/day. Further investigations in Norwegian waters have supported this provided that the thickness of the homogeneus surface layer does not exceed a certain depth.

Around the middle of March the global radiation in Reykjavík surpasses 150 ly/day on the average and in the last week of March a mean level of 230 ly/day is reached (Anon. 1976-1981). Thus, by the end of March the blooming is well established in the inner shallow part of Faxaflói (Fig. 2) but only there. At locations within the same depth range off the south coast the blooming starts later and the culmination does not take place until early May (Fig. 2). The reason for the slow increase in the production in the nearshore area off the south coast as compared to Faxaflói may probably be attributed to the topography of the south coast which is unsheltered and faces the open ocean directly. There the wind stress is liable to cause less stable conditions than in the more sheltered Faxaflói where stratification is normally less variable (Thórdardóttir and Stefánsson 1977). The late blooming in offshore waters (Fig. 2) may be explained by the late establishment of a thermocline

Variations in the magnitude of surface radiation do not seem to be an important cause of growth fluctuations visulized in Fig. 4. On the other hand, it appears that both variations in stability and grazing are important. In 1981 more frequent observations were carried out in the area south of Iceland than in any other year discussed in this paper. As an example we will consider changes at the 2nd station from land on the section crossing the Selvogsbanki (Fig. 6; for location see Fig. 1). It will be seen that a thermocline did not form at this station until after the middle of May. Prior to May 21 the stability was thus dependant upon the distribution of fresh water. In this way a stratification had developed in late April due to a lowering of salinity, but this condition was only temporary, as the water column became homogeneous again by May 10 as the result of vertical mixing. The relatively high productivity on April 25 was in accordance with the stratification established at that time, but was appreciably lowered, when the water column became vertically mixed on May 10. It is notable, however, that in spite of the formation of a thermocline by May 25, the phytoplankton growth did not accelerate, even though there was an ample supply of nutrients. This apparent discrepancy can be attributed to grazing as the result of a marked increase in the density of zooplankton after the middle of May (Fig. 6).

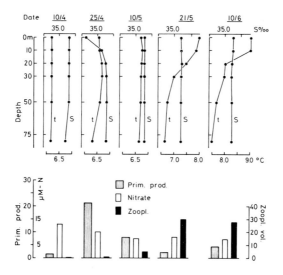

Fig. 6. Vertical distribution of salinity and temperatute (°C), nitrate (µM) 0-20 m primary production (mg C m^{-3} h^{-1}) 0-10 m, and zooplankton (ml per 21 m^3) 50-0 m, at the 2nd station from land on the section over Selvogsbanki 1981. For location see Figure 1.

From the foregoing and other examples it is evident that the offshore extension of low-salinity surface water supplied by runoff may constitute an important process for initiating the development of an early phytoplankton bloom in this area. Such a stratification in early spring caused by a vertical salinity gradient may, however, be easily disturbed or even broken down due to the action of winds. The effect of wind in this respect may thus be twofold, viz. to bring lower salinity surface water offshore and distribute it over relatively large areas, and to create turbulent mixing whereby the water column may become vertically homogeneous (Thórdardóttir 1976).

Results from a buoy station located a few miles west of the Selvogsbanki section (Fig. 7) combined with information on the wind regime (Fig. 8) provide a further insight into the effects of winds on the distribution of fresh water and the resulting impact on the outcome of primary production during the spring of 1981.

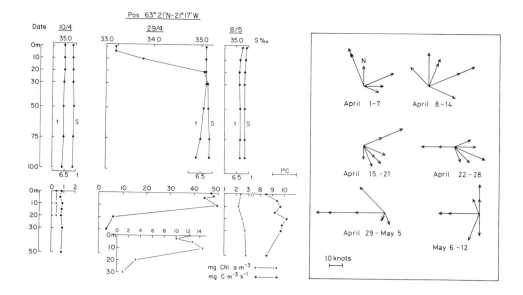

Fig. 7. Vertical distribution of temperature (°C), salinity, primary production (mg C m^{-3} h^{-1}) and chlorophyll a (mg m^{-3}) at a buoy station on Selvogsbanki (63°21′N, 21°17′W) on April 4, April 29 and May 8, 1981.

Fig. 8. Daily means of wind direction and velocity (in knots) at Vestmannaeyjar (Westman Islands) from April 1 - May 12 1981.

Three more or less distinct wind patterns can be indentified in the period April 1 - May 12: 1) rather strong winds from SW to SE during April 1-14, 2) moderate winds from NW dominating from April 15-28 and 3) winds from E in the period April 29 - May 12, with high velocities for several days during the first week of May.

During the first of these three periods the wind would tend to accumulate the fresh water close inshore and preserve unstratified conditions farther offshore in the coastal area. This would explain the slow development of plant growth at the buoy station by April 10. On the other hand, the moderate westerly winds, presumably not strong enough to cause upwelling near the shore, would favour the outward extension of low salinity water. This in turn would lead to an increased stratification at the buoy station and an outburst of blooming as indeed observed near the end of April. However, it appears that the stormy weather during the first week of May broke down the stratification and led to an uniform mixing down to the bottom. This would explain the even vertical distribution of both productivity and chlorophyll a within the uppermost 50 m on May 8.

Similar conditions as found at the buoy station on May 8, 1981, were observed over extensive areas off the south coast in early May of that year (Fig. 4), viz. relatively high level of productivity in homogeneous water. Such a large plant stock could hardly have developed in turbulent water, and presumably was the outcome of plant growth during late April, when northwesterly winds most likely brought low salinity water considerable distances away from the coast. In this connection it may be pointed out that at the mouth of the bay Faxaflói primary production did not increase appreciably in early spring of 1966, although the critical depth (Sverdrup 1953) exceeded the depth to the bottom. Also there, the development of a moderate stratification was needed before the onset of an appreciable primary production could take place (Thórdardóttir and Stefánsson 1977).

Not only variations in the direction and velocity of winds will have an important effect of the timing of a substantial bloom in the waters south of Iceland. Also, variation from year to year in runoff will play a significant role, and the time of spring floods may vary considerably. Thus the years 1977 and 1979 were both very cold, especially 1979, and the spring floods did not occur until May 17 and May 31, respectively. On the other hand, the spring flood started as early as March 30 in 1981. This suggests that the retardation of spring blooming in the years 1977 and 1979, as compared to 1981, may be explained by the smaller fresh water runoff in spring during these years.

It is seen that zooplankton densities were relatively small in the area south of Iceland in the spring of 1979 (Fig. 9). It appeared in appreciable quantities only in small isolated patches inside a low salinity "pool", and in Faxaflói. These patches coincided with high primary production at the time in question (Fig. 4). On the other hand, the zooplankton densities were considerably greater in 1981. In early May it was most apparent in the eastern part of the area where vigorous plant growth was established already in late April (Fig. 4) and in late May the highest densities were found in the middle and western part. This year, 1981, belongs to the 3 years (the other two are 1973 and 1976) with the highest zooplankton density off the SW coast in spring during the period 1971-1981 (Astthórsson et al. 1983). In these three years the plant production started relatively early due to a wide distribution of fresh water, causing stratification of the surface layers (Thórdardóttir 1976, Friðgeirsson et al. 1978). The fresh water runoff and its horizontal distribution, largely governed by the

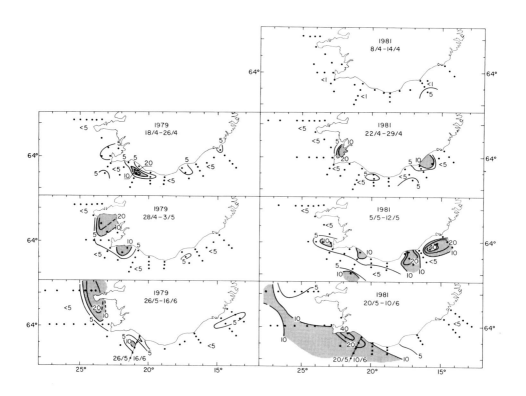

Fig. 9. Zooplankton volumes (ml per 21 m^3) sampled at different times with a Henson net from 50 m to the surface during spring of 1979 and 1981.

wind regime, thus appears to be a process of major importance for both primary and secondary production in the shelf area south of Iceland.

ACKNOWLEDGEMENT

I wish to thank my colleagues of the Marine Research Institute, Reykjavík, for making available data on temperature, salinity, nitrate and zooplankton. Thanks are also due to the Icelandic Meteorological Office and the National Energy Authority for kindly providing information on wind and springflood respectively. I am grateful to my co-workers at the botanical section of this institute who have assisted at all stages in the preparation of the paper. Last but not least I want to thank professor Unnsteinn Stefánsson who kindly read the paper and made many valuable suggestions.

REFERENCES

Anon. *Vedráttan*, ársyfirlit (yearly summary) 1976-1981.

Ástthórsson, Ó.S., I. Hallgrímsson and G.S. Jónsson 1983. Variations in zooplankton densities in Icelandic waters in spring during the years 1961-1982. *Rit Fiskideild.* 7:73-113.

Bainbridge, V. and B.J. McKay 1968. The feeding of cod and redfish larvae. *Spec. Publ. int. Comm. Northw. Atl. Fish.* 7. Norwestlant Part I:187-217.

Friðgeirsson, E. 1984. Cod larvae sampling with a large pump off SW-Iceland. In the Propag. of Cod Gadus Morhua L. Ed. by E. Dahl et al. *Flödev. Rapp. ser.* 1:317-333.

Friðgeirsson, E., S. Einarsson, E. Hauksson, J. Ólafsson and Th. Thórdardóttir, 1979. Environmental conditions and spring spawning off south and southwest Iceland 1976-1978. *ICES/ELH SYMP*. DA:5.

Hermann, F. and H. Thomson 1946. Drift bottle experiments in the Northern North Atlantic. *Medd. Dan. Fisk. Havunders.*

Ser. Hydrogr. 3,4.

Malmberg, S.A. 1968. Beinar straummælingar á hafi úti. Straummælingar í Faxaflóa 12-13/8 1966. *Náttúrufraedingurinn, 37:* 64-76.

Sakshaug, E. 1976. Dynamics of Phytoplankton blooms in Norwegian fjords and coastal waters. In *Freshwater on the sea* (Skreslet, S. et al. eds.). The Association of Norwegian Oceanographers, Oslo. 139-143.

Smayda, T. 1959. The seasonal incoming radiation in Norwegian and Arctic waters, and indirect methods of measurements. *J. Cons. Perm. Int. Explor. Mer. 24*:215-220.

Steeman Nielsen, E. 1952. The use of radioactive carbon (C-14) for measuring organic production in the sea. *J. Cons. Perm. Int. Explor. Mer. 18*:117-140.

Stefánsson, U. and G. Guðmundsson, 1978. The freshwater regime of Faxaflói, Southwest Iceland, and its relationship to meteorological variables. *Est. Coast. Mar. Scienc. 6*:535-551.

Sverdrup, H.U. 1953. On conditions for the vernal blooming of phytoplankton. *J. Cons. Perm. Int. Explor. Mer. 18*:287-295.

Thórdardóttir, T. 1976. The spring primary production in Icelandic waters 1970-1975. *ICES C.M.* no L 31.

Thórdardóttir, T. and U. Stefánsson, 1977. Productivity in relation to environmental variables in the Faxaflói region 1966-1967. *ICES C.M.* no. L 34.

PRODUCTION, GRAZING AND SEDIMENTATION IN THE NORWEGIAN COASTAL CURRENT *

Rolf Peinert
Institut für Meereskunde
Düsternbrooker Weg 20, D-2300 Kiel 1
FRG

* Contribution no. 5 of the Joint Research Programme 313 (SFB 313), University of Kiel

ABSTRACT

The relationship between pelagic system structure and sedimentation was studied during spring and summer in the Norwegian Coastal Current off the Lofoten Islands. Sedimentation was monitored directly with free drifting sediment traps and indirectly on the basis of a nutrient budget. The seasonality of processes leading to a loss by sinking or retention by recycling of essential elements in the pelagic system is discussed: Direct sedimentation of phytoplankton cells and phytodetritus following nutrient depletion impoverishes the pelagic system during spring when the food web is poorly developed. Copepod grazing retains essential elements, as faecal pellets are apparently recycled. Rapidly sinking euphausiid faeces, however, dominate sedimentation during summer. By inducing patchiness in spring and providing new nutrients during summer, the Norwegian Coastal Current will be well suited to exploitation by large, motile grazers such as euphausiids.

INTRODUCTION

Pelagic system structure in temperate and boreal latitudes is subject to seasonal changes which can be described as a succession from "new" production to "regenerated" production systems, following the terms of Margalef (1978), Eppley and Peterson (1979) and Smetacek (1984). In the Norwegian Coastal Current (NCC), accordingly, the spring bloom (primary production based on new nutrients) is followed by a steady-state system (regenerated nutrients) in summer stratified waters.

Freshwater discharge here enhances vertical stability, induces patchiness, can provide nutrients by entrainment and thus has a direct impact on pelagic system structure. The temporal and spatial patterns of production, grazing and sedimentation in the NCC are most likely linked to the time and space scales of the hydrographical structures. In this study new and regenerating systems, as encountered in the NCC, are described with special emphasis on loss or retention of essential elements.

METHODS

Data were collected on the north Norwegian shelf in spring 1983 along a transect and at three stations (A,B and C) and in summer 1983 at a frequently visited station located near B (Fig.1). Salinity profiles, as measured with a CTD-probe, were kindly made available by R.Wittstock, Institute for Applied Physics, Kiel University. Standard methods were used for collecting particulate and dissolved substances (Niskin-water bottles, planktonnets: 55μm and 300μm meshsize).

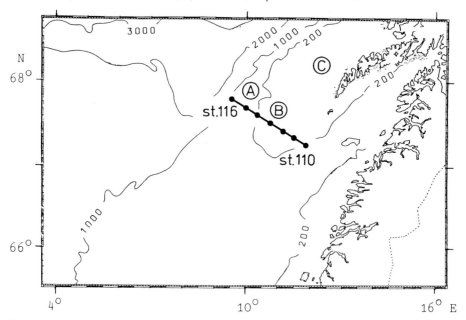

Fig.1. The investigation area in the Norwegian Coastal Current: transect across the shelf (st. 110-116), stations A,B and C with additional deployment of free drifting sediment traps.

Nutrients were measured with an autoanalyzer according to Graßhoff (1976). Particulates were filtered on Whatman GF/C filters (poresize 1μm) after prestraining through a 300μm gauze. Free drifting sediment traps (Zeitzschel et al. 1978), suspended at 60m depth for 24hr-intervals were deployed five times in spring and summer, respectively. Trap samples were suspended in artificial seawater, subsamples were also filtered on GF/C filters and analyzed for organic carbon (CHN-Analyzer, Hewlett Packard 185b) and chlorophyll $\underline{a}$. The equations of Jeffrey and Humphrey (1975) were used for calculating the chl.$\underline{a}$-content of the water column. As the preservative chloroform in the sediment traps converts chl.$\underline{a}$ into phaeopigment $\underline{a}$ (without any further breakdown according to Hendrikson, 1975), the equations of Lorenzen (1967) were used to estimate chl. $\underline{a}$ and phaeopigment $\underline{a}$ in the sedimented matter. Both were added together and are referred to as chlorophyll $\underline{a}$ equivalents. Phytoplankton and protozooplankton cell numbers in trap and water samples were counted under an inverted microscope (Utermöhl 1958) and converted to carbon using appropriate factors based on the respective plasma volumes. Primary production was measured by the ^{14}C-method ($\underline{in\ situ}$ simulated; Steemann-Nielsen 1958). Zooplankton was counted at the species level under a stereo microscope. Rough estimates of zooplankton grazing rates were obtained from calculated biomasses (using conversion factors from the literature) by assuming constant daily rations of 100% (protozoans) and 30% (mesozooplankton; mainly copepods) of body carbon.

RESULTS

The spring bloom
Salinity distribution showed the typical pattern with strongly stratified waters on the shelf and vertical homogeneity in the Atlantic water mass beyond the shelf break (Fig. 2a). Spring bloom development had not yet begun in the Atlantic water masses, as indicated by high nutrient reserves and low chlorophyll concentrations. In the NCC itself, however, large

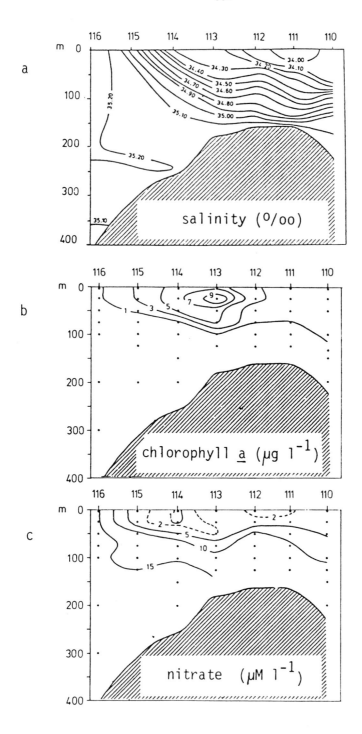

Fig.2. Isopleths for salinity (a), chlorophyll a (b) and nitrate (c) on a transect across the shelf.

variation in nutrient and chlorophyll concentrations showed the presence of different developmental stages of the spring diatom bloom not strictly coupled to salinity distribution on the smaller scale: Near the coast the bloom peak had passed (low chlorophyll and low nitrate concentrations), whereas on the middle of the shelf chlorophyll concentrations were still higher (Fig. 2b,c). Thus, within a distance of 60 nm, several weeks' difference in bloom development was present.

A rough nitrogen budget, comparing the winter nutrient load with measured concentrations (nutrients and particulates) was carried out applying the following equation:

$$L\% = \frac{(TIN_w + PON_w) - (TIN + PON + MZN)}{(TIN_w + PON_w)} \times 100$$

with L% : Nitrogen loss as percentage of nutrient concentration prior to the spring bloom
TIN_w: winter concentration of total inorganic nitrogen (NO_3, NO_2, NH_3 in µM l^{-1})
PON_w: winter particulate organic nitrogen load (1.4 µM l^{-1} assumed, as measured well below the productive zone)
TIN, PON: measured concentrations of total inorganic nitrogen and part. org. nitrogen (µm l^{-1})
MZN : Metazooplankton-nitrogen, as calculated from biomass estimates, assuming a C:N-ratio of 6:1 (atoms)

Føyn and Rey's (1981) finding that low salinity waters in the NCC in winter contained less nitrate than Atlantic water masses was taken into consideration when roughly estimating winter nutrient concentrations by extrapolating from values below the nitricline (Table I).

The two pools of reactive nitrogen not quantitatively measured in this study are the dissolved organic nitrogen (DON) and the biomass of large, migratory macrozooplankters such as euphausiids. Transfer to the former pool via phytoplankton exudation and sloppy feeding of zooplankton is unlikely to account for more than 20% of primary production (Larsson and

Hagström 1982). Incorporation into the macrozooplankton biomass (primarily euphausiids) will at most amount to a few percent. Thus, the losses of nitrogen from the upper 100m (Fig. 3) can be primarily attributed to sedimentation of particles.

Table I. Estimated winter nitrate concentrations in different water masses on the north Norwegian shelf.

Station	116	115	114	113	112	111	110
NO_3 (µM l^{-1})	16	15	13	13	13	10	10

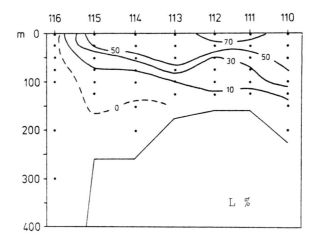

Fig.3. Isopleths for nitrogen losses (L%), as percentage of pre-bloom nutrient concentrations on a transect across the shelf.

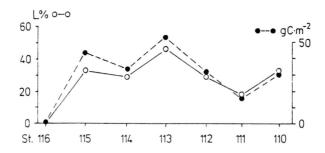

Fig.4. Nitrogen losses (L%), as percentage of pre-bloom nutrient concentrations for 100m water columns (left scale); estimated corresponding carbon losses (gC/m2) prior to the survey on a transect across the shelf (right scale).

Losses higher than 70% were estimated to have occurred in the coastal surface layer. In the mid-shelf region, layers well below the euphotic zone (approx. 40m according to Secchi-depth readings) had also suffered considerable nitrogen losses. In the unstratified high salinity Atlantic water, however, no losses had occurred. Results of calculations integrating over a 100m water column and then converting percentage nitrogen losses into absolute carbon losses, assuming a C:N-ratio of 6,6:1 by atoms (Redfield et al. 1963), are given in Fig. 4. These calculations yield extremely high losses due to sinking ranging up to 50g C m^{-2} for the period from spring bloom initiation to the date of the survey and show a factor of two variation for the stations on the shelf. Such a calculation, although being heavily biased, indicates that intensive sedimentation events must have taken place prior to the survey. Differences between the stations A, B and C (not presented here) support the view that sedimentation during spring is as heterogeneous in time and space as spring bloom development itself.

The pelagic system in spring and summer

The spring and summer situations on the Norwegian shelf are compared in terms of standing stocks of phytoplankton and zooplankton, primary production, grazing and sedimentation by means of a schematic representation (Fig. 5, 6). A characteristic spring situation (new production) was found near the shelf break (st. A). Because of the transitory character of the bloom and the prevailing patchiness related to bloom development, this is a momentary picture only. Phytoplankton biomass here was much higher than zooplankton standing stocks (mainly adults and copepodites of _Calanus finmarchicus_), which rarely exceeded 1gC m^{-2} (100m water column) at any station. Nutrients at this station were still far from being depleted (10 µM NO$_3$ l^{-1} in surface layers) and calculated nitrogen losses were still negligible (L% < 10%). Primary production exceeded by far the estimated heterotrophic removal by grazing (Fig. 5).

Table II. Daily carbon sedimentation ($mgC\ m^{-2}\ d^{-1}$) in spring and summer in the Norwegian Coastal Current (from 5 trap deployments)

	SPRING		SUMMER	
	average	range	average	range
part. org. carbon	50	30-117	30	17-66
phytoplankton-C	4	1-11	3	0.1-7
faecal matter-C	8	0.5-27	44	29-77

Measured sedimentation rates at this station at the shelf break, as well as at the others on the shelf proper, were low: Mean daily vertical carbon fluxes were 50 $mgC\ m^{-2}\ d^{-1}$ (Table II). These low recorded sedimentation rates do not necessarily contradict the high loss rates calculated from the nutrient budget. Observations from other areas (Smetacek 1980, v.Bodungen et al. 1981, Billett et al. 1983) have shown that spring bloom sedimentation is a short-term event during which a large proportion of the cells sink out en masse. The likelihood of encountering such an intensive sedimentation event, in the patchy and highly dynamic environment characteristic of the NCC in spring, with a few short-term trap deployments is rather low. The traps, however, provide useful information on the quality of sedimenting particles. Major constituents of sedimented matter were "fresh" amorphous phytodetritus (C:Chl.a-ratio $\leq$ 100:1) and, to a lesser extent, intact phytoplankton cells dominated by diatoms, and also faecal matter of euphausiids (Table II). The presence of euphausiids, whose biomass could not be determined quantitatively is thus documented by their faeces in the sediment traps. Copepod faecal pellets, however, did not contribute significantly to the vertical flux of particulates.

During summer an inverse biomass relationship was present throughout, with the mesozooplankton standing stock (dominated by adult Calanus finmarchicus and also Limacina retroversa) being much higher than the low phytoplankton biomass (Fig. 6). Protozoans, however, had not increased accordingly. High nutrient reserves below the thermocline (40m, > 5µM $NO_3\ l^{-1}$) were found on a transect located near that of the spring

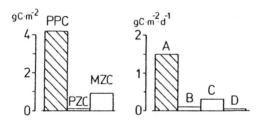

Fig.5. Schematic relation between standing stocks and flux rates in spring on the north Norwegian shelf (100m water column). Left: phytoplankton (PPC, shaded), protozooplankton (PZC) and mesozooplankton (MZC). Right: primary production (A, shaded), estimated daily grazing rates of protozooplankton (B) and mesozooplankton (C), measured daily sedimentation rates (D).

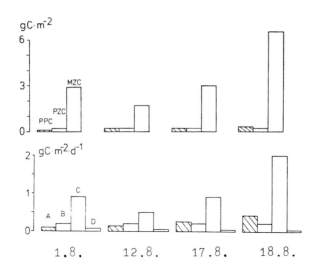

Fig.6. Standing stocks and flux rates (100m water columns) at 4 stations (near B in fig.1) in summer on the north Norwegian shelf. Upper: phytoplankton (PPC, shaded), protozooplankton (PZC) and mesozooplankton (MZC). Lower: primary production (A, shaded), estimated daily grazing rates for protozooplankton (B) and mesozooplankton (C), measured daily sedimentation rates (D).

cruise: apparently the lowered nutrient concentrations found at a depth of 50m on the mid-shelf during spring had been replenished in the intervening period by water transport. The low primary production recorded in summer could not have met the calculated heterotrophic demand. Further, measured sedimentation rates were of the same order of magnitude as during spring. The composition of the sedimented matter, however, showed major differences as compared to spring: euphausiid faeces almost entirely dominated trap collections and contained large numbers of fragments of tintinnid shells (Parafavella sp.), indicative of a food chain from phytoplankton to protozoans and euphausiids during this season. Copepod biomass was 3-6 times as large as during spring; their faecal pellets, however, again only contributed insignificantly to vertical carbon flux. Phytoplankton (no diatoms collected) was also only of minor importance (Table II).

DISCUSSION

In spring blooms, as "new-production systems" characterized by an imbalance of primary production and remineralisation, sedimentation is the fate of a large portion of the phytoplankton biomass with nutrient depletion being one major triggering mechanism. This is a well known fact from shallow water systems (Smetacek 1980, v.Bodungen et al. 1981, Peinert et al. 1982) and there is increasing evidence that a sedimentation pulse of freshly produced matter takes place following the spring bloom in open oceanic waters as well (Deuser and Ross 1980, Billett et al. 1983, Lampitt et al. in press). With regard to the fate of essential elements, it is appropriate to categorize processes within the surface pelagic system according to their impact on the fate of the primary produced matter and thus to distinguish between "process chains" leading to loss ("sedimentation chains") or to retention of these elements ("retention chains"). The processes leading to build-up of large phytoplankton biomass, particularly in the case of diatoms, and their eventual mass sedimentation is an example of such a "sedimentation chain". The grazing activity of macrozooplankters (euphausiids, salps; Schnack in press, Alldredge 1984) and herbivorous fish such as anchoveta (Staresinic et al. 1984) with their fast sinking fecal matter also enhance losses via sedimentation. Grazing by protozoans and mesozooplankters such as copepods, whose excretion products are to a great extent recycled in the pelagic food web, in contrast, are "retention chains", as they conserve biogenic elements within the surface system. Thus, typical spring bloom dynamics are favoured, when the size of the overwintering copepod standing stock is small and when its growth behaviour is not adjusted to cope with rapid spring phytoplankton growth. In this conceptual framework, the seasonal succession from "new production systems" (spring bloom) to "nutrient regenerating systems" (summer) results from the interaction between different sedimentation and retention chains, each of them enfolding on its own characteristic time and space scale.

The impact of freshwater outflow on marine systems such as the NCC in terms of sedimentation and retention of particulates is difficult to assess but some speculations can be offered. It is for instance likely, that the time and space scales of the physical, salinity-related structures have a twofold influence. The first being effective at the base of the food web, inducing patchiness and a phase-shifted development of the spring bloom, as well as potentially enhancing nutrient availability during summer by entrainment. Apart from the direct effects on phytoplankton, it can be assumed that conditions in the NCC are conducive to large macrozooplankters living in swarms, such as euphausiids, if the scales of phytoplankton patchiness are compatible with their horizontal migratory capabilities.

In shallow coastal waters where copepod grazing pressure is low because of small overwintering standing stocks, heavy sedimentation of diatoms which deplete the water column of essential elements in spring is the rule. In deeper waters, maintaining larger overwintering copepod stocks, grazing pressure is more intense during bloom growth and hence losses due to sedimentation of phytoplankton cells is reduced. Of course, copepods will also contribute to losses (e.g. by vertical migration or being predated by carnivores) but their overall impact, compared with an ungrazed diatom bloom, will be to retain essential elements in the surface layer. In case new nutrients are not available during summer stratified conditions, the size of the overwintering copepod standing stock and its feeding activity during the spring bloom will hence control the amount of essential elements retained within the productive layer. This conditions the future feeding grounds for the copepods as well as those for the euphausiids. The latter, however, "undermine" the retention activity of the copepods because of their fast sinking faecal matter. In this context, a possible introduction of new nutrients during summer makes the NCC particularly well suited for exploitation by swarms of large, motile grazers as the losses engendered by their feeding would be replenished eventually, at the same time ensuring copepod food resources. Further, because of the close proximity of water masses with different temporal development - NCC water and Atlantic water - such motile macrozooplankters can move to new feeding grounds after depletion of the old. Lateral displacement of the NCC-Atlantic front by regulation of freshwater discharge could thus have an effect on the patterns of interlocking sedimentation and retention chains on the Norwegian shelf.

REFERENCES

Alldredge, A.L. 1984. The quantitative significance of gelatinuous zooplankton as pelagic consumers. in: The flows of energy and materials in marine ecosystems, M.J.R. Fasham (ed.), Plenum Press, N.Y., pp 407-434.

Billett, D.S.M., R.S. Lampitt, A.L. Rice, R.F.C. Mantoura 1983. Seasonal sedimentation of phytoplankton to the deep-sea benthos, Nature 302, 520-522.

Bodungen, B.v., K.v.Bröckel, V. Smetacek and B. Zeitzschel 1981. Growth and sedimentation of the phytoplankton spring bloom in the Bornholm Sea (Baltic Sea). Kieler Meeresforsch., Sonderheft 5, 49-60.

Deuser, W.G. and E.H. Ross 1980. Seasonal change in the flux of organic carbon to the deep Sargasso Sea. Nature 283, 364-365.

Eppley, R.W and B.J. Peterson 1979. Particulate organic matter flux and planktonic new production in the deep ocean. Nature, 282, 677-680.

Føyn, B.R. and F. Rey 1981. Nutrient distribution along the Norwegian Coastal Current. in: The Norwegian Coastal Current, R.Saetre and M. Mork (eds.), Bergen, pp 629-639.

Graßhoff, K. (ed.) 1976. Methods of seawater analysis. Verlag Chemie, Weinheim, 317 pp.

Hendrikson, P. 1975. Auf- und Abbauprozesse partikulärer organischer Substanz anhand von Seston- und Sinkstoffanalysen. Diss.Univ.Kiel, 160pp.

Jeffrey, S.W. and G.F. Humphrey 1975. New spectrophotometric equations for determining Chlorophylls a, b, c_1 and c_2 in higher plants, algae and natural phytoplankton. Biochem. Physiol. Pflanzen 167, 191-194.

Lampitt, R.S. 1985. Evidence for the seasonal deposition of detritus to the deep-sea floor and its subsequent resuspension. Deep-Sea Res., in press.

Larsson, U. and A. Hagström 1982. Fractionated phytoplankton primary production, exudate release and bacterial production in a Baltic eutrophication gradient. Mar. Biol. 67, 57-70.

Lorenzen, C.J. 1967. Determination of chlorophyll and phaeopigments: spectralphotometric equations. Limnol. Oceanogr. 12, 343-346.

Margalef, R. 1978. Life-forms of phytoplankton as survival alternatives in an unstable environment. Oceanol. Acta 1, 493-504.

Peinert,R., A.Saure, P.Stegmann, C. Stienen, H. Haardt and V. Smetacek 1982. Dynamics of primary production and sedimentation in a coastal ecosystem. Neth. J. Sea Res. 16, 276-289.

Redfield, A.C., B. Ketchum and F.A. Richards 1963. The influence of organisms on the composition of sea water. in: The Sea, 2, M.N. Hill (ed.), Wiley, N.Y., 26-77.

Schnack, S. 1985. A note on the sedimentation of particulate matter in Antarctic waters during summer. Polar Biol., in press.

Smetacek, V. 1984. The supply of food to the benthos. In: M.J.R. Fasham (ed.), Flows of energy and materials in marine ecosystems: theory and practice. Plenum Press, 517-548.

Smetacek, V. 1980. Annual cycle of sedimentation in relation to plankton ecology in western Kiel Bight. Ophelia 1, 65-76.

Staresinic, N., Hovey Clifford, C. and E.M. Hulburt 1984. Role of the southern Anchovy, *Engraulis ringens*, in the downward transport of particulate matter in the Peru coastal upwelling. in: Coastal upwelling: its sedimentary record. E.Suess and J. Thiede (eds.), Plenum Press, N.Y.

Steemann-Nielsen, E. 1958. Experimental methods for measuring organic production in the sea. Rapp.P.-v.Cons. int. Explor. Mer. 144, 38-46.

Utermöhl, H. 1958. Zur Vervollkommnung der quantitativen Phytoplankton-Methodik. Mitt. Int. Ver. Limnol. 9, 1-38.

Zeitzschel, B., P. Diekmann and L. Uhlmann 1978. A new sediment trap. Mar. Biol. 45, 285-288.

ADVECTION OF CALANUS FINMARCHICUS BETWEEN HABITATS IN NORWEGIAN COASTAL WATERS.

S. Skreslet and N.Å. Rød
Nordland College
N-8001 Bodø, Norway

ABSTRACT

The exchange of the copepod Calanus finmarchicus between the Vestfjord and the Norwegian coastal current was studied by net sampling and hydrography. The species was probably recruited to wintering habitats in the Vestfjord from reproduction habitats in the frontal zone of the Norwegian coastal current. The wintering generation spawns in the Vestfjord in late winter, but their offspring are flushed out by freshwater outflow, possibly recruiting to the spawning habitat of a shelf break front to the west of the Lofoten Islands. Offspring from this habitat seems to accumulate at a thermal front farther offshore. C. finmarchicus is to some extent transported by water of coastal origin, into the central Norwegian Sea.

INTRODUCTION

Calanus finmarchicus is the dominant copepod in Norwegian coastal waters (Wiborg 1954). It stays in deep basins during winter, mainly as copepodid stage V, i.e. the last stage before moulting to sexually mature males and females. The Vestfjord, a wide bay between the Lofoten Islands and the mainland (Fig. 1), is a major wintering habitat for the species. Sømme (1934) suggested that C. finmarchicus is stationary in the Vestfjord only during mid winter, and that the stock is dispersed by surface currents to shelf waters during spring. On the other hand, Wiborg (1954) found that the average size of copepodid V which accumulate near the Lofoten Islands during summer, may indicate a stationary stock. The present investigation assesses the exchange of C. finmarchicus between the Vestfjord and the

Norwegian coastal current which flows essentially to the north (Sætre & Ljøen 1971). However, during summer, a seaward vector (Haakstad 1977) feeds coastal water into the central Norwegian Sea (Sætre 1966), due to freshwater outflow. Thus, we have also studied how the dispersal of C. finmarchicus is influenced by dynamical features of the Norwegian coastal current.

METHODS

Retention of C. finmarchicus in the Vestfjord was studied from February to December 1977, by zooplankton sampling in wintering habitats (Stations B and F) identified by Sømme (1934) and where coastal water flows into the Vestfjord (Sts C, D and E) according to Sundby (1978). Sts G and H were established in a geostrophic current advecting local runoff to the west. Station I was established to the north of the Lofoten Islands (Fig. 1). On each station five replicate vertical tows with a 180u, 0.1 sq.m. Juday net were taken from about 10 m above the bottom to the surface, and from 50 m depth to the surface. Temperature and salinity was recorded in standard depths by Nansen bottles and reversing thermometers. Currents were recorded by an Aanderaa RCM-4 current meter in 15 m depth on Stn D, to assess the flow of coastal water into the Vestfjord.

To study the production of C. finmarchicus in shelf habitats which exchange water with the Vestfjord, a 130 n.mile transect of 14 sampling stations to the north-west of the Lofoten Islands, was sampled from 2330hr on 2 July to 0400 on 4 July 1983. A modified Bongo 200 net was towed over the distance of 10 n. miles between each station, to outrule patchiness (cf. Wiebe 1971). The sampling occurred at about 6 m depth, at a speed of 5 knots. To avoid clogging by large samples and allow for the high speed, the net openings were fitted with hemispherical bulbs with apertures of 5.0 and 2.5 cm, allowing for estimated volumes of 36 and 9 cu.m. to be filtered through nets with 180 and 60u mesh, respectively. These samples were on each station checked against one vertical tow from 200 m depth, or from the bottom over shallow shelf areas, by a 180μ, 0.1 sq.m. Juday net. Water samples were at 5 n. miles intervals, including the Juday stations, pumped from 2 m depth for measurements of temperature and salinity.

Water samples were analysed for salinity by a laboratory conductivity meter (C.S.I.R.O.). Plankton samples were stored in 4o/o formaldehyde and worked up in a few months. Large samples were subdivided by a Wiborg-Lea plankton divider. Numbers of C. finmarchicus copepodids collected in 1977 were based on direct counts. This was also the case with copepod nauplii sampled by the 60u Bongo net in 1983, but they were not identified to species and moulting stage. The remaining groups of zooplankton sampled in 1983 were quantified by a "short-cut" method where 100 specimens were identified and copepdids of C. finmarchicus grouped according to moulting stage. Total numbers were calculated from direct counts of one C. finmarchicus stage, obtained from the whole sample or a subsample. Rare species not present in the "short-cut" sample were recorded during this procedure.

The cephalothorax length of C. finmarchicus was measured by stereo microscope and stage micrometer in up to 110 specimens. A chi-square test for two independent samples (Siegel 1956) was

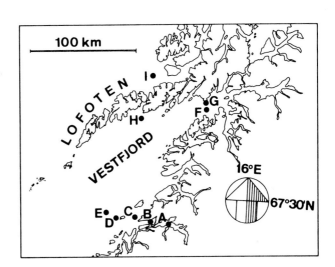

Fig. 1.
The Vestfjord, north Norway, with sampling stations 1977.

used to test the significance of class differences in copepodid stage distribution between stations. For this purpose five sub-samples, 1/10 of the total volume of each sample, were lumped to outrule patchiness.

RESULTS

C. finmarchicus wintering in the Vestfjord in 1977 were copepodids III-VI. In February average concentrations of about 40000 per sq.m. sea surface were recorded in tows from 350 and 500 m depth to surface at Sts B and F, respectively. Concentrations were lower than 7000 per sq.m. sea surface at Sts A, C, D and E, where tows were made from 500, 200, 190 and 190 m depth, respectively. Tows from 50 m depth to the surface at all these stations, and the shallow Sts G, H and I, featured concentrations lower than 500 per sq.m. sea surface.

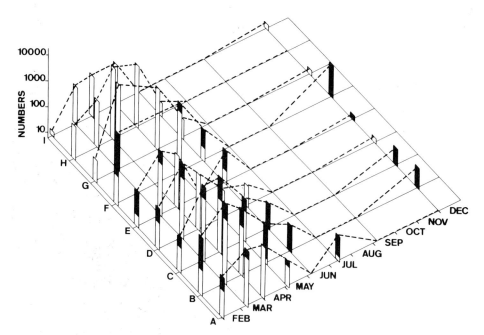

Fig. 2. Average number (log scale) of female C. finmarchicus in vertical Juday tows from 50 depth to the surface (open part of columns) and from bottom to surface (entire columns) at stations in the Vestfjord 1977.

Table I. Average number per sq.m. sea surface, of different C. finmarchicus copepodid moulting stages on four sampling stations in the Vestfjord 1977. Results from the chi-square tests compares stage distributions on Stn D with Stn F.

		Copepodid moulting stage								
St	Date	I	II	III	IV	V	IV	df	X2	p
F	3Feb				3980	28240	5240			
D	8Feb				720	6160	720	2	6.6	.05
D	29Mar					140	840			
F	2May	3060	1620	600	20	100	3140			
D	10May	3080	6400	8060	100	180	40	3	479.8	.001
F	7Jun	1960	14460	21640	4900	1080	120			
D	31May	1420	5960	12420	5240	2040	40	4	121.7	.001
F	13Jul		20	740	9640	2620				
D	12Jul	20	100	200	900	240		2	242.4	.001
F	8Sep			180	4260	3500				
D	31Aug				700	600		1	0.1	.8
F	29Nov				1480	5320	340			
D	5Dec				360	900		1	1.9	.2

Concentrations of females culminated at Sts A-E in March, and then declined. They were still high at Sts F-I in May, but were few at all stations in June, and in July only present at Stn A. None were found in September, but a few occurred in December (Fig. 2).

A new generation was present in May and gradually moulted into higher stages, but did not develop mature females in July-September. Proportionally, there were more copepodids in low stages at Stn F than at Stn D in May and June, but in July there was a predominance of advanced stages at Stn F (Table I).

Copepodids stage V had in February shorter cephalothorax than those sampled during summer. Those with a cephalothorax length of about 3 mm present in samples from May to July, were not represented in later samples, and the minimum size decreased from June to September. Specimens sampled from bottom to surface, were consistently larger than those sampled from 50 m depth to the surface (Fig. 3).

There were noticable reductions in surface salinities from 31 May to 12 July at Stn C, from 7 June to 13 July at Stn F, from 10 March to 4 April at Stn H, and from August-September to November-December at all Vestfjord stations. Surface temperatures in March were close to 4 C at Stn E, and lower at the other stations.

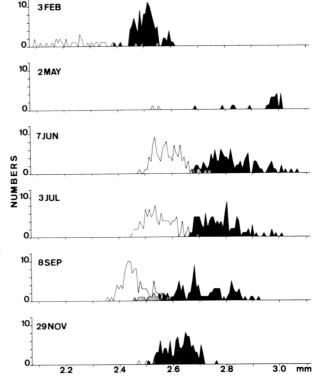

Fig. 3.
Frequency distribution of cephalothorax length of C. finmarchicus copepodid V taken in Juday tows from 600 m depth to surface (filled) and from 50 m depth to surface (open) at Stn F, Vestfjord 1977. Stipled: Overlap.

The net water transport at 15 m depth, Stn D, equals a distance of 1000 km from 15 February to 1 June 1977 (Fig. 4).

In the transect made to the north-west of the Lofoten Islands from 2 to 4 July 1983, surface salinities increased from less than 34 So/oo near-shore to 35 So/oo at the shelf break, and stayed high farther out. The temperature at the same depth, was higher than 10 C from the coast to 80 n.miles off-shore, while it decreased to 10 C or lower farther into the Norwegian Sea (Fig. 5).

Female *C. finmarchicus* were present in horizontal Bongo tows, in a zone between 30 and 50 n. miles off-shore, in numbers close to 100 per cu.m. The number between 120 and 130 n. miles off-shore was 470 per cu.m. (Fig. 5, Table II). The cephalothorax length of females from the two areas varied between 2.3 and 3.2 mm and there were no significant differences between the size frequency distributions. Females were absent or present in just low numbers in all other Bongo tows. They were also present in low numbers in vertical Juday tows 30 to 90 n.miles off-shore (Fig. 5).

C. finmarchicus copepodids I-III were numerous in vertical Juday tows at 90 and 100 n.miles off-shore, but also present in numbers higher than 150 per cu.m. in both Bongo and Juday tows farther into the Norwegian Sea, and in the near-shore zone inside 50 n.miles. Copepod nauplii were very numerous in both Bongo and Juday samples in the 30-50 n.mile zone, but were also

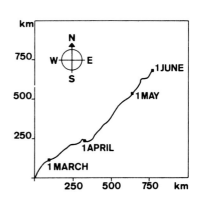

Fig. 4.
Progressive vector diagramme for water transport in 15 m depth at Stn D, Vestfjord 1977.

quite numerous in Juday samples 90 and 100 n.miles off-shore (Fig 5). Early copepodids dominated among C. finmarchicus sampled 40-50 n.miles off-shore, while advanced copepodids dominated 120-130 n.miles off-shore. Other copepod species were few in both areas (Table II).

Larvae of the family Galatheidae (Decapoda Anomura) were present in Bongo samples from the near-shore zone inside 30 n.miles and from 110-120 n.miles off-shore. Four specimens selected at random from each zone, all belonged to the species Munida sarsi, and were in their zoea stage III or IV.

DISCUSSION

The observation that specimens of C. finmarchicus copepodid V, were consistently smaller in tows from 50 m depth, than from

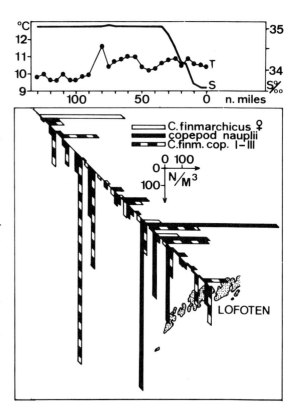

Fig. 5. Salinity and temperature at 2 m depth (upper graph) and average density of copepod nauplii, females and copepodids I-III of C. finmarchicus sampled by 10 n.mile Bongo tows (horizontal columns), and Juday tows (vertical columns) from 200 m depth or the bottom to the surface (lower graph), in a transect from Gimsøy in Lofoten to N 70° 14' E 08°58' in 1983.

Table II. Average number of copepods per cu.m. of seawater filtered by 180µ Bongo net at 6 m. depth over 10 n.miles at 40-50 (A) and 120-130 n.miles (B) to the northwest of the Lofoten Islands, 2-4 July 1983.

	A	B
C. finmarchicus females	88	470
-+- copepodid V	69	224
-+- -+- IV	178	213
-+- -+- I-III	275	96
Copepod nauplii	772	92
Copepodids of indet. species	19	53

the bottom (Fig 3), indicates that individuals near the surface avoided the net when towed from large depths, probably due to disturbances from the winch wire. This feature questions the representativity of samples from vertically towed nets, as it will obviously vary with the seasons, depending on the vertical distribution of the plankton. We therefore choose to interpret our Juday samples with care.

High concentrations of C. finmarchicus in deep pelagic habitats at Sts D and F in February, and low concentrations near the surface, support Sømme (1934) that the wintering stock occupies particular parts of the Vestfjord region. The copepodids moulted to sexual maturity in late winter (Fig. 2) and by May a new generation was evident (Table II). The uniform disappearance of females from May to July, may be explained by mortality, but seems contradicted by Sømme (1934) who kept spring spawning females alive in captivity until early August. It may at least partly be due to the extensive seaward advection of surface water along the Lofoten Islands (Sundby 1978) where local runoff lowered surface salinities in April and probably increased the advection. Flushing due to freshwater discharge from the mainland lowered surface salinities during summer, and may be the reason why the new generation did not form a new stock of summer spawners in the Vestfjord (Table II).

Copepodids stage V had in May and June larger body size than that of the parent generation in February (Fig 3). As moulting stage size in C. finmarchicus is negatively correlated with temperature (Matthews 1966), it is probable that the larger individuals of the new generation hatched in cold fjord water early in the season. This fits with the surface temperatures in the Vestfjord, being lower than 4 Centigrades in March, but considering the extensive surface advection into the Vestfjord at Stn D in this period (Fig. 6), a substantial part of these copepodids may have originated from equally cold fjords farther south.

The appearance of smaller specimens in July (Fig. 3), indicates recruitment of copepodids hatched after the seasonal increase in temperature had begun and, may be, advection from warmer habitats in the Norwegian Sea (Cf. Østvedt 1955, Wiborg 1955, Lie 1968). We can not exclude that the smallest copepodids present near the surface at Stn F in June, were retained and contributed to the wintering stock present in December, but find it difficult to accept this explanation. A retention of recruits ought to result in a predominance of older copepodids on the wintering habitats than in localities where they enter the Vestfjord, for instance at Stn D where coastal water runs into the area (Sundby 1978). In July there was a predominance of significantly more advanced stages at Stn F which could indicate a retention of copepodids, but it was not manifest in September (Table II). The disappearance of copepodid V with a cephalothorax longer than 2.8 mm, from September to November, was associated with lowered sea surface salinities, probably due to a rather weakly understood annual exchange of water between fjords and the Norwegian coastal current at this time of the year (Haakstad 1979). This exchange probably also introduced the wintering stock which thus seems to be established in the Vestfjord not before late autumn. Their small size (Fig. 3) indicate ontogeny in warm water. The observation that C. finmarchicus spawns vigorously in the outer margin of the Norwegian coastal current in June-July (Ruud 1929), may indicate that the wintering stock in the Vestfjord, has originated from such a spawning habitat, somewhere along the Norwegian shelf.

The presence of females and copepodids I-III of *C. finmarchicus*, and high numbers of copepod nauplii, in the haline front over the shelf break outside the Lofoten Islands in early July 1983 (Fig. 7), replicate the observation by Ruud (op.cit.) and indicate that a summer spawning habitat of this species may have a wide extension along the coast. The few females and low spawning activity over the shelf itself is in accordance with Wiborg (1954) who sampled frequently on a near-shore station throughout two summers. This indicates that the spawning habitat of the haline shelf break front is ecologically different from the central core of the coastal current.

An additional, thermal front was present 90 n. miles off-shore. It was associated with high numbers of copepod nauplii and *C. finmarchicus* copepodids I-III sampled by vertical Juday tows (Fig. 7). Few specimens were taken in the Bongo tows in this area, probably due to sampling made during day-time when the plankton had migrated into deeper water. Lack of females at the thermal front and low abundance of the species between the two fronts, makes us suggest that this front recruited young stages from the haline shelf break front by fast advection.

The absence of significant size differences between females from the shelf break front and those sampled 130 n. miles off-shore, may indicate a common origin. Larvae of *Munida sarsi* which is a common benthic decapod in Norwegian fjords (Brinchmann 1936) were present among the latter females as well as over the shelf, and make us believe that the plankton at the outermost station were of coastal origin.

Transport of coastal water into the central Norwegian Sea has been demonstrated by Halldal (1953) and Sætre (1966). Wiborg (1955) found that neritic zooplankton was distributed in a plume-like fashion from the Lofoten Islands into the Norwegian

Sea, and suggested that small C. finmarchicus found in Greenland waters originated from the southern Norwegian Sea, or from the coast of western or northern Norway. Thus, there is some evidence that the progression of the "biological spring" from the eastern to the western Norwegian Sea, as demonstrated by Pavshtiks and Timokhina (1972), may be due to advection. The seaward advection may be particular during large freshwater outflow, and less important during other seasons (Cf. Haakstad 1977).

ACKNOWLEDGEMENTS

The Norwegian Fisheries Research Council financed our work in 1977 (Project No I 718.03). We thank Magne Haakstad for good advice on hydrography and all assistants for keen support.

REFERENCES

Brinchmann, A. 1936. Die nordischen Munida-arten und ihre Rhizochephalen. Bergens Mus. Skr. 18, 1-111.
Haakstad, M. 1977. The lateral movement of the coastal water and its relation to vertical diffusion. Tellus 29, 144-150.
Haakstad, M. 1979. Reexchange of water between the Norwegian coastal current and the fjords in autumn. Nordland College, Section of Natural Science, Rep. No. 4/79.
Halldal, P. 1953. Phytoplankton investigations from weather ship M in the Norwegian Sea, 1948-49. Hvalrådets Skr. 35, 1-191.
Lie, U. 1968. Variations in the quantity of zooplankton and the propagation of Calanus finmarchicus at station "M" in the Norwegian Sea, 1956-1966. Fiskeridir. Skr. Ser. Havunders. 14 (3), 121-128.
Matthews, J.B.L. 1966. Experimental investigations of the systematic status of Calanus finmarchicus and C. glacialis (Crustacea: Copepoda). pp 479-492 in H. Barnes, Some contemporary studies in marine science. Allen- Unwin Ltd., London.
Pavshtiks, E.A. and A.F. Timokhina 1972. History of investigations on plankton in the Norwegian Sea and the main results of Soviet investigations. Proc. R. Soc. Edinburgh (B) 73, 267-278.
Ruud, J.T. 1929. On the biology of copepods off Møre 1925-1927. Rapp. P. V. Reun. Cons. Int. Explor. Mer 56 (8), 1-84.
Siegel, S. 1956. Nonparametric statistics for the behavioral sciences. McGraw-Hill Kogakusha, London. 312 pp.

Sundby, S. 1978. In/out flow of coastal water in Vestfjorden. ICES CM 1978/C:51, 17 pp (Mimeo).

Sætre, R. 1966. De øvre vannlag på værskipstasjon M. M.Sc. thesis, University of Bergen, 62 pp.

Sætre, R. and R. Ljøen 1971. The Norwegian coastal current, pp 514-535 in Proceeding POAC Conference, The Technical University of Norway, Trondheim.

Sømme, J.D. 1934. Animal plankton of the Norwegian coast waters and the open sea I. Production of Calanus finmarchicus (Gunner) and Calanus hyperboreus (Krøyer) in the Lofoten area. Fiskeridir. Skr. Ser. Havunders. 4, 1-163.

Wiborg, K.F. 1954. Investigations on zooplankton in coastal and offshore waters of western and north-western Norway. Fiskeridir. Skr. Ser. Havunders. 11 (1), 1-246.

Wiborg, K.F. 1955. Zooplankton in relation to hydrography in the Norwegian Sea. Fiskeridir. Skr. Ser. Havunders. 11 (4), 1-66.

Wiebe, P.W. 1971. A computer model study of zooplankton patchiness and its effects on sampling error. Limnol. Oceanogr. 16, 29-38.

Østvedt, O.J. 1955. Zooplankton investigations from weather ship M in the Norwegian Sea, 1948-1949. Hvalrådets Skr. 40, 1-93.

THE ECOLOGICAL IMPACT OF THE EAST GREENLAND CURRENT ON THE NORTH ICELANDIC WATERS

Svend-Aage Malmberg
Marine Research Institute
Reykjavík, Iceland

ABSTRACT

Weight and growth of cod and capelin of different year classes in Icelandic waters are compared with variations of hydro-biological conditions in North Icelandic waters. The main results show a postive relation of these fish stocks to hydrographic conditions. The growth of capelin seems to be related to the hydrographic conditions but the growth and maturity of cod related to the stock size of capelin and/or to the occurence of arctic water in the North Icelandic shelf area.

INTRODUCTION

The current system in the Iceland Sea north and east of Iceland is more or less fed by waters from the warm and saline Irminger Current (t>4°, S>35.0) and the cold East Greenland Current (t<0°, S<34.0), resulting in the arctic or polar current named the East Icelandic Current (t<0°,<S 34.8; Fig. 1; Knudsen 1898, Kiilerich 1945, Stefánsson 1962, Swift and Aagaard 1981). The East Icelandic Current, an ice-free arctic current in the period from approximately 1930 to the year 1963 later advanced south and eastwards and developed into a polar current in 1964-1971, transporting and maintaining drift ice (Malmberg 1969, 1972, 1984a).

When summarized, the results of hydrobiological investigations in North Icelandic waters show that Atlantic water (t>4°, S>35.0) dominated in spring during the period 1924-1964, in 1972-1974, in 1980 and in 1984-1985 (Table 1, Figs. 2 and 3). The late sixties (1965-1971), however, as well as shorter periods thereafter (1975, 1979) were characterized by polar influence. These changes were manifested by the appearance of sea ice (Sigtryggsson 1972) and had biological implications in North Icelandic waters, viz. low nutrient supply, reduced spring production, low zooplankton concentrations and changed migration pattern of herring and even the collapse of herring fisheries in Icelandic waters (Thórdardóttir 1977, 1980, Jakobsson 1980, Ástthórsson et al. 1983, Malmberg 1984b). The changes were also linked to atmospheric pressure variations as well as to variations in the physical marine environment throughout the northern North Atlantic (Dickson et al. 1975, Malmberg and Svansson 1982).

In 1981 the hydrobiological conditions in North Icelandic waters were extremely unfavourable, of neither Atlantic nor polar character but with very homogenous water of an arctic character (t= 1-3°, S~34.8; Fig. 3). These arctic conditions may possibly be related to the so-called "mid-seventies salinity anomaly" (Dooley et al. 1984, Dickson et al. 1984a,b), i.e. a decrease in salinity observed in the northern North Atlantic in the late seventies. It may either be directly connected to an increased outflow of polar water from the European Arctic and Subarctic Seas during the ice-years in the late sixties, or indirectly through atmospheric variations, traced from the Iceland-Greenland Channel (1968) to Labrador (1972) and from there back to the coasts of Europe (1976) and into the European Subarctic (1979-1981; see also Aarkrog et al. 1983). Whether or not this is the case, the hydrobiological conditions in North Icelandic waters were most unfavourable in 1981-1983 (Malmberg 1984b).

It is well known that the main spawning grounds of Icelandic capelin and cod are found in the warm water south of Iceland, whereas the nursery and feeding grounds are mainly found north of Iceland. The purpose of this paper is to precent an overwiew of the hydrographic conditions in North Icelandic waters in the period 1958-1984, and the biological impact of the environmental variations.

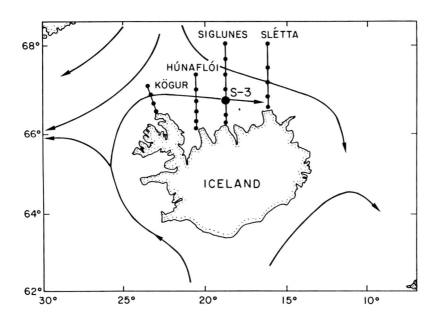

Fig. 1. Main ocean currents and location of standard hydrographic sections in Icelandic Waters.

Table 1. Indices of hydrobiological conditions in North Icelandic Waters 1920 - 1984

Year	Sea Ice	t°C	S	Phpl.pr.	Zoopl. conc.	0-gr. capelin	0-gr. cod	3 year cod
1920-57	0							
1958	0							
1959	0							
1960	0							
1961	0	5	35.0	11	58			289
1962	0				79			254
1963	0				37 }55			271 }285
1964	0				46			326
1965	3				8			172
1966	1				7			252
1967	1-2	0-3	34-34.5	5	4 }23			185
1968	3				22			177 }198
1969	2-3				7			135
1970	2			11	18	2 }7	848 }550	300
1971	1-2			19	23	12	214	168
1972	0			22	34	52	36	263
1973	0	5	35.0	7 }18	21 }21	46 }52	757	427 }278
1974	0			25	9	57	30	144
1975	1-2	2		10	11	46	73	222
1976	0	4		27	21	39	2015	249
1977	1-2	2	34.2-34.8	8	40 }23	19 }30	306	139 }173
1978	0	4		21	26	29	334	138
1979	2-3	1		9	22	25	345	115
1980	0	5	35.0	25	19	20	507	240
1981	0			28	12	6	19	170
1982	0	1-3	34.5-34.8	21	6 }9	5 }10	4 }30	(100 }153
1983	0				8	18	66	(190)
1984	0	3-4	34.8		29	17	826	

Table 1. Indices of hydrobiological conditions in North Icelandic waters in 1920-1984.

Key:
Sea ice: 0 ice-free, 1 insignificant ice, 2 moderate ice.
t°C: Characteristic temperature in the upper layers off Siglunes in spring.
S: Characteristic salinity in the upper layers off Siglunes in spring.
Phpl.pr.: Index of primary production in spring on sections Kögur, Siglunes and Slétta (Fig. 1).
Zoopl.conc.: Index of zooplankton concentration in spring on sections Kögur, Húnaflói, Siglunes and Slétta (Fig. 1).
0-gr. capelin/cod: Abundance indices of 0-group capelin and 0-group cod in North Icelandic waters in August.
3 year cod: Recruitment of 3 year old cod in Icelandic waters according to VPA.
t°C∼5° and S∼35.0: Atlantic water; 1920-1964, 1972-1974, 1980.
t°C 0-4° and S 34.0-34.8: Polar water; 1965-1971, 1975-1979.
t°C 1-3° and S 34.5-34.8: Arctic water; 1981-1983.

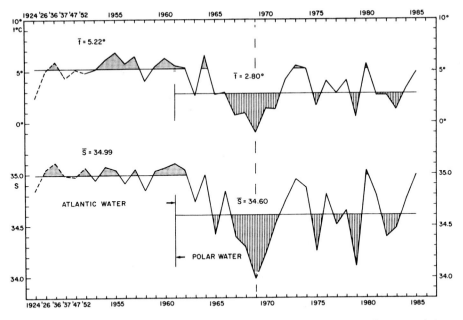

Fig. 2. Temperature and salinity at 50 m on a hydrographic station in North Icelandic waters (S-3; for location see Fig. 1) in May-June 1924, 1926, 1936, 1937, 1947 and 1952-1985.

MATERIAL

Information on the hydrobiological observations made in annual spring surveys into North Icelandic waters is given in Table 1. The 0-group estimates are from surveys in August in these same waters, and the estimates on recruitment of 3 year old cod are based on VPA from catches all around Iceland. The information on sea ice are from the Icelandic Meteorological Office (Anon. 1983), and other data are from the Marine Research Institute (Anon. 1984, 1985). The temperature and salinity data are based on one characteristic station in North Icelandic waters (S-3; 66°32'N, 18°50'W), whereas the primary production and zooplankton concentration estimates are based on all observations available from the spring surveys on four standard sections in North Icelandic waters (Kögur, Húnaflói, Siglunes and Slétta; Fig. 1). Information on weight by age group and maturity of the Icelandic cod and weight by age group of the Icelandic capelin is based on standard fishery-biological investigations in Icelandic waters recently reported (Anon. 1984, 1985, Vilhjálmsson 1983).

RESULTS AND DISCUSSION

The data used in the overview shown in Table 1 are certainly of varying reliability. Nevertheless, some large-scale features can be detected. Thus favourable hydrobiological conditions in North Icelandic waters prior to 1965 coincided with small extension of drift ice. Recruitment of 3 year old cod in Icelandic waters in 1961-1964 was also well above the average of 220×10^6. In 1965-1971 sea ice and hydrobiological conditions in North Icelandic waters were unfavourable (polar water, low production of phytoplankton in 1965-1969 and low zooplankton concentrations in 1965-1971), and recruitment of 3 year old cod was on the average below 200×10^6, with only two years (1966 and 1970) above the average value. In 1972-1974 sea ice and hydrobiological conditions were again more favourable or almost like those prior to 1965, with inflow of Atlantic water into North

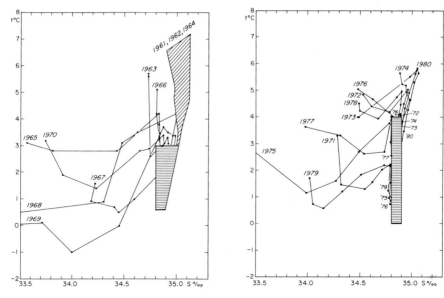

Fig. 3a. Temperature – salinity relationships in May-June 1961-1980 at a station in North Icelandic waters (S-3; for location see Fig. 1). Shaded areas cover observations in 1961, 1962 and 1964 as well as those from other years with salinities between 34.8 and 34.9 and temperatures between 0.6° and 3° or 0° and 4°C. Depth of observation points are 0, 20, 50, 75, 100, 150, 200, 300 and 400 m (Malmberg and Svansson 1982).

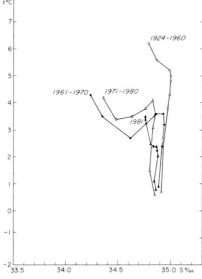

Fig. 3b. Mean temperature – salinity relationships in May-June 1924-1960, 1961-1970 and 1971-1980 and in 1981 at a station in North Icelandic waters (S-3; for location see Fig. 1; Malmberg and Svansson 1982).

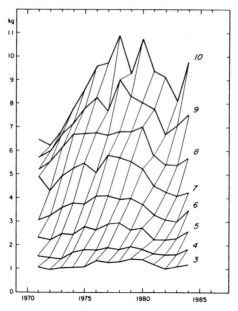

Fig. 4. Mean weight (kg) of 3-10 years old cod in Icelandic waters in 1971-1984 (Anon. 1984, 1985).

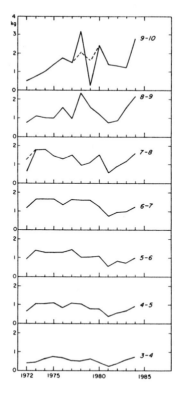

Fig. 5. Mean increase in weight (kg) of 3-10 years old cod in Icelandic waters from year to year in 1971-1984.

Icelandic waters and indication of high production of phytoplankton (1972, 1974) and large zooplankton concentrations (1972, 1973), as well as relatively high 0-group indexes for capelin (1972, 1973, 1974) and cod (1973) and also for 3 year old cod (1972, 1973). In the following period (1975-1980) cold and warm years alternated and these were also reflected in the production of phytoplankton but not in the zooplankton concentration (see Fig. 8). A decrease in the 0-group indices of capelin was observed in these years but not in the 0-group cod. A decrease also appeared in the recruitment of 3 year old cod which showed an average value below 200×10^6. Then in 1981-1983 a cold arctic water mass dominated in North Icelandic waters (t =1-3°, S~34.8), seemingly followed by very low values for at least zooplankton concentrations and 0-group estimates, but not for the production of phytoplankton. It must be noted that only indices of total amount of phytoplankton production and zooplankton concentration are shown in Table 1 and Fig. 8 but not other specifications such as division in cold and warm water species.

The results in Table 1 may explain some of the effects of environmental variables on the living conditions in North Icelandic waters; is it the temperature alone that is critical, the food availability or other more complex conditions? It has been shown (Jónsson 1965) that a variation of 1.0°C may lead to a variation of about 4 cm in length of cod of the same year class in Icelandic waters. However, the availability of food must also be taken into consideration. The Icelandic cod showed about 25% decrease in weight by age group 1981-1983 (Fig. 4; Anon. 1984, 1985) and a delay of maturity according to age (Fig. 9) which may possibly be related to temperature or hydrometeorological conditions and/or food availability. It will be seen (Fig. 4) that 10 year old cod from the year classes 1968-1970 had gained the most weight during the period 1971-1984 (over 10 kg) but those from the year classes 1961-1962 had gained the least (about 6 kg). The latter group seems to have been stunted by the effects of the ice-years in North Icelandic waters of the late sixties but the former benefited

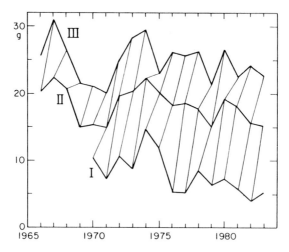

Fig. 6. Mean weight (g) of 1-3 years old capelin in Icelandic waters in 1966-1983 (Vilhjálmsson 1983).

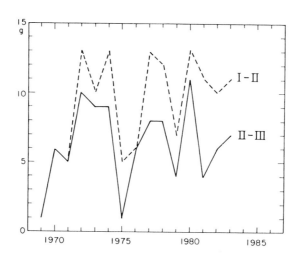

Fig. 7. Mean increase in weight (g) of 1-3 years old capelin in Icelandic waters from year to year in 1969-1983.

from the improved hydrographic conditions in these waters in the seventies. Younger fish, 3-8 years old, also showed a relatively high weight in 1975-1980. After that the weight decreased in 1981-1983 but increased in 1984, which is in general agreement with the hydrographic conditions observed in these years in North Icelandic waters. The question is whether this is due to temperature alone or more complex hydrobiological conditions and/or to food supply.

The 3-7 years old cod (Fig. 5) showed a rather constant growth rate in the seventies, followed by a distinct decrease in the early eighties. The older fish had a more variable growth, with an increase from year to year up to 1978-1980, followed by a decrease in 1981-1983 and an increase in 1984. It is noteworthy that all year classes or ages had a growth minimum 1981, but for other years no obvious relation was found between growth and hydrobiological conditions in North Icelandic waters, not even in 1975 and 1979, when polar water was observed in these waters. The minimal growth in 1981 leads to speculation about the condition of the Icelandic capelin stock, which was at a minimum in 1981-1982 (Vilhjálmsson 1983), since cod feed to a high degree on capelin (Pálsson 1983).

The weight data for capelin from 1966-1968 are questionable (Fig. 6), but year classes from 1969-1971 showed a lesser weight than those of 1972-1975, but after that the growth fluctuated, showing a slight decrease towards the end of the period. This has been compared with temperature in North Icelandic waters (Vilhjálmsson 1983) and hydrographic conditions in general (Malmberg 1984b). The greatest weights of the capelin were observed when Atlantic water reached North Icelandic waters and the lowest when polar or arctic water was found. Whether hydrographic conditions and/or food supply are responsible is still questionable.

The growth rate of 1-3 years old capelin (Fig. 7) was high during 1972-1974, 1976-1978 and in 1980 but low in 1969-1971, 1975, 1979 and in 1981-1982. The correlation with

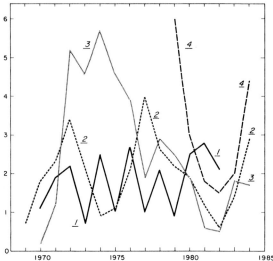

Fig. 8. Indices of primary production of phytoplankton (1), and of zooplankton concentration (2) in spring and 0-group estimates of capelin (3) in August in North Icelandic waters in 1970-1984, and abundance estimates of the spawning stock of capelin in winter 1979-1984 (4) measured in 100.000 tons.

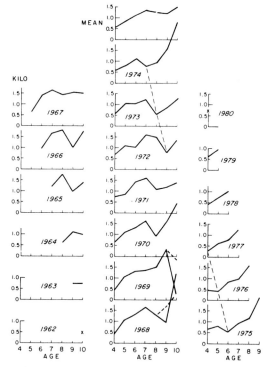

Fig. 9. Mean increase in weight (kg) of Icelandic cod from year to year for all year classes 1962-1980 and for each year class as well.

hydrographic conditions is striking. According to Figs. 2 and 7 nearly 0.1 in salinity and 0.5-0.6° in temperature correspond to about 1 g in weight of capelin, or 10 g to approximately 1 in salinity and 5-6° in temperature. These temperature and salinity values distinguish between Atlantic and polar water in North Icelandic waters, and are thus crucial for the food conditions in the area.

The biological data on primary production and zooplankton (Fig. 8) may not show any obvious correlation to growth of capelin. Notable are the low values found after 1980, both in zooplankton, 0-group capelin and the spawning stock of capelin, compared with findings in the seventies, when the Icelandic capelin fisheries reached values of 400-500 thousand tons during the winter season. The small stock size of capelin after 1980 and the minimum growth rate for cod in 1981 indicate that the hydrographic conditions in North Icelandic waters were not the only cause for this growth minimum, but rather the conditions of the capelin stock as food supply for cod. On the other hand the hydrographic conditions were directly or indirectly responsible for the growth rate of the capelin, but overfishing, as well as hydrographic conditions, responsible for the small stock size which led to a fishing ban in 1982 (Vilhjálmsson 1983).

The conditions in 1981-1983 may also have caused the delay observed in maturity of the Icelandic cod after 1980 (Anon. 1984, 1985). The general maximum in growth found at the age of 7 years for the year classes from 1967-1973 (Fig. 9) followed by a decrease, may be connected with the maturity (Jónsson 1983). This was interrupted by the conditions in 1981 for the 1974-1976 year classes delaying their maturity, and possibly also for the year classes prior to 1967 because of the conditions during the ice years in the late sixties. The favourable hydrographic conditions found in 1984 (and also 1985) and the increased stock size of capelin, may result in an improvement in the cod stock, both as regards rate of growth and maturity.

CONCLUSION

From the discussion above it can be concluded that the ecological impact of the East-Greenland Current on Icelandic coastal waters was not only apparent during the ice years in 1965-1971, but possibly also in 1976 on the spawning grounds south of Iceland (the mid seventies anomaly), and possibly on the nursery and feeding grounds north of Iceland in 1981. The abrupt changes in the hydrobiological conditions in North Icelandic waters after 1965 have had far-reaching adverse consequences as to the abundance and growthrate of capelin and to the growthrate and maturity of cod. The question arises how long we have to live with these conditions and whether we can hope for a return of the stable conditions prior to 1965. We have once more had to face the limits to growth in the sea around Iceland due to adverse hydrographic situation and we have seen that the living conditions are subject to certain environmental preconditioning.

Acknowledgement

The author expresses his thanks to colleagues at the Marine Research Institute for useful discussion and the data provided, to Hrefna Einarsdóttir and Kristín Jóhannsdóttir for typing the manuscript and Sigthrúdur Jónsdóttir for drawing the figures and many good advice, and to professor Unnsteinn Stefánsson who kindly read the paper and made many valuable suggestions.

REFERENCES

Aarkrog, A., H. Dahlgaard, L. Hallstadius, H. Hansen, E. Holm 1983: Radiocaesium from Sellafield effluents in Greenland waters, Nature, 304, 5921, 49-51.

Anon. 1983: A report on choice of industrial sites (in Icelandic). The Icelandic Goverment 5,7, Sea Ice:32.

Anon. 1984: The State of Marine Stocks in Icelandic Waters and Fishing Prospects for 1984 (in Icelandic), Hafrannsóknir, 28.

Anon. 1985: The State of Marine Stocks in Icelandic Waters and Fishing prospects for 1985 (in Icelandic), Hafrannsóknir, 31.

Astthórsson, Ó.S., I. Hallgrímsson and G.S. Jónsson 1983. Variations in zooplankton densities in Icelandic waters in spring during the years 1961-1982, Rit Fiskideildar, 7, 2:73-113.

Dickson, R.R., H.H. Lamb, S.A. Malmberg and J.M. Colebrook 1975: Climatic reversal in the northern North Atlantic, Nature, 256:479-482.

Dickson, R.R., J. Blindheim 1984a: On the abnormal hydrographic conditions in the European Arctic during the 1970's, Rapp. P.-v. Réun. Conc. int. Explor. Mer, 185:201-213.

Dickson, R.R., S.A. Malmberg, S.R. Jones and A.J. Lee 1984b: An investigation of the earlier Great Salinity Anomaly of 1910-1914 in waters west of the British Isles, ICES C.M. 1984/GEN:4. Minisymposium (mimeo).

Dooley, H.D., J.H.A. Martin and D.J. Ellett 1984: Abnormal hydrographic conditions in the north-east Atlantic during the nineteen-seventies, Rapp. P.-v. Réun. Cons. int. Explor. Mer, 185:179-187.

Jakobsson, J. 1980: The North Icelandic herring fishery and environmental conditions, 1960-1968, Rapp. P.-v. Réun. Conc. int. Explor Mer, 177:460-465.

Jónsson, E. 1982: A survey of spawning and recruitment of Icelandic Cod, Rit Fiskideildar, 6,2:1-45.

Jónsson, J. 1965: Temperature and growth of cod in Icelandic waters, Int. Comm. Northwest Atl. Fisheries, spec. publ., 6, 537-539.

Kiilerich, A.C. 1945: On the hydrography of the Greenland Sea, Meddr. om Grönland, 144 (2)

Knudsen, M. 1898: Hydrografi, Den danske Ingolf Expedition, 1(2), Copenhagen.

Malmberg, S.A. 1969: Hydrographic changes in the waters between Iceland and Jan Mayen in the last decade, Jökull, 19:30-43.

Malmberg, S.A. 1972: Annual and seasonal hydrographic variations in the East Icelandic Current between Iceland and Jan Mayen, in "Sea Ice Proc. Int. Conf." Reykjavík. Ed. by T. Karlsson. National Res. Council, 4:42-54.

Malmberg, S.A. 1984a: Hydrographic conditions in the East Icelandic Current and sea ice in North Icelandic waters, 1970-1980, Rapp. P.-v. Réun. Cons. int. Explor, Mer, 185: 170-178.

Malmberg, S.A. 1984b: A note on climate, hydro-biological investigations and fish recruitment in Icelandic waters in 1958-1983, ICES C.M. 1984/L:17 (mimeo).

Malmberg, S.A. 1985: Hydrographic conditions in Icelandic waters in May/June 1981, 1982 and 1983, Annls. biol., 40 (in press).

Malmberg, S.A. and A. Svansson 1982: Variations in the physical marine environment in relation to climate, ICES C.M. 1982/GEN:4. Minisymposium (mimeo).

Pálsson, Ó.K. 1983: The feeding habits of demersal fish species in Icelandic waters, Rit Fiskideildar, 7, 1:1-60.

Sigtryggsson, H. 1972: An outline of sea ice conditions in the vicinity of Iceland, Jökull, 22:1-11.

Stefánsson, U. 1962: North Icelandic waters, Rit Fiskideildar, 3, 296 pp.

Swift, J. and K. Aagaard 1981: Seasonal transitions and water mass formation in the Iceland and Greenland Seas, Deep-Sea Res., 20A (10):1107-1129.

Thórdardóttir, T. 1977: Primary production in North Icelandic waters in relation to recent climatic changes, in Polar Oceans", 655-665. Ed. by M.J. Dunbar. Proc. Oc. Congress. Montreal 1974.

Thórdardóttir, T. 1984: Primary production north of Iceland in relation to water masses in May-June 1970-1980, ICES C.M. 1984/L:20 (mimeo).

Vilhjálmsson, H. 1983: Biology, abundance estimates and management of the Icelandic stock of capelin, Rit Fiskideildar, 7, 3:153-181.

OCEANOGRAPHIC FUNCTIONS OF FRESHWATER DISCHARGE AND CONSEQUENCES OF CHANGE

Group work

INTRODUCTION

The organisers felt that some efforts should be devoted to disciplinary issues. Accordingly, the participants were invited to convene according to their own choice for discussions in one of the three groups: 1) Physical oceanography, 2) Phytoplankton dynamics, 3) Zooplankton and fish.

Each group was supposed to discuss and report their conclusions on two themes:
A. What is the function of freshwater outflow?
B. What are the consequences of change?

PHYSICAL OCEANOGRAPHY

Chairman: P.LeBlond
Rapporteur: T.McClimans

Effects of freshwater runoff.

This group decided to address the fundamental tendencies associated with freshwater runoff under the following headings:

1. Static vertical stability

The addition of freshwater to the sea will in general increase the vertical stability although to a degree which will depend on the mixing characteristics of the location of the source. As consequences:

- There will be an inhibition of vertical exchange of momentum, turbulent energy and scalar properties, tending to reduce the coupling between the upper and lower parts of the water column. This effect will affect the wind response of the ocean.
- Changes in stability affect the internal time-dependent response characteristics of the water column, determining internal water properties at all frequencies (including tides), acting as a wave guide for the transmission of local wind effects to remote places.

2. Dynamics

Freshwater floats on the sea and tends to flow down on the surface pressure gradient from its source, being therefore transported seawards. Secondary effects such as mixing or the Coriolis force will induce circulations in the vertical (the classical estuarine system) on horizontal planes and will affect the structure of the flow, restricting it to a coast as in buoyancy-driven coastal flows.

3. Energetics

The energy source to mix the water dolumn in the face of increased stability is provided
- near the river input by shear-generated entrainment
- further downstream by the wind, instabilities and tidal mixing processes.

In some special cases, man-made energy sources may be important (submerged inputs, shipping).

4. Fronts

Fronts are sharp horizontal gradients, here basically in salinity which arise because of
- pulsations of the source - as in tidally pulsed river plumes,
- mixing gradients - as in shelf edge fronts, where sharp changes in tidal mixing occur,
- wind or current generated convergences.

Frontal zones are broader and are more typical of the outer edges of coastal currents. They may include multiple fronts and are subject to a variety of lateral flow instabilities which depend on their scales and stratification as well as on bathymetry.

5. Ice

Ice formation will generally be affected by the presence of fresh water. Because of increased stability and, in the presence of sufficient dilution for the existence of a temperature of maximum density, ice formation will be easier in fresher, more stratified water.

Consequences of runoff changes

Again, we have identified general tendencies. Specific effects depend on local conditions and on the relative influence of all factors involved. General conclusions are therefore not to be drawn for specific applications.

1. Mobility

More runoff implies generally increased stability and hence
- less restricted exchange
- increased frequencies of internal oscillations.

2. Vertical exchange

The energetics of vertical exchange are to be considered in two zones:
- In the near zone (~10 river mouth widths away from the source) entrainment increases proportionately to the rate of outflow: near surface concentrations are therefore not affected. The thickness of the surface layer will however be increased.
- Downstream, mixing by the wind and other effects is dominant and the additional thickness of fresher water will be eroded by these normal oceanic mechanisms.

3. Advection

Greater runoff will spread brackish water over broader areas, although the increase in area is not simply related to then runoff. Fronts will move offshore and their stability will be affected. This is an area of active research where no firm conclusions can be drawn at this stage.

4. Ice

More ice will form with increased winter runoff. How much more will depend on the relative importance of several factors.

5. Variability

Even in regulated areas, enough natural variability may remain that conditions may be as variable as before regulation. Near the river outlet a sudden change of power production may lead to greater variability than natural conditions.

PHYTOPLANKTON DYNAMICS

Chairman: J.-C.Terriault
Rapporteur: B.R.Heimdal

Ecological effects of freshwater runoff in relation to phytoplankton dynamics

The working group dealt with the effects of freshwater runoff in relation to environmental factors affecting the distribution and growth of marine phytoplankton as documented in many previous publications (e.g. Sakshaug 1976; Skreslet 1981) and this workshop.

The addition of freshwater to the sea will in general increase the stability of the upper layer (as defined by the Physics group). In fjords and estuaries, the discharge sets up an estuarine circulation, with an outflowing brackish layer compensated by an inflow at some deeper layer, resulting in an import of nutrients to the euphotic zone. large freshwater outflow causes rapid seaward transport of brackish water, and high flushing rates which will prevent the phytoplankton from establishing large populations near the river mounth. Maximum phytoplankton abundance may therefore be located in areas farther seaward, and it must be expected that much of the primary production stimulated by the freshwater runoff occurs outfjord in such cases. The influence of the runoff depends not only on the oceanographic conditions of the sea and on the possibilities of dispersion and mixing of different water masses. Depending on the nature of the discharge, essentially different water qualities may be brought into the sea. At the same time, the light conditions are influenced both quantitatively and qualitatively by the material carried into the recipient and the distribution of this material. Thus, the runoff may significantly affect both the community structure and the productivity of the phytoplankton.

The annual changes in phytoplankton abundance and community structure are also heavily affected by grazing, causing large fluctuations in biomass, growth rate and nutrient supply. What happens to the zooplankton, and also to the bacteria, during freshwater runoff is therefore important to the total pelagic system. As outlined by the Zooplankton group, freshwater runoff may effect herbivorous zooplankton directly or by dispersion of food patches.

Consequences of changes in freshwater runoff

Conclusions from the working group support the view that changes in freshwater outflow largely influence the alternating freshwater, brackish water and marine conditions in the recipient area. At the same time, changes in phytoplankton abundance and community structure, which occur with some degree of predictability in temperate water, must be expected, although the plankton may be as variable as before. Decreased discharge may lead to reduced productivity in some areas and govern biological processes in others. Increased winter runoff may also influence the local climate, i.e. fog and ice formation, and possibly the starting point of the spring bloom.

Reference

Sakshaug, E. 1976. Dynamics of phytoplankton blooms in Norwegian fjords and coastal waters. Pp. 139-143 in: S.

Skreslet, R. Leinebø, J.B.L. Matthews and E. Sakshaug (eds). <u>Fresh water on the Sea.</u> The Association of Norwegian Oceanographers, Oslo.

Skreslet, S. 1981. Information and opinions on how freshwater outflow to the Norwegian Coastal Current influences biological production and recruitment to fish stocks in adjacent seas. Pp. 712-748 in: R. Sætre and M. Morsk (eds). <u>Proceedings from the Norwegian Coastal Current Symposium. Geilo, 9-12 September 1980.</u>

ZOOPLANKTON AND FISH

Chairman: J.Lenz
Rapporteur: S.Kaartvedt

The discussion went along two lines which are reflected by the two parts of our report. The first one was the attempt to identify causal relationships between environmental factors and biological key species,e.g. Calanus, and commercially important fish stocks (capeline, herring, sprat, cod) such as the spawning event, their food chain relationships and the interlinkage with freshwater run-off. The second approach consisted in listing up well documented observations on the impact of freshwater discharge in fjords, estuaries and marginal seas on fish and lobster stocks. - Causal

Causal relationships

The discussion of causal relationships between physical and biological processes, and in food webs, rose more questions than answers:
- Copepod reproduction
Which factors are governing the spawning period of Calanus after overwintering: temperature, day light length or phytoplankton standing stock? Are advection processes significant in bringing up the adults from their overwintering places in deep water masses and in transporting them into

Table I References to evidence with regard to effects of natural and regulated freshwater outflow on marine invertebrates and fish, grouped according to habitats under influence

	Estuaries and fjords	Coastal currents	Marginal seas
Natural variability			
Zooplankton	Parsons et al 1969 Strømgren 1974, 1976 Beyer 1976 Fosshagen 1980	Malmberg 1985	
Fish	Bakken 1975 Stevens 1977	Skreslet 1976, 1981 Sutcliffe et al 1977, 1983 Koslow 1984 Pati 1984	Sutcliffe 1972, 1973 Bugden et al 1982 Doubleday & Beacham 1982 Sinclair et al 1985
Commercial invertebrates		Ruello 1973 Driver 1976 Glaister 1978 Meeter et al 1978	Sutcliffe 1972, 1973 Bugden et al 1982 Sinclair et al 1985
Regulation			
Zooplankton	Aleem 1972 Sands 1984		
Fish	Aleem 1972 Ochman & Dodson 1982		Tolmazin 1979
Commercial invertebrates	Aleem 1972 Stora & Arnoux 1983	da Silva 1985	Panchian 1980

phytoplankton-rich areas ?

- Spawning of fish

Fish populations are known to have specified spawning grounds and rather precise spawning periods. Here again we have the same question as to the inducing factors for spawning. Are copopod and fish spawning independently influenced by the same or different environmental factors or are they in one way or the other linked together ?

Impact of regulation of freshwater runoff

Factors like temperature, salinity, stratification, current systems and input of detritus are influenced by freshwater discharge. These factors all have impact on marine life, and thus changes in the freshwater discharge will probably lead to some biological responses.

Impact on zooplankton and fish may be direct, or biological responses may originate from effects on their food organisms, competitors or predators. A lot of speculation on mechanisms and responses may undertaken, but to surpass pure theoretical and nonquantitative statements a search for documented covariations between freshwater discharge and biological variables is needed. Existing information (Table I) is inconclusive, but document that the impact of freshwater discharge on different marine ecosystems vary.

References

Aleem, A.: 1972. Effect of river outflow management on marine life. Mar. Biol. 15: 200-208.

Bakken, E. 1975. Utbredelse og mengde av årsyngel brisling i Vest-Norge høsten 1974. Fiskets Gang. 61:67-73.

Beyer, F. 1976. Influence of freshwater outflow on the distribution and production of plankton in the Dramsfjord. Pp. 165-171 in Skreslet S., R. Leinebø, J.B.L. Matthews and E. Sakshaug (eds). Fresh water on the sea. The Association of Norwegian Oceanographers, Oslo.

Bugden, G.L., B.T. Hargrave, M.M. Sinclair, C.L. Tang, J.C.

Therriault and P.A. Yeats. 1982. Freshwater runoff effects in the marine environment: The Gulf of St. Lawrance example. Can.Tech.Rept.Fish.Aquat. Sci. 1978: 1-89.

da Silva, A.J. 1985. River runoff and shrimp abundance in a tropical coastal ecosystem - The example of the Sofala Bank (Central Mozambique). This Volume.

Doubleday, W.G. and T.Beacham 1982. Southern Gulf of St. Lawrence cod - a review of multi-species models and management advice. InM.C. Mercer (ed). Multispecies approaches to fisheries management advice. Spec. Publ.Can.J.Fish.Aquat.Sci. 59.

Driver, P.A. 1976. prediction of fluctuations in the landings of Brown Shrimp (Crangon crangon) in the Lancashire and Western Sea fisheries district. Estuarine and Coastal Marine Science 4: 567-573.

Fosshagen, A. 1980. How the zooplankton community may vary within a single fjord system. Pp. 399-405 in Freeland, H.J., D.M. Farmer and C.D. Levings (eds). Fjord oceanography. Plenum Publishing Corp., New York.

Glaister, J.P. 1978. The impact of river discharge on distribution and production of the School Prawn Metapenaeus macleayi (Haswell) (Crustacea: Penaeidae) in the Clarence River Region, Northern New South Wales. Aust.J. Freshwater Res. 29: 311-323.

Koslow, J.A. 1984. Recruitment patterns in northwest Atlantic fish stocks. Can.J.Fish.Aquat.Sci. 41: 1722-1729.

Malmberg, S.A. 1985. The ecological impact of the East Greenland Current on the North Icelandic Waters (This Volume).

Meeter, D.A., R.J. Livingston and G.C. Woodsum. 1979. Long-term climatological cycles and population changes in a river-dominated estuarine system. Pp. 315-338 in Livingston, R.J. ed. Ecological processes in coastal and marine systems. Plenum Press, N.Y.

Ochman, S. and J.Dodson. 1982. Composition and structure of the larval and juvenile fish community of the Eastmain River and estuary, James Bay. Nat.Can. 109: 803-813.

Pandian, T.J. 1980. Impact of dam-building on marine life.

Helgolander Meeresunters. 33: 415-421.

Parsons, T.R., R.J. LeBrasseur, J.D. Fulton, and O.D. Kennedy. 1969. Production studies in the Strait of Georgia. Part II. Secondary production under the Fraser River plume, February to May, 1967. J.exp. mar.Biol.Ecol. 3: 39-50.

Pati,S. 1984. Observations on the relation between rainfall and the inshore fishery long the Orissa coast. J.Cons.int.Explor.Mer. 41: 145-148.

Ruello, N.V. 1973. The influence of rainfall on the distribution and abundance of the School Prawn Metapenaeus macleayi in the Hunter River region (Australia). Mar.Biol. 23: 221-228.

Sands, N.J. 1984. Zooplankton in Skjomen before and after control of freshwater runoff 1970-72 and 1977-79. Rådgivend utvalg for fjordundersøkelser, Skjomenprojektet, Rapport nr. 5. 95 pp.

Skreslet, S. 1976. Influence of freshwater outflow from Norway on recruitment to the stock of Arcto-Norwegian cod (Gadus mordua). Pp.233-237 in Skreslet S., R. Leinebø, I.B.C. Matthews and E. Sakshaug (eds). Fresh water on the sea. The Association of Norwegian Oceanographers, Oslo.

Skreslet, S. 1981. Information and opinions on how fresh- water outflow to the Norwegian coastal current influences biological production and recruitment to fish stocks in adjacent seas. Pp.712-748 in Sætre, R. and M. Mork (eds). The Norwegian coastal current. University of Bergen.

Stevens, D.E. 1977. Striped Bass (Morone saxatilis) year class strength in relation to river flow in Sacramento-San Joaquin estuary, California. Trans.Am.Fish.Soc. 106:34-42.

Strømgren, T. 1974. Zooplankton investigations in Skjomen 1969-73. Astarte 7: 1-15.

Strømgren, T. 1976. Relationship between freshwater supply and standing crop of Calanus finmarchicus in a Norwegian fjord. Pp. 173-177 in Skreslet S., R. Leinebø, J.B.L. Matthews and E.Sakshaug (eds). Fresh water on the sea. The Association of Norwegian Oceanographers.

Stora, G. and A.Arnoux 1983. Effects of large freshwater diversions on benthos of a Mediterranean lagoon.

Estuaries, 6: 115-125.

Sutcliffe, W.H. 1972. Some relations of land drainage nutrients, particular material and fish catch in two eastern Canadian bays. J.Fish.Res.Bd.can., 29: 357-362.

Sutcliffe, W.H. 1973. Correlations between seasonal river discharge and local landings of American lobster (Homarus americanus) and Atlantic halibut (Hippoglossus hippoglossus) in the Gulf of St. Lawrence. J.Fish.Res.Bd.Ca., 30: 856-859.

Sutcliffe, W.H., K. Drinkwater and B.S. Muir 1977. Correlations of fish catch and environmental factors in the Gulf of Maine. J.Fish.Res.Bd.Can., 34: 19-30.

Sutcliffe, W.H., R.H. Loucks, K.F. Drinkwater and A.R. Coote 1983. Nutrient flux onto the Labrador Shelf from Hudson Strait and its biological consequences. Can.J.Fish.Aquat.Sci., 40: 1692-1701.

Tolmazin, D. 1979. Black Sea - dead sea? New Scientist, 84: 767-769.

INTERDISCIPLINARY ASSESSMENT OF ECOLOGICAL PROBLEMS RELATED TO FRESHWATER OUTFLOW IN COASTAL MARINE ECOSYSTEMS

Group work

INTRODUCTION

One working session was devoted to identification of interdisciplinary problems and selection of appropriate methods. This was due to the recognition that a systems approach is needed, involving close cooperation between oceanographic disciplines, and development of holistic concepts. Three working groups were established, to deal with 1) Estuaries and fjords, 2) Marginal seas, 3) Coastal currents.

ESTUARIES AND FJORDS

Chairman: C.Hopkins
Rapporteur: P.Tett

The working group dealt mainly with ecological processes in marine embayments receiving a significant input of freshwater and in which exchange with the sea is restricted to a small part of the total periphery. Most are long in relation to their width and are dominated by longitudinal salinity gradients. Tidal currents often play an important role in horizontal exchange and vertical mixing. Where river flows are relatively large significant vertical gradients of salinity, and associated 2-layer ('estuarine') circulations are found. These physical transports import and export plant nutrients, plankton, organic detritus, and inorganic particles, depending on the relative internal and external concentrations of these substances and on the sinking rates of the particles. Depending on the nature of the catchment, runoff may import inorganic nutrients, particulate inorganic material, or dissolved and particulate organic material. All these inputs and outputs may significantly affect the local ecosystem, and all depend, sometimes in non-linear fashion, on the rate of runoff.

Questions relating to production are central to an understanding of the ecology of estuaries and fjords. It is important to quantify :
- The total amount and the spatial and community distribution of autochthonous primary production (e.g. planktonic, benthic or littoral; in microalgae, macroalgae or higher plants, at the head, middle or mouth of the estuary or fjord, at the surface of the water column or in the pycnocline), and its seasonal variation.
- The spatial, community and temporal variation of secondary and higher levels of production and of consumption of organic material at these trophic levels, including the organims of the detritus food chain.
- The nature, amount and seasonal variation of the main allochthonous sources of organic material consumed in or passing through the estuary or fjord as detritus, plankton or nekton.
- The balance between import and export of particulate and dissolved organic material and of organic and inorganic nitrogen and phosphorus.
- The way in which all the above vary with changes in runoff.
These questions are susceptible to practical and theoretical investigation.

With regard to practical investigation it was considered that systems belonging to several categories should be investigated : the overall strategy would be that of 'compare and contrast'. One division between categories is that between estuaries and fjords : for present purposes an estuary is defined with regard to the water between certain minimum and maximum isohalines; the numeric value of these isohalines differs between locations but in general both refer to regions of discontinuity, such as at the front between an estuarine plume and the sea. The estuarine community includes the pelagic organisms in this water, and the benthos and littoral that are regularly exposed to it. Most fjords include an estuarine circulation but many have in addition underlying seawater which may exhibit a complex circulation or a pattern of stagnation and intermittent renewal. In the fjordic category it is probably necessary to distinguish between small, shallow-silled `polls` and larger, deeper fjords proper.

Practical investigations of fjords and estuaries should pay particular attention to (a) the role of very small algae (picoplankton); (b) the role of planktonic and benthic microheterotrophs of the detritus food chain; (c) input and sedimentation of inorganic material and its effect on benthic metabolism and community structure in relation to organic sedimentation and (d) detailed analysis of food-chain mediated or physically determined links between estuarine/fjordic variability and fish populations. Many estuaries and fjords are physically complex and biologically heterogenous systems, and particular thought should therefore be given to appropriate sampling strategies.

With regard to theoretical investigation it was suggested that a continuum approach should be adopted, and that a search should be made for a few high-information-content biological parameters which could be related across a range of fjords and estuaries to physical parameters such as Richardson numbers. These quantify the relationship between stabilizing buoyancy input (mostly provided by freshwater) and the destabilizing velocity shear. One suitable biological parameter is the ratio of local primary production to total consumption of organic material; another might be the ratio of the mean growth rate of an organism to a turnover rate of the fjord or estuary.

The principal purpose in understanding the ecosystem of fjords and estuaries is to provide practically relevant information to users of these environments and thus avoid possible conflicts between their natural conservation and their use for recreation, aquaculture, waste disposal, shipping, and so on. The present problem is that literature reviews of production, systematics, physical and chemical data sets often result in attempts to compare studies carried out by with different techniques. Thus, misintrepation is easy and can lead to poor advice to managers. The methodology exists for carrying out the investigations proposed here. Our central argument is thus for a programme of properly comparable studies of a range of estuaries and fjords.

MARGINAL SEAS

Chairman: W.Leggett
Rapporteur: B.Cote

The definition of marginal sea used by the working group is the following: a semi-enclosed body of water with a mouth dimension on the order of several internal Rossby radii allowing a significant exchange with the ocean. Examples of marginal seas include the gulf of St.Lawrence, Hudson Bay, the North Sea, the Bay of Fundy, Ungava Bay and the Barents sea. The Baltic sea and the Black Sea was not considered a marginal sea on account of their mouth dimension.

In seeking what physical, chemical and biological processes might be most affected by variations in river runoff, the group was led to formulate a series of unanswered questions in each of the fields. These questions are enumerated below:

Physical processes

1. Vertical mixing processes
- To what extent does the stabilizing effect of freshwater, influence nutrient dynamics both on temporal and spatial scales? For instance, what is the influence of fresh water on the nutrient pool in winter?, what is the effect of variable runoff on initial utilization of nutrient by the spring bloom?, what is the effect of variable stratification on summer regeneration of nutrients?
- Localized effects may also be important, for instance to what extend does vertical stratification induced by freshwater

inflow lead to the generation of internal tides at the head of channels (e.g. Laurentian channel in the gulf of St.Lawrence)?
- What is the contribution of freshwater induced vertical stability to the rate of heating, and the strength and depth of the thermocline.
- What is the influence of vertical stabilization on turbidity and light conditions?
- What effect will changing stratification have on wind induced upwelling? This could be either a positive or negative effect depending on the situation.

2. Lateral transport processes.
- What influence do differences in volume of freshwater inflow have on:
a. the strength and stability of marginal sea "coastal currents" (i.e. currents with a width on the order of one internal Rossby radius)
b) the manifestation of secondary flow at the deep interface of the "coastal current" (i.e. does it exist and what is the coupling)
- What is the magnitude of exchange between the marginal sea and the ocean, in terms of a) deep residual current and b) coupling between circulation and ocean water?
- What influence will freshwater inflow have on the formation, strength and localization of fronts?
- How does the location and intensity of freshwater sources affect circulation and mixing?
- What will be the influence of freshwater induced circulation on residence times and pattern of transport of a) biological organisms and b) sediment and organic material?

Chemical processes
(i.e. nutrients, pollutants, sediments, suspended- dissolved material)

- What is the importance of silicate deficiency in source waters on the productivity of marginal seas?
- Will freshwater inflow/density shifts induce phase shifts

(dissolved/suspended) in nutrients and heavy metals?
- What are the interacting influences between stability and nutrient input on oxygen deficits?

Biological processes

1. Phytoplankton
- How does the timing and size of the spring bloom vary in relation to freshwater induced changes in winter and/or spring nutrient pool?
- What will be the impact of freshwater induced vertical stratification and stabilization on phytoplankton production and species composition (fronts, mixing depth, nutrient exchange, turbidity)?

2. Bacteria and benthic systems
- What is the relative contribution of transported (riverine) vs produced organic material to bacterial and benthic systems?
- What will be the impact of freshwater induced changes on the magnitude of sedimentation to the benthos?
- What influence will freshwater inflow have on a) the production of neritic (small) vs oceanic (large) plankton particles and on b) the transfer to benthic vs palagic food chains? Also what effect will it have on the diversity of organisms?

3. Zooplankton and ichtyoplankton
- What is the effect of freshwater induced horizontal transport processes on population distribution (specially longer living zooplankton species and fish species with longer larval stages)?
- What is the effect of vertical stratification on vertical migration and food chain interactions?

General processes

- What is the impact of freshwater induced changes in the timing of biological production cycles on the coupling between

trophic levels (temporal match/mismatch)?
- What is the impact of freshwater induced changes in vertical stability on the temporal and spatial coupling between palagic and benthic systems?
- What is the contribution of freshwater inflow to marginal seas fisheries production?

Recommendations

1. Physical oceanography

It is felt that vertical stability is moderately well understood and can be modelled in a general sense. More effort should therefore be spent on the dynamics of circulation.

2. Chemical oceanography

The impact of freshwater inflow on the temporal and spatial dynamics of "nutrients" (i.e. nutrients, pollutants, sediments, etc.) in marginal seas merits further study.

3. Biological oceanography

A simple model is proposed, dealing with phytoplankton production, that can be verified rather easily. The two main premises of the model are (1) that runoff has a stabilizing effect on the water column and (2) that entrainment due to freshwater inflow is negligible once landed in the marginal seas. According to the model there is a positive correlation between stratification and runoff, a negative correlation between nutrient concentration and stratification (the slope would be a function of the particular case study), a positive correlation between phytoplankton production and nutrients and consequently a negative correlation between phytoplankton production and runoff.

Sampling would require surveying the whole area under study at the appropriate frequency and integrating this information over the seasonal cycle. This sampling scheme would require several years of study. In the case of the German Bight and the North Sea some data are already available.

The committee considered examining the higher trophic

levels, however, given the sampling difficulties involved and the lack of information at the primary production level it was felt that most effort should be expanded in verifying the proposed model.

COASTAL CURRENTS

Chairman: J.H.Simpson
Rapporteur: T.Koslow

By a coastal current we mean a buoyancy driven flow confined to the coast by the action of the Coriolis forces. We recognize two forms: plume-like flows from a single source and flows from distributed sources.

Coastal currents are widespread, persistent consequences of river discharge from land. Ten currents of this type can be identified in the North Atlantic region alone. They represent major structural features of coastal zones and, as such, may constitute significant controls on the coastal ecosystem. Coastal currents exhibit spatial and temporal variability on a variety of scales.

Physical considerations

To the physical oceanographer these systems present a number of challenging questions, e.g.
- To what extent is the buoyancy input responsible for driving the flow?
- What is the influence of other forcing agents such as bottom friction, ambient currents and longshore windstress?
- How is the flow affected by bottom topography?
- What is the secondary transverse circulation?
- How much cross-frontal mixing is effected by frontal eddies and other instabilities?
- How does the downstream extent of the current respond to the changes in source buoyancy flux?
- What is the role of a coastal current in propagating distur-

bances alongshore?

Biological considerations

Areas of high biological productivity often coincide with the location of coastal currents. As a consequence of this, the group suggests the following statement: "Coastal currents operate as unidirectional transport systems for biomass".

The group identified the following questions to be addressed when validating this statement:
- Does buoyancy input create stability early in the season and thus advance the timing of primary production? We need to know how the timing of the spring bloom is related to development of stability from whatever mechanism. We already have a fair amount of information on this but more detail is needed concerning the critical depths and requirements of dominant species (this is not specifically a coastal current exercise).
- Are communities in coastal currents dependent on terriginous nutrient supply? This is an important but purely descriptive question tackled by estimating budgets (supply/demand).
- Does advection from fjords provide seed organisms in spring? The biological studies must be conducted at the species (population) level and be run in conjunction with plume studies.
- Does advection from oceanic water significantly alter community structure? Transect profiles of community structure, conducted principally in summer when biological tracers of oceanic water can be expected to be most prominent, should reveal admixture of expatriate forms.
- Is primary production enhanced in frontal areas? On the basis of observation (including remote sensing) conditions for primary production should be investigated in sections perpendicular to the coastal current/ocean water front. Convoluted fronts may increase the potential for production. From concurrent studies of natural supply, zooplankton, biomass etc. the proportion of "new" and "regenerated" primary production should be assessed, in order to estimate vertical

loss from the system.

- Are biological systems ("cycles") compatible with both northward and southward transport? There seem to be striking contrasts between the ecology of the Norwegian and Labrador currents, and successful fish stocks in the western area seem to depend on areas of retention. Comparative studies may indicate causal relationships. We note the significant difference in the potential for passively drifting plankton and migratory nekton capable of returning to a spawning area.

- Is there an effect of horizontal displacement on stock/recruitment relationships in coastal current systems? First of all stock/recruitment relationships need to be confirmed in situations large enough and sufficiently enclosed for the effects of shear to be negligible. Then one may be able to apply the general conclusion to observed recruitment in the coastal current to see whether recruitment efficiency has been affected.

The group also discussed the question of biological variability and of enhanced production of gelatinous plankton where stable conditions and a rich food supply may cause this to happen.

A foundation experiment for the interdisciplinary study of a coastal current

The objective of this suite of naturally supportive experiments should be to determine the distribution and dynamics of physical and biological processes that occur within and characterize some representative coastal current. Our thinking revolved around the analogy of a coastal current to a "leaky pipe", albeit a pipe subject to flow reversals and some fluctuations.

The current transport and exchange of materials with deeper and with offshore waters are effectively addressed only be an extensive physical oceanographic experiment which will involve several cross-sections instrumented with recording current meters and other oceanographic sensors. To provide the necessary temporal coverage, observations should continue for at least a year.

The complexities of the offshore frontal zone of a coastal current can be daunting, but for the purpose of this experiment, it need only be parameterized as bulk horizontal exchange rates. One particularly revealing measurement toward this end could be performed by use of a "purposeful tracer", such as SF_6, to characterize the advection, mixing, and dilution of the coastal current.

Among the tools which will probably find application during this experiment are SST images from space, tracked buoys, and HF-radar as well as conventional current and hydrographic measurements. The construction of diagnostic circulation models will be required to account for topographic, wind-induced and other sources of variability.

Biological oceanographic investigations of the coastal current are recommended, such as C^{14} production, chlorophyll a, size spectra, and the like. A coastal current experiment of this nature is probably not the appropriate setting for studying development and physiological ecology. At this stage, these are better conducted in smaller sites and applied to coastal currents through modelling and confirmatory experiments.

A primary emphasis is put on cross-sections of productivity and biomass. The objective of those sections is to determine the origins, advection, and fate of primary and secondary production. The need to integrate biological and physical oceanographic fluxes cannot be over-emphasized, given the dynamic nature of these regions.

Alongshore fluxes are of interest and importance, such as the advection of reduced carbon into down-stream communities, and the fate of dispersion life-history stages.

The separation in space of temporal sequences in biology, e.g. zooplankton peaks which lag phytoplankton peaks, demands a well defined advective and dispersive physical model in order to interpret biological observations. Typical of the measurements required to capture the appropriate flux estimates is an array of sediment traps to capture downward fluxes that occur along the length of the coastal current.

In general, the approach we recommend is a "bottom-up" as opposed to a "top-down" investigation, since we reason that the effort expanded in thoroughly characterizing the "leaky pipe" of a coastal current will be repaid by the subsequent realism and generality of future modelling and experimental design.

ON THE ROLE OF FRESHWATER OUTFLOW ON COASTAL MARINE ECOSYSTEMS -- A WORKSHOP SUMMARY

K. F. Drinkwater
Department of Fisheries & Oceans
Marine Ecology Laboratory
Bedford Institute of Oceanography
Canada, B2Y 4A2

INTRODUCTION

The papers presented at the workshop and found within this volume indicate the profound effect river input to the oceans has on the physical, chemical and biological processes in coastal waters. Freshwater induces important circulation patterns, effects vertical stability, modifies mixing and exchange processes, and influences nutrients and primary production. Particular organic and inorganic compounds, as well as organisms, carried seaward in rivers, are incorporated into marine food chains. Physical and biological characteristics of coastal waters reflect the seasonal and interannual variability of the incoming rivers. This includes the fisheries as interannual fluctuations in the yields of certain commercial fish species are found to covary with runoff. These freshwater effects are not limited to an area close to the river mouth but can extend over a thousand kilometers in the case of large rivers.

On the one hand the role of freshwater outflow and its influence on the environment and biota of the coastal regions is complex and the details not well understood; on the other hand, the requirement for hydroelectric power, flood control, and water for agriculture has resulted in heightened awareness by oceanographers of the possibly important effects to the marine ecosystem that may result from increased freshwater regulation. Not only do hydroelectric projects result in major alterations to the natural seasonal discharge cycle, but in the case of freshwater diversion schemes, even the total annual discharge is altered. Such changes can result in significant and sometimes detrimental effects to large areas of adjacent caostal waters (e.g., Tolmazin, 1985). The increasing number of large-scale water management schemes being planned emphasizes the immediate need to address further the environmental effects of freshwater outflow. It was for this purpose that the worskhop was convened.

The main objectives of the workshop were two-fold: first, to review and critically assess the present understanding of the importance of freshwater runoff (including its seasonality, interannual variability and regulation) on the productivity and population dynamics of coastal marine ecosystems; and, secondly, to identify critical gaps in our knowledge and suggest fruitful areas for future research.

The following is a brief summary of the papers presented at the workshop and the discussion which followed them. It is based primarily upon a summary talk given at the time but for completeness has been expanded to include certain issues raised during the working group discussions. I have attempted to cover the major points and recurring themes that arose during the meeting but in the final analysis the paper represents a personal account of the proceedings. I shall begin, as the workshop began, by discussing the role of freshwater on the physical environment.

PHYSICAL OCEANOGRAPHIC EFFECTS

The addition of freshwater to the ocean creates a succession of physical regimes each with their characteristic dynamics, circulation and spatial scales. As a river enters the sea it initially spreads over saltwater forming a highly stable but shallow brackish plume of freshwater, known as a river plume. Mixing between the near freshwater and saltwater occurs at the edges of the plume. The spatial scales are typically a few kilometres. Eventually the mixing produces a brackish surface layer. If it is confined within a narrow estuary or fjord it flows seaward with a necessary compensating landward flow in the subsurface layers. Estuaries and fjords have length scales of 10 to 100 kilometres and widths an order of magnitude smaller. Upon reaching the shelf the surface layer, sometimes referred to as an estuarine plume, turns to the right (in the northern hemisphere) under the influence of the Coriolis force. It then flows in a narrow band parallel to shore and is called a coastal current. It has along-shelf length scales of 100 to 1000 kilometres and widths on the order of an internal Rossby radius (10 to 20 kilometres). Marginal seas adjacent to coastal currents may exhibit some estuarine characteristics, but the mean flows are generally weaker than coastal currents. Examples of marginal seas include the North Sea, the Gulf of St. Lawrence and the Bering Sea. Their spatial scale is generally on the order of hundreds of kilometres. Not all of the described regimes occur for a given river. Intense tidal mixing may eliminate the river plume. Rivers that enter directly onto the shelf proceed from river plumes to coastal currents directly. Small rivers will not form coastal currents. The physical dynamics and the role of freshwater varies between regimes. Even within a given regime the relative importance of freshwater outflow differs depending upon other conditions such as the tides, winds, topography and offshore forcing. Therefore, while there is sufficient knowledge of coastal dynamics to describe the general effects of varying river runoff, application to specific regions requires detailed knowledge of the relative importance of these other factors.

It was generally agreed that one of the major effects of freshwater, is on the vertical stability of the water column. Freshwater addition increases stability, thereby suppressing vertical mixing and momentum transfer between the surface and subsurface waters. This acts to suppress nutrient enrichment of euphotic zone. Freshwater-induced mixing through vertical velocity shear, i.e., vertical entrainment, only occurs near river mouths. Beyond approximately 10 river-mouth widths downstream, further mixing of freshwater and saltwater is dominated by other processes such as tides, winds and internal waves. Increased runoff tends to decrease salinity and to extend the area of brackish waters but the relationship is not a simple function of runoff.

A second major effect is the generation of horizontal circulation from freshwater induced horizontal density differences, such as the coastal current. It was noted that coastal currents form an almost continuous cyclonic circulation pattern around the perimeter of the North Atlantic. A similar situation occurs in the North Pacific. Two recently identified coastal current systems were described during the workshop. These were the Scottish Coastal Current off northern Scotland believed to be driven, in part, by the low salinity outflow from the Irish Sea and supplemented by rivers from the Scottish coast (J. Simpson),* and the Vancouver Island Coastal Current off the British Columbia coast of western Canada whose source is as yet undetermined but is thought to be linked to the Fraser River outflow, direct discharge from local rivers off Vancouver Island or both (P. LeBlond). The importance of other factors besides freshwater were clearly shown in these studies. Topographic steering and tidal mixing play influential roles in the Scottish Coastal Current while interactions with wind-

---

*Note that undated references refer to workshop participants induced upwelling appear to be of major importance in the Vancouver Island Coastal Current. These contrast with the Norwegian Coastal Current which is dominated by large scale instabilities producing swirls and eddies (T. McClimans). A major question concerning coastal currents is the extent to which they are buoyancy driven. Even in the Norwegian Coastal Current which has by comparison has been relatively well studied, we still do not know the answer. Another question that has received even less attention is the cause of their demise.

The effects of freshwater are not confined to the surface layers. An example of runoff controlled exchange of deep bottom water in a shallow-silled Scottish Loch was described by A. Edwards. During high runoff conditions low density water forces higher salinity coastal water further offshore and covers the shallow sills. This prevents renewal of deep water by high density saline waters. A well-known example of subsurface effects is the landward flow which occurs in the lower layers of estuaries and fjords to compensate the seaward surface drift. Observations on the Middle Atlantic Bight off the east coast of the United States show near-bottom transport toward the mouths of estuaries as far offshore as 70 km (R. Garvine). Further investigations are required to determine the forcing mechanism and what role, if any, freshwater runoff plays.

The last decade has seen great advancements in our understanding and ability to model coastal dynamics using analytical, physical and numerical methods. Some model results presented at the workshop included river plume dynamics (R. Garvine), the instabilities of the Norwegian Coastal Current (T. McClimans), and the topographic influences on the Scottish Coastal Current (J. Simpson). It was reported that Norwegian Coastal Current models have accurately predicted the formation of current instabilities and eddy shedding.

The importance of the h/u^3 tidal mixing parameter of Simpson and Hunter (1974) in predicting stratification and understanding the horizontal distribution of biological properties was noted. Although originally developed for regions where density stratification is due to solar heating, the parameter is also useful in estuarine regions (Bowman and Esaias, 1981).

BIOLOGICAL EFFECTS

Freshwater effects on biological processes in coastal waters are generally indirect occurring through the influence of runoff on the physical and chemical environments. While the present understanding of phytoplankton dynamics are sufficient to provide some insights into the role of freshwater runoff, at the higher levels of the food chain we are not as far advanced. Indeed our lack of basic knowledge governing the distribution, production, survival and growth of fish larvae impedes efforts to determine the precise role freshwater may play in interannual recruitment variability. I shall return to this topic after discussing the effect of runoff on primary production.

The necessary requirements for phytoplankton growth are vertical stability, sufficient light and available nutrients. All of these factors can be influenced by freshwater outflow. The response is not simple, however, as primary production may be enhanced or suppressed depending upon several conditions. Light reduction from heavy silt loading, prevention of nutrient addition to the surface layer by strong vertical stability, or rapid advection such that the flushing time exceeds the time scale of phytoplankton growth will act to reduce production. On the other hand, in river plumes where freshwater and saltwater are actively mixed, primary production is usually high. Fronts associated with estuarine plumes and coastal currents were also noted as regions of increased production. In regions where there are sufficient surface nutrients vertical stability imposed by runoff helps to maintain phytoplankton in the euphotic zone thereby enhancing production. As with the physical oceanography, while the general effects may be understood the specific response to varying runoff depends upon other factors.

Increased primary production also occurs through direct nutrient input from rivers. Examples were presented from the Bay of Brest on the west coast of France (B. Queguiner), and from the Dutch (G. Franz) and German (U. Brockman) coasts in the southern North Sea. The high nutrient loadings in the rivers emptying into these regions result from a combination of industrial, agricultural and urban wastes. In the Bay of Brest phytoplankton blooms are generated by pulses of nutrients discharged by the rivers. High tidal mixing and rapid exchange between the Bay and adjacent shelf waters prevents eutrophication. In the southern North Sea, however, eutrophic conditions are observed during the growing season. Off the German coasts these can be destroyed by intense mixing under storm conditions. Along the Dutch coast eutrophication is believed to have resulted in a shift from a diatom to a flagellate-dominated population.

The timing of the phytoplankton production can also be affected by runoff. Increased vertical stability from high runoff reduces vertical exchange as discussed under physical oceanographic effects. In the St. Lawrence Estuary this prevents early seeding of the surface mixed layer by phytoplankton cells resulting in a delayed spring bloom (J. -C. Therriault). In other regions such as the Icelandic Coastal Current (T. Thordardottir) and the Norwegian Coastal Current (R. Peinert) vertical stability imposed by freshwater produces an earlier spring bloom than in adjacent oceanic waters.

A further effect of freshwater runoff on phytoplankton is its influence on species composition. Several examples were given during the talks. These included differences in species due to seasonality of runoff (in the Bay of Brest, B. Queguiner), between different physical regimes (in the St. Lawrence Estuary, J. -C. Therriault; in Chesapeake Bay, H. Marshall) and from variability in river-borne nutrient supply (along the Dutch coast, G. Franz). A detailed examination of species related effects was presented by V. Smetacek. He expressed the view that species are selected for specific environmental regimes. Besides temperature, salinity and nutrient tolerances, the selection process must include the ability of the organism to maintain a residual population, often in presence of horizontal advection. Certain estuarine phytoplankton species form aggregates or become coated with mineral particles. These 'drop out' out of the surface layer due to increased specific gravity. The mass sinking does not represent mortality, although most are eventually consumed, but rather part of a life cycle which includes a resting stage in deep waters or on the seabed. These resting plankton then act as seed cells for the surface layer at a later time (as described by J. -C. Therriault for the St. Lawrence Estuary and mentioned earlier). Those species that become covered with particulates and sink out have important implications on the sedimentation rates. This in turn acts to enhance production downstream by increasing light levels. These effects indicate inter-relationships and feedback systems between physical and biological processes. In recent years, many studies of phytoplankton have not collected species information. It was felt that much can still be learned from such data and that changes in the species composition may be used to indicate changes in the physical or chemical environments.

Freshwater effects on zooplankton are less clearly defined than in the case of phytoplankton. Some effects were noted, however. High zooplankton abundance was measured off the Dutch coast in association with high phytoplankton biomass caused by high nutrient discharges from rivers (G. Franz). The possible role of advection on the copepod *Calanus finmarchicus* off western Norway was described (S. Skreslet). Transport by the Norwegian Coastal Current and local freshwater-induced circulation in Vestfjord at different stages in their life cycle carry them onto their spawning and feeding grounds. The role of freshwater on the species composition of zooplankton was discussed by V. Smetacek.

Results from a simulation model on the response of phytoplankton and zooplankton in an estuary to changes in the physical, chemical and biological environments were described by T. Parsons. Under certain conditions increased light extinction coefficients can increase zooplankton production. This occurs because the zooplankton can follow more closely phytoplankton production. At lower extinction coefficients phytoplankton production proceeds more rapidly and zooplankton cannot respond quickly enough. This results in a rapid depletion of nutrients and a sinking out of the surface layer by the phytoplankton. Nutrient concentrations were shown to play a role in the form of detrital material, phytodetritus being dominate under high nutrient conditions and zoodetritus under low nutrient conditions. Initial conditions of zooflagellates were shown to determine zooplankton production. Such modelling attempts are extremely useful, especially in exploring non-linear effects. These can sometimes produce counter intuitive results as revealed by the prediction of increased zooplankton production with increased extinction coefficient. It must be remembered, however, that such models are only as good as their assumptions and the dynamics upon which they are built.

Our ability to determine the effects of freshwater runoff on larval fish, and hence adult fish, is hampered by the lack of understanding of the factors governing larval survival and year-class strength. While both food availability and predation are known to play major roles, the details remain unresolved (W. Leggett). Recent work has suggested that the size of food particles are as important as the quantity available. The role of freshwater remains unclear in light of the many uncertainties.

Studies that show runoff and fish covary are often cited as evidence for cause and effect, albeit in some indirect manner. The problems with correlation analysis are well known (see J. da Silva). It was also pointed out that the sign of the correlation can be 'justified' or 'explained' biologically whether it is positive or negative because of the diverse nature of freshwater effects on the biota. The variability in runoff is often similar to other environmental variables. An example of strong covarability between the St. Lawrence runoff and both geostrophic winds and large scale pressure patterns was given (A. Koslow). It suggests that St. Lawrence runoff-fish correlations (e.g., Sutcliffe et al., 1977) may not indicate a freshwater influence but rather the effect of a different environmental variable on fish. This again points to the necessity of understanding the relative importance of freshwater outflow on the physical environment. In defence of correlation analysis they are often consistent with biological information. For example, in the Sutcliffe et al. (1977) study the highest correlations occur for a lag time between runoff and fish that generally equals the mean age at commercial size, a result in agreement with the general belief that survival is most strongly influenced by events in the first year of life of the fish. This does not resolve the previous concerns raised by A. Koslow, however.

One particular correlation study mentioned by M. Sinclair involves a lagged relationship between the St. Lawrence River and lobster catches in the Gulf of St. Lawrence. Originally published by Sutcliffe (1973), it was slightly modified and shown to be predictive (Sheldon et al. 1982). Further updating to 1984 indicates the relationship continues to give good predictions of lobster landings (M. Sinclair). It was suggested that this region would be excellent for detailed studies of the mechanisms of lobster survival to learn what role freshwater plays.

A food chain hypothesis is often inferred from runoff-fish correlations. It implies a simple transfer of energy up through the food chain initiated by river effects on phytoplankton usually through nutrient availability. An alternative viewpoint was expressed that for the higher levels of the food chain, such as zooplanktors and fish, advection and diffusion from strategic areas at the larval stages are the important factors determining survival (M. Sinclair). This fundamental question as to the importance of trophic linkages was vigourously debated at the meeting but remains unresolved.

FRESHWATER REGULATION

Only one paper dealt exclusively with the effects of runoff regulation on the coastal environment. The study presented by S. Akenhead was conducted on the Labrador Coast in eastern Canada. Construction of a hydroelectric project in the early seventies on the Churchill River had been blamed for failure of a nearby cod fishery. The fishery was located about 40 kms south of Hamilton Inlet, a narrow, long (180 km) embayment into which the Churchill River empties. The regulated flow produced a seasonal shift in discharge rate with a 30% decrease in spring peak flows (from 4500 to 3500 m^3s^{-1}) and a tripling of the winter minimum (from 550 to 1600 m^3s^{-1}). Temperature and salinity data collected in the inlet during pre- and post-dam conditions did not support the idea that regulation produce any environmental changes. However, no definite conclusion can be reached as to the dams effect because of the limited temporal coverage of the data. The effects of regulation of the rivers entering the Gulf of St. Lawrence were presented by M. Sinclair. Although there is persuasive evidence that inter-annual variability in runoff has an impact on fisheries production within the Gulf of St. Lawrence it was concluded that regulation of the St. Lawrence system of rivers does not seem to have had a measurable impact on fisheries production. If there has been an effect it has been masked by the natural inter-annual variability of runoff and the effects of fishing. Most coastal regions exhibit large natural variability in freshwater under normal conditions. This points out the necessity of having to collect long-time series of data to be able to demonstrate the effect or lack of one due to regulation.

Further discussion on the effects of regulation are given in the working group reports.

CONCLUDING REMARKS

The papers presented at the workshop reflect the diverse role freshwater outflow plays in the coastal regions. They also reveal the wide range of time and space scales over which such effects occur. Detailed mechanisms are generally unknown and specific effects require knowledge of other important physical and biological factors. Much further work is required before a comprehensive understanding of role of freshwater is achieved. This is especially true at the higher levels of the food chain where even general effects cannot be readily established. The task is not easy nor will it be accomplished quickly. Major advancements, it was felt, would require a multidisciplinary approach to the problem.

REFERENCES

Bowman, M.J., and W.E. Esaias. 1981. Fronts, stratification, and mixing in Long Island and Block Island Sounds. J. Geophys. Res. 86: 4260-4264.

Sheldon, R.W., W.H. Sutcliffe, Jr., and K.F. Drinkwater. 1982. Fish production in multispecies fisheries, p. 28-34 In Multispecies approaches to fisheries management advice. Can. Spec. Publ. Fish. Aquat. Sci. 59.

Simpson, J.H., and J. R. Hunter. 1974. Fronts in the Irish Sea. Nature 250: 404-406.

Sutcliffe, W.H., Jr. 1973. Correlations between seasonal river discharge and local landings of American lobster (Homarus americanus) and Atlantic halibut (hippoglossus hippoglossus) in the Gulf of St. Lawrence. J. Fish. Res. Board Can. 30: 856-859.

Sutcliffe, W.H. Jr., K. F. Drinkwater, and B.S. Muir. 1977. Correlations of fish catch and environmental factors in the Gulf of Maine. J. Fish. Res. Board Can. 34: 19-30.

Tolmazin, D. 1985. Changing coastal oceanography of the Black Sea. I: Northwestern Shelf. Prog. Oceanog. 15: 217-276.

RECOMMENDATIONS FOR JOINT INTERNATIONAL RESEARCH COOPERATION

Plenary adoptions

Chairman: T.Parsons
Rapporteur: G.Fransz

INTRODUCTION

Based on information from the working groups, a plenary session was launched to discuss the feasibility and organisational aspects of joint international research cooperation. After a general discussion of possible modes of cooperation and presentation of national projects of possible interest, three proposals for continuing cooperation were put forward. These were discussed in separate parallel sessions by three as hoc working groups. They drafted terms of reference which were discussed and adopted by the pleanry assembly of workshop participants.

PROPOSAL 1. A STUDY OF THE COASTAL CURRENTS OF ICELAND AND FACTORS LEADING TO BIOLOGICAL PRODUCTION AND LARVAL COD SURVIVAL

K.Drinkwater, T.Koslow, R,Sætre, J.H.Simpson and T.Thordardottir.

To assess the desirability and feasibility of an international interdisciplinary experiment to study the Icelandic Coastal Current nd its ecological impact, the group identifies the following steps: a) A comparative study of fresh-water induced coastal currents, with the objective of site selection (1986). b) To assemble the available data base, including I-R (infra-red) imagery, and access to existing knowledge of the buoyancy-driven flows around Iceland and its importance to fisheries (1986). c) Detailed plans of design, responsibility analysis, ship schedules, permissions, etc. for an experiment in 1988 (1987).

PROPOSAL 2. A REVIEW OF DIFFERENT ESTUARIES IN TERMS OF THE FLUX OF ORGANIC MATERIALS; THE NEED FOR FURTHER MEETINGS TO DISCUSS DIFFERENCES AND CONCEPTS IN ESTUARINE ECOSYSTEMS.
C.Hopkins, W.Leggett, S.Skreslet. H.Svendsen, P.P.Tett and J-C.Therriault.

The group draws attention to the lack of cohesive, holistic quantifications in estuaries and fjords and recommends the following steps of progress:
a. To establish and fund a procedure for carrying out and publishing a review of existing data on estuaries and fjords in order to search for empirical relationships between physical, chemical and biological components.
b. To consider the need for further meetings to discuss the findingns of this review and its implications for future field, laboratory and modelling studies related to these ecosystems.

PROPOSAL 3. COOPERATIVE INVESTIGATIONS IN THE NORTH SEA AND NORWEGIAN CURRENT WITH PARTICULAR REFERENCE TO FRESHWATER FLOW ON NUTRIENT AND PRODUCTION CYCLES.
U.Brockmann, B.Cote, A.Edwards, H.G.Fransz, B.R.Heimdal and T.McClimans.

The group concludes that the main question of interest for a future international cooperation would be: what are the effects of the fresh water entering the North Sea, with special emphasis on toxic blooms - exchange between coastal currents and fjords - inputs (physical, chemical and biological) to the marginal sea. The chosen object is the North Sea with emphasis on the Norwegian Coastal Current (low in nutrients) and the Germanic Bight (high in nutrients), providing a comparative case for the Gulf of St. Lawrence (variable nutrients).

RELATIONS TO NATIONAL AND INTERNATIONAL RESEARCH ORGANISATIONS

The procedures for international cooperation are somewhat uncertain. National programs and/or programs within a nation should be coordinated through the national committees (Norwegian Oceanographic Committee, for example). International programs staged by ICES are of interest (SCAPINS, for example). In the least, inclusion of biological and chemical analysis in on-going physical programs (Norwegian Coastal Current studies, for example), directed to the above goals, would be a cost-effective way of paving the road towards a more systematic program.

J.B.L.Matthews will bring the recommendations with regard to cooperation in the marginal North Sea to the attention of ICES with the objective to put them under the umbrella of SCAPINS (projects on Summer Circulation And Production In the North Sea).

FUTURE COOPERATION

With respect to further coordination of the proposed activities it was decided that different time scales and geographic conditions would favour independent operation of the working groups, which does not require mutual coordination. However, K.Drinkwater, T.McClimans and P.P.Tett will provisionally act as correspondents for the three ad hoc working groups which are supposed to follow up on their initiatives. S.Skreslet will distribute a newsletter with information he receives about the progress in working group activities, among the members of the present workshop.

PARTICIPANTS

CANADA

Akenhead, S.
Boulva, J.
Drinkwater, K.F.
Cote, B.
Koslow, J.A.
LeBlond, P.H.
Leggett, W.C.
Parsons, T.
Sinclair, M.
Therriault, J-C.

FRANCE

Queguiner, B.

GERMANY

Brockmann, U.H.
Lenz, J.
Peinert, R.
Smetacek, V.

ICELAND

Malmberg, S.Å.
Thordardottir, T.

NETHERLANDS

Fransz, H.G.

NORWAY

Haakstad, M.
Heimdal, B.R.
Hopkins, C.
Kaartvedt, S.
McClimans, T.
Skreslet, S.
Solemdal, P.
Svendsen, H.
Sætre, R.

PORTUGAL

Silva, A.J. da

UNITED KINGDOM

Edwards, A.
Matthews, J.B.L.
Simpson, J.H.
Tett, P.P.

U.S.A.

Garvine, R.W.
Marshall, H.G.

SUBJECT INDEX

Acartia, 88, 91, 99
Actinocyclus, 95
Advection
   114, 119, 189, 195, 375, 407
American shad, 123
Anchovy, 120
Arctic characteristics
   East Iceland Current, 390
Arcto-Norwegian cod, 7
Asterionella glacialis
   322, 324
Australian Gulf, 35

Baltic Sea, 98
Barents Sea, 8
Bay
   anchovy, 121
   of Brest, 219
Belt Sea, 98
Benthos
   sedimentation of food, 91
Biddulphia, 95
Biomass
   dispersion, 4
Brackish plumes, 47

C:Chl.a ratio, 368
C:N ratio, 366

Calanus, 93
   finmarchicus, 114, 367, 375
      advection, 375
      spawning habitat, 385
      wintering habitat, 383
Capelin, 119
   growth, 389
Centropages, 99
Cerataulina pelagica
Ceratium, 90
   furca, 98
   tripos, 98
Chaetoceros, 89, 96, 97, 324
Chesapeake Bay, 36, 88
   phytoplankton, 319
   Plume, 48, 325
Chlorophyceae, 322, 325
Chlorophyll
   seasonal cycles, 245
   specific extinction, 164
Chrysophyceans, 325
Ciliates, 99
Circulation
   two-layer, 18
Classification
   estuaries, 15
   fjords, 18
   mixing processes, 21
   physical regimes, 14
   plumes, 48
   shelves, 20

Coastal Current
  ecology, 92
  experiment (CCE), 311
  fronts, 385
  Icelandic, 346
  laboratory model, 82
  meanders, 82, 302
  Norwegian, 7, 82, 92
  of Mozambique, 333
  Scottish, 295
  Vancouver Island, 310
Coccolithophores, 325
Cod
  Arcto-Norwegian, 7
  Icelandic
    growth, 389
  Labrador, 153
  larvae, 122
  recruitment, 8, 151, 394
  St. Lawrence, 6, 148
Coherence spectrum, 123
Compensation currents
  estuarine, 18
  plume systems, 60
Connecticut River
  Plume, 50
Copepod
  abundance, 243
  community structure, 99
    related to flushing
      108, 375
  grazing, 90, 361
Coriolis effect
  estuaries, 15
  plumes, 49
Coscinodiscus, 95, 324
Coupling
  primary/secondary production

Critical period, 117
Cryptophyceans, 325
Cs radioisotope, 297
Cyanobacteria, 322

Deep water renewal, 188
Density current, 196
Detritus food chain, 419
Diatom
  abundance, 258
  biomass, 245
  blooms, 87
  tychopelagic, 94
Diffusion, 115, 119, 195
Dinoflagellates, 90, 98, 258
Dispersion
  biomass, 4
  zooplankton, 375
Dynamic processes, 67

Early life history of fish,
  117, 139
  mesoscale features, 261
East Greenland Current, 390
East Iceland Current, 390
Ecological
  effects of
    advection, 114
    climate, 8
    freshwater regulation,
      10, 92, 149
    geochemicals, 5
    natural variations in
      freshwater outflow
      87, 97, 108, 141, 259
      262

Ecological
 effects of
  "phasing functions" on
   "forcing functions" in
    numerical modeling
    161
 variability, 86
  in copepod communities, 99
  north Icelandic waters
   389
  south Icelandic waters
   345
Ekman transport, 123
El Nino, 153
Emiliana huxleyi, 323, 324
Enclosures,
 rearing of larval fish, 122
Entrainment velocity, 164
Environmental variations
 biological impact, 391
Estuarine circulation, 197
Estuarine production
 response to
  "phasing" functions, 173
Estuary
 classification, 15
 plume, 48, 56
Euglenophyceans, 325
Euphausiids, 93, 361
Eutrophication, 247

Faecal pellet recycling, 361
Faxafloi, 345
Feeding
 exogenous, 19
 grounds
  Icelandic fish stocks
   402

Fish larvae
 in shelf habitats, 122
 survival, 118, 141, 268
Fisheries related to
 freshwater outflow
  8, 139, 261
Fjord
 hydrography, 37, 183, 195
 stagnant water, 195
Flagellates, 245
Flemish Cap, 283
Flux
 nutrient rates, 201
 phytoplankton in lochs
  205
 salt and heat, 283
Food
 availability, 117, 268
 chain, 9, 141, 419
 web, 3, 100, 410,
Food chain hypothesis, 141
Freshwater
 outflow effects on
  Bay of Brest
  biological processes
   408, 410, 422, 424, 429,
   432
  Calanus, 410
  chemical processes, 421
  coastal current flow
   301, 309, 424
  cod recruitment, 151
  copepod reproduction, 410
  coupling between
   trophic levels, 423
  Dutch waters, 241

Freshwater
- outflow effects on
  - ecological processes
    10, 87, 108, 141, 259, 375, 417,
  - estuaries, 161, 417
  - fish spawning, 412
  - fisheries
    8, 139, 261, 330
  - flushing, 188, 375
  - fjord circulation
    19, 183, 196
  - fjords
    13, 67, 183, 195, 205
  - food chains, 141, 268
  - German Bight, 231
  - ice, 407
  - lobster landings, 148
  - marginal seas
    231, 241, 251, 261
  - marine
    - ecosystems, 417, 429
    - production, 85
  - match/mismatch, 423
  - north-west Atlantic
  - Norwegian Coastal Current
    361, 375
  - nutrient loading, 219
  - oceanographic fronts
    47
  - physical oceanograpic processes, 406, 420, 424, 430
  - phytoplankton dynamics
    408
  - primary production
    358, 409

Freshwater
- outflow effects on
  - recruitment to stocks
    9, 139, 261, 330
  - salinity and temperature
    north-west Atlantic, 271
  - Scottish
    Coastal Current, 295
    sea-lochs, 195, 205
  - secondary production, 359
  - shelf waters
    271, 283, 295, 309, 319, 329, 345, 361, 375, 389
  - shellfish poisoning, 359
  - shrimp abundance, 329
  - Sofala Bank, Mozambique
    329
  - south Iceland waters, 345
  - St. Lawrence
    Estuary, 251
    Gulf, 139, 261
  - static vertical stability
    405, 408
  - stratification, 345
  - vertical mixing, 407, 420
  - zooplankton production, 107
- regulation, 10, 92
  - effects on
    - advection, 407
    - fish, 412
    - fisheries, 437
    - fjord circulation, 183
    - ice, 408
    - Lake Melville, 183
    - Nile Estuary, 92

Freshwater
  regulation
    effects on
      oceanographic variability, 408
      phytoplankton
        communities, 247
        dynamics, 409
      sea surface temperature 149
      shrimp abundance, 333
      vertical water exchange 407
      water mobility, 407
      zooplankton, 412
      river plumes, 48
  residence time, 187
Fronts
  as zooplankton habitats 113, 385
  Norwegian coastal current 385
  of plumes, 54, 262
Froude numbers
  two-layer models, 18

Gaspe Current, 262
Geochemicals,
  ecological effects of, 5
German Bight, 98
Gulf
  Australian, 35
  of Maine, 8
  of St. Lawrence, 6, 261
Grazing, 354
  by copepods, 90
Growth, 111

Haddock, 120
Herring
  larvae, 123
Hudson
  River plume, 56
Hydro-biological variations
  North Icelandic waters, 389
Hypothesis of
  run-off driven food chain 141
  population persistence, 155
Ice, 394, 407, 408
Iron, 241
Internal waves, 16,

Katodinium rotundatum, 89
Kelvin waves
  in laboratory models, 67
  in shelf waters, 48
Kelvin numbers
  in brackish plumes, 49, 59
Kiel Bight, 89
Knight Inlet, 34

Laboratory modeling, 67
Lake Melville fjord, 183
Larval survival, 117
Lateral transport, 421
Leptocylindrus
  danicus, 322, 324
  minimus, 322
Light extinction
  effects on zooplankton, 167
Limacina retroversa, 368
Lined sole, 121
Loch
  Ardbhair, 208
  Creran, 208

Loch
  Etive, 195, 208
  Spelve, 208
  Striven, 209

Macrophytes, 95
Maturation, 111
Meganyctiphanes norvegica, 117, 131
Metapenaeus monoceros, 329
Microzooplankton, 123
"Mid-seventies salinity anomaly", 390
Mississippi, 15
  River Plume, 48
Mixed layer depth, 164
Mixed Upper Layer Ecosystem Simulation (MULES), 161
Mixing
  classification, 21
  definition of, 13
  diagram, 222
  effects on plankton, 5, 88
  estuaries, 27
  fjords, 30
  kinetic energy, 22
  related to
    freshwater outflow
      405, 417, 420, 424, 429
  shelves, 34
Modeling
  conceptual, 108
  numerical
    37, 64, 161, 243, 303
  laboratory, 67

Mortality, 111, 291
  effects of container size 129
Mozambique, 329
Munida sarsi
  zoea, 382
Mysids, 87

Nano-
  flagellates, 90, 258
  plankton, 321
NCC ratio, 93
Net plankton, 321
Nile River
  effects of regulation, 92
  estuarine ecology, 92
  Plume, 92
Nitrate
  198, 221, 233, 242, 355, 368
Nitrogen budget, 365
Nitzschia, 96
  pungens, 322
North Sea
  oxygen depletion, 237
  plankton, 241
Norwegian
  Coastal Current
    7, 82, 93, 361, 375
  fjord ecology, 91
  Sea, 375
Numerical models
  estuarine plankton ecology
    161, 243
  fjord circulation, 37, 214
  loch phytoplankton dynamics, 205
  plume return flow, 64

Numerical models
  topographic influence on
    coastal current, 303
Nursery grounds
  Icelandic fisk stocks, 402
Nutrient
  budget, 361
  input by rivers, 231
  regeneration, 88
  uptake 226

O-group estimates, 394
Oxygen
  consumption rate
    in fjord bottom water,
    195
  depletion, 237

Paracalanus, 99
Parafavella, 368
Parameters
  circulation, 17
  stratification, 17
Partial renewals of
  deep fjord water, 201
Patchiness, 121
Penaeus indicus, 329
Phaeocystis pouchetii, 93
Phosphate, 198, 231, 242
Physiological adaptation
  of phytoplankton, 226
Phyto-/zoodetritus ratio
  correlation with nitrate,
    172
Phytoplankton
  biomass change, 205
  exudation, 165

Phytoplankton
  life history strategy, 96
  nutrient uptake, 221
    195
  production, 143
  sedimentation, 90, 361
  seeding, 96
  spatio-temporal distribution, 251
  stock size control, 88
picoplankton, 419
Plaice, 118
Plume, 47, 80
  "biophysics", 113
  brackish, 47
  ecology, 92, 97
  laboratory model, 78
Polar characteristics
  East Iceland Current, 390
Predator- prey relations, 117
  macroinvertebrate predators,
    129
  prey
    densities,117
    size, 125
  Spectral analysis, 123
Prorocentrum marie-lebouriae,
  89
Primary production, 345
  coupling to
    secondary production, 161
  estimates, 394
Protogonyaulax tamarensis
  98, 259
Pseudocalanus, 99
Pteropods, 93
Pycnoclines
  primary and secondary, 197

Red tide, 98
Redfish larvae, 291
Regeneration
  of plant nutrients, 88
Reproduction, 111
Rhizosolenia
  delicatula, 322
  fragilissima, 322
Richardson number
  flux formula, 23
  relation to stability, 26
River
  plume, 48, 70, 92, 235, 259
  regulation, 10, 92, 405, 436

San Francisco Bay, 87
Scale
  space, 6
  time, 6
  variability, 86
Scottish Coastal Current, 295
Sea bream, 121
Secondary production
  coupling to primary production, 161
Sediment traps, 361
Sedimentation
  of clay by copepods, 100
  of euphausiid faeces, 361
  of phytoplankton, 90
  related to
    pelagic system structure 361
Sewage effluents, 242, 325
Shelf break front, 385
Shellfish poisoning, 259
Shrimp abundance, 329
Silicate, 198, 221, 231, 242

Skeletonema, 96
  costatum
    87, 89, 322, 323, 324, 325
Sofala Bank, 329
South Atlantic Bight, 34
Spawning grounds
  Icelandic fish stocks, 402
Spectral analysis
  prey-predators, 123
Spring bloom, 345, 361
St. Lawrence
  cod stock, 6
  Estuary, 96, 123, 251
  Gulf, 6
    fisheries production, 139, 261
  River
    plume, 48
    spring discharge, 261
Stagnant water in fjord, 195
Stoichiometric relations
  in nutrient regeneration, 202
Strait of
  Georgia, 311
  Juan de Fuca, 313
Stratification
  effects on phytoplankton 88, 355
  in estuaries, 15
Striped bass, 87, 118
Survival
  curve, 118
  fish larvae, 118, 291

Thalassionema nitzschioides, 322

Thalassiosira, 95, 322
  descipiens, 87
  eccentrica, 87
Thermal front
  Norwegian Coastal Current 385
Tidal forcing
  effects on water exchange 226
  long-period, 8
Time/space resolution, 118
Tintinnid shells, 368
Tracer
  Cs radioisotope, 297
Trophic interactions, 6
  time and space dimensions, 6
Turbulence
  kinetic energy, 23
  sources of energy, 25

Upwelling, 122

Vertical migration, 111

Water exchange
  effects on zooplankton 114, 125, 375
  estuary-shelf 36
  fjord-shelf 37
  shelf-ocean, 38
  tidally induced, 226
Wind effects
  on primary production, 356

Zaire River Plume, 97

Year-class strength 117, 143, 268
Yolk-sac
  absorption, 120
  nutrition, 119

Zooflagellates
  importance to zooplankton, 172
Zooplankton
  ecology 6, 90, 263, 268, 358
  patches, 121
  production estimates, 394
  response to
    allochtonous substrate, 167
    light, 167
    primary production, 358
    zooflagellate abundance, 172

NATO ASI Series G

Vol. 1: **Numerical Taxonomy.** Edited by J. Felsenstein. 644 pages. 1983.

Vol. 2: **Immunotoxicology.** Edited by P.W. Mullen. 161 pages. 1984.

Vol. 3: **In Vitro Effects of Mineral Dusts.** Edited by E. G. Beck and J. Bignon. 548 pages. 1985.

Vol. 4: **Environmental Impact Assessment, Technology Assessment, and Risk Analysis.** Edited by V.T. Covello, J.L. Mumpower, P.J.M. Stallen, and V.R.R. Uppuluri. 1068 pages. 1985.

Vol. 5: **Genetic Differentiation and Dispersal in Plants.** Edited by P. Jacquard, G. Heim, and J. Antonovics. 452 pages. 1985.

Vol. 6: **Chemistry of Multiphase Atmospheric Systems.** Edited by W. Jaeschke. 773 pages. 1986.

Vol. 7: **The Role of Freshwater Outflow in Coastal Marine Ecosystems.** Edited by S. Skreslet. 453 pages. 1986.

DEPT. Oceanography ✓
O.N. 9894
PRICE
ACCN No. UC868